BIOLOGY
at a Glance

Third edition

Judy Dodds
Brampton College, London

Illustrations by
Annette Whalley
and
Cactus Design

MANSON
PUBLISHING

CONTENTS

BIOLOGY IN THE NEWS

Mandy Allwood's Octuplets

After taking fertility drugs, Mandy Allwood found herself pregnant with octuplets. Going against doctors' advice, she refused to abort any of them, reportedly did a deal with the News of the World, and then miscarried all eight.

THE SECRET OF LIFE

The DNA code is finally cracked

STILL THE SLAUGHTER GOES ON

The foot and mouth outbreak continues

Test Tube babies

Frozen Assets: Diane Blood

Diane Blood made history in 1997 by giving birth to a child years after the father died of bacterial meningitis. Doctors took some of her husband's sperm as he lay dying and froze it.

SCIENTISTS ARE RACING TO CREATE A GENETICALLY MODIFIED 'SUPER MOSQUITO' THAT WILL DESTROY MALARIA INSTEAD OF SPREADING IT

Gene code opens new fields of medicine

Giant bat triumphs over fruit farmer...

A north Queensland farmer has been ordered to stop electrocuting thousands of giant bats that were feasting on his fruit crops

Super-dingoes

Australian sheep farmers are being terrorised by packs of highly aggressive dogs, resulting from mating between dingoes and domestic dogs gone wild. The new crossbreed recently killed a young boy.

White House targeted in anthrax terror campaign

Human remains identified as the last Romanovs, by DNA testing

GENETIC SUPER BABIES STORM

Key discovery raises spectre of designer children with high IQs

NEW SARS TRAVEL ALERT

The WHO has confirmed that the total deaths from Sars has now passed 500. Sars (Severe Acute Respiratory Syndrome) has killed people in China, Hong Kong, Singapore, Canada and Taiwan. A new alert has been issued advising against non-essential travel to many areas of China as alarm grows over the spread of the disease.

First Human Embryo Cloned

Maize crop modified by GM genes

Questions:

1. Use the internet to write one page about one of the topics in the news. List your sources at the end.
   (A useful web site is www.bbc.co.uk/genes)
2. Collect articles relating to biology over the last few weeks. Stick them on a page in a similar way. Why did you choose these articles?
3. Why are people concerned about cloning?

CELLS

CELLS A cell is the basic unit of life. All living organisms are made of cells.

Animal and plant cells share many features but there are differences.

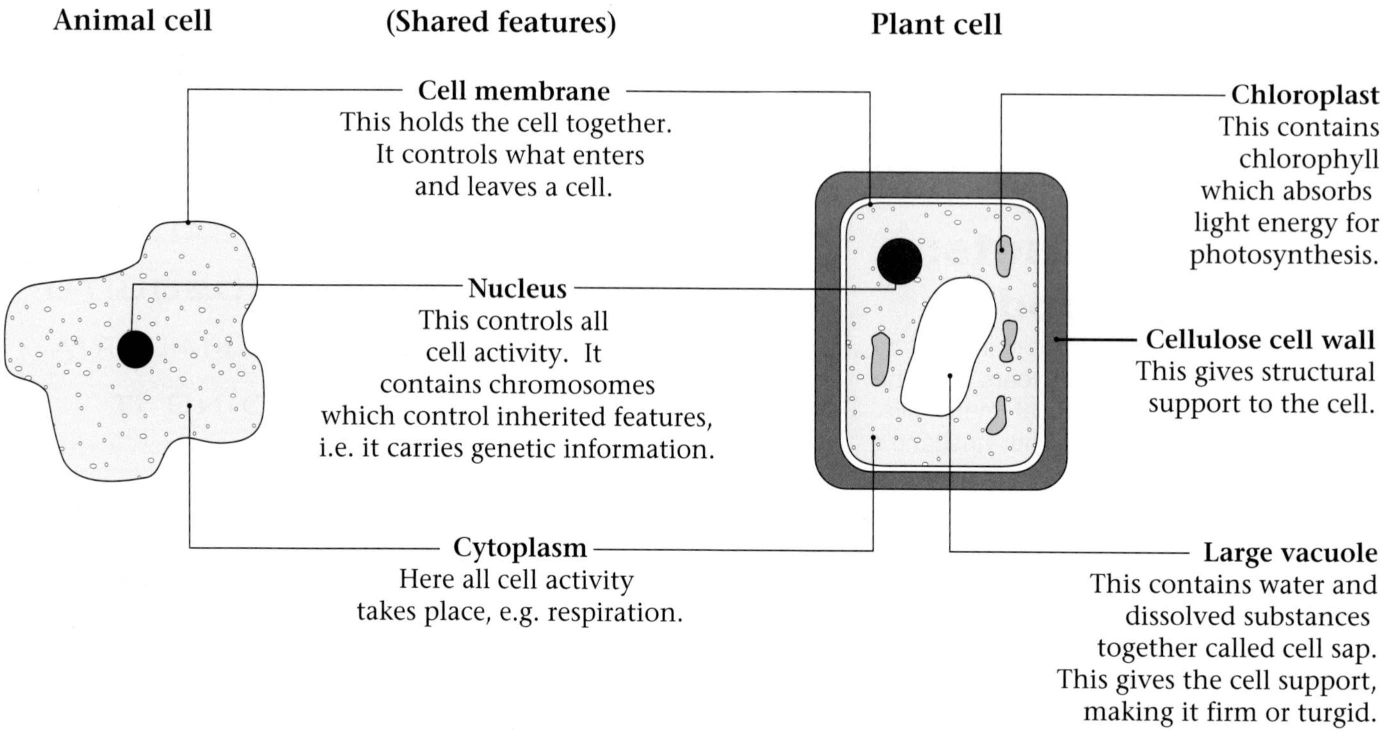

Animal cell
(Shared features)
Plant cell
Cell membrane
This holds the cell together. It controls what enters and leaves a cell.
Nucleus
This controls all cell activity. It contains chromosomes which control inherited features, i.e. it carries genetic information.
Cytoplasm
Here all cell activity takes place, e.g. respiration.
Chloroplast
This contains chlorophyll which absorbs light energy for photosynthesis.
Cellulose cell wall
This gives structural support to the cell.
Large vacuole
This contains water and dissolved substances together called cell sap. This gives the cell support, making it firm or turgid.

In addition, cells have little organelles called mitochondria which are the site of aerobic respiration, and ribosomes, where proteins are made in the process called protein synthesis.

	Animal cells	Plant cells
Features in common	Have a nucleus. Have a cell membrane. Have cytoplasm.	Have a nucleus. Have a cell membrane. Have cytoplasm.
Differences	Do not have a cell wall. Do not have chloroplasts. Do not have a large vacuole.	Have a cell wall made of cellulose. Have chloroplasts. Have a large vacuole filled with cell sap.

The size of a cell is limited by the distance over which diffusion is efficient.

Questions:

1. State two differences between animal and plant cells.
2. What is the function of the cell membrane?
3. Which three features do animal and plant cells share?
4. When plant and animal cells are placed in water, most animal cells will burst, whereas plant cells will not. Explain this difference.
5. Where does photosynthesis take place in a plant cell?

VARIETY OF CELLS

Animal cells

1. Red blood cell

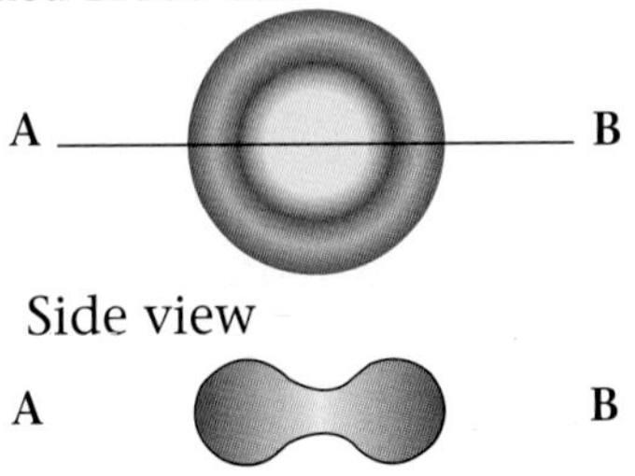

This has no nucleus.
It contains **haemoglobin** which combines with oxygen. Red blood cells carry oxygen round the body.

2. Cheek cell

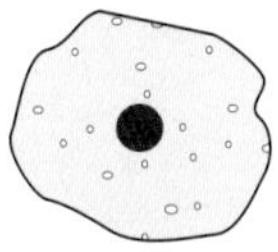

A simple unspecialised animal cell.

3. Sperm cell

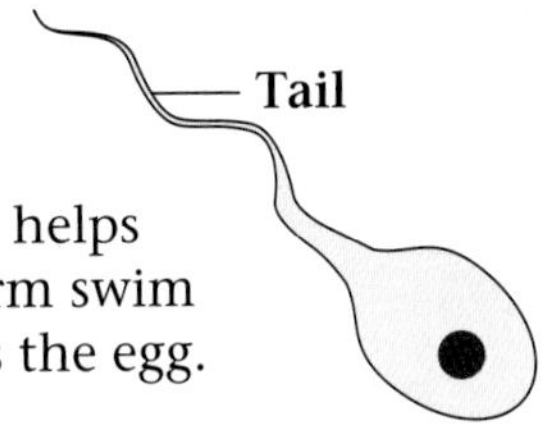

The tail helps the sperm swim towards the egg.

4. Egg (ovum)

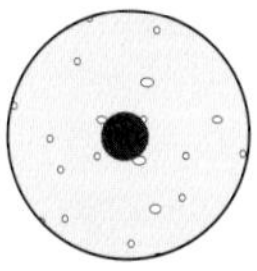

Food can be stored in this large cell.

5. Intestine cell

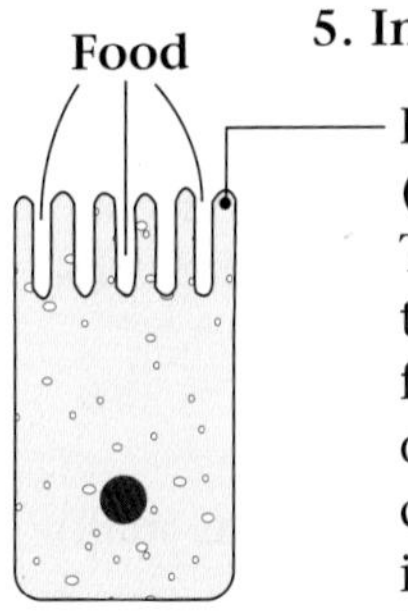

Folds (microvilli)
These increase the surface area for absorption of food. The cells are found in the small intestine.

6. Tracheal cell

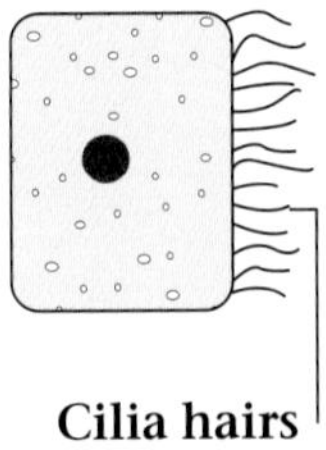

This **ciliated cell** is found in the trachea where the cilia hairs sweep up mucus and any bacteria and dust that enter, helping to keep the lungs clean.

7. Neurone (nerve cell)

These long cells quickly carry messages round the body.

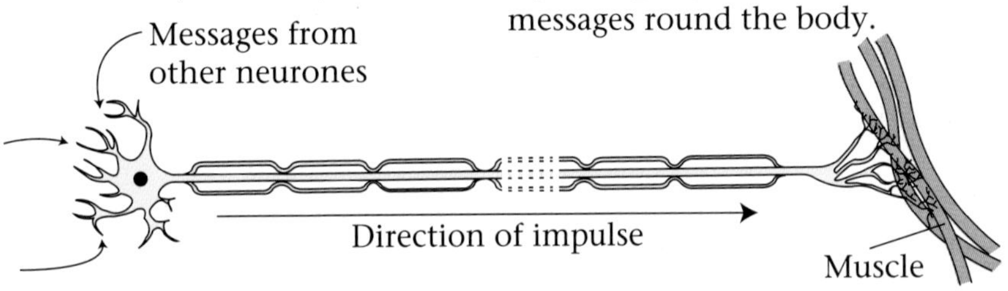

Plant cells

1. Palisade cell (in leaf)

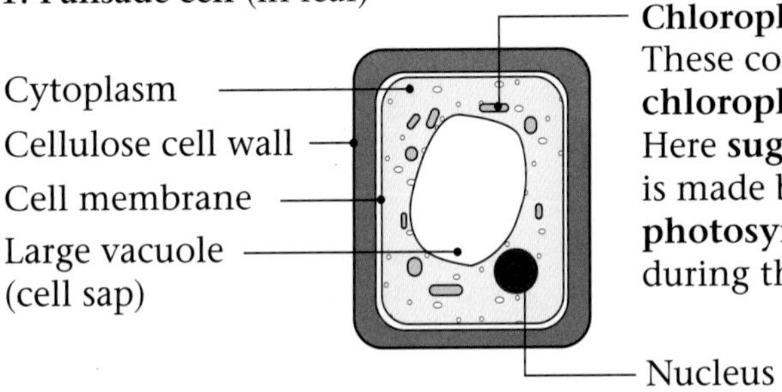

Chloroplasts
These contain **chlorophyll.** Here **sugar** is made by **photosynthesis** during the day.

2. Root hair cell

This **root hair cell** has a large surface area for anchorage and absorption of water and minerals. Lots of water can enter quickly through the large surface area. There are no chloroplasts in root cells as there is no light in the soil.

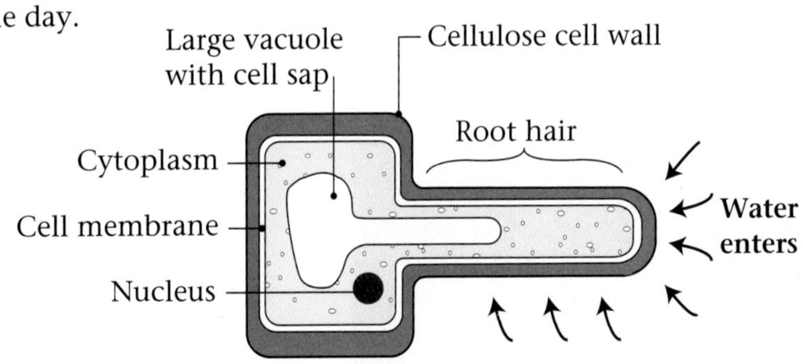

Simple organism, e.g. ***Amoeba***

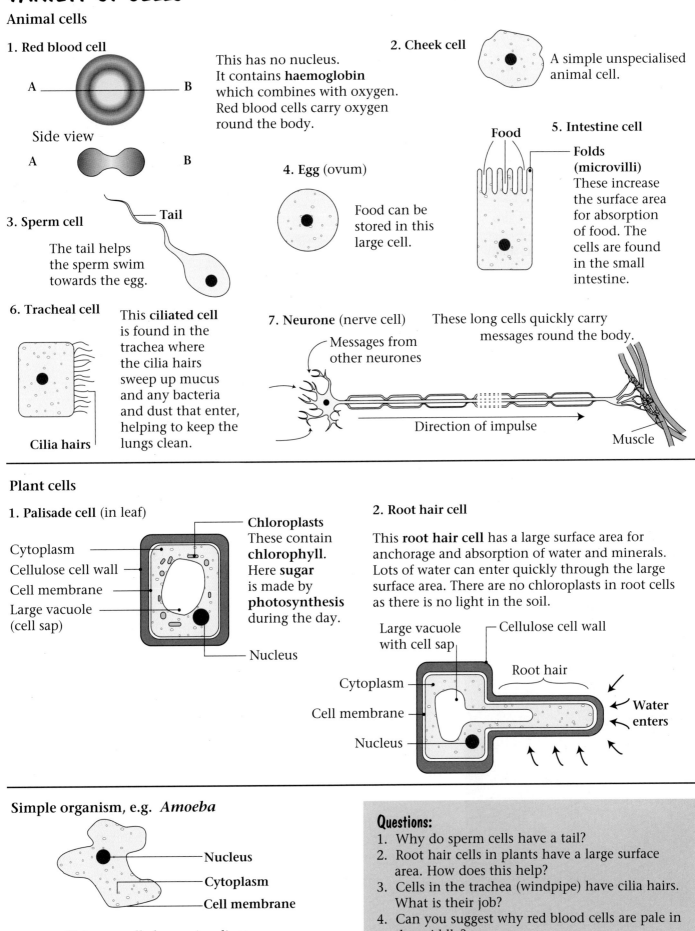

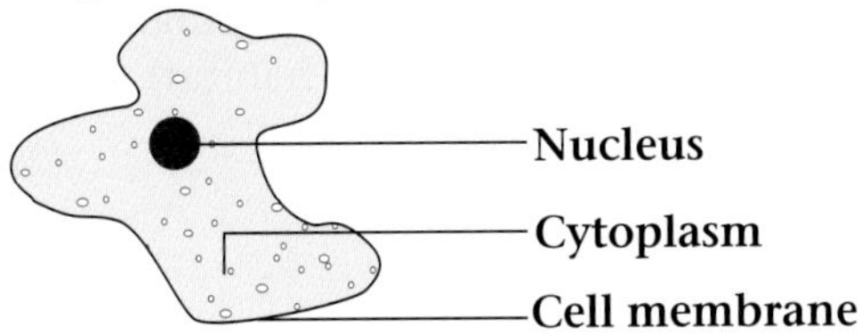

This one-celled organism lives in freshwater ponds.

Questions:

1. Why do sperm cells have a tail?
2. Root hair cells in plants have a large surface area. How does this help?
3. Cells in the trachea (windpipe) have cilia hairs. What is their job?
4. Can you suggest why red blood cells are pale in the middle?
5. Why must the ovum be larger than the sperm cell?

LEVELS OF ORGANISATION

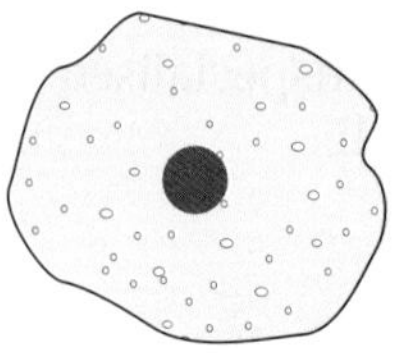

Cells

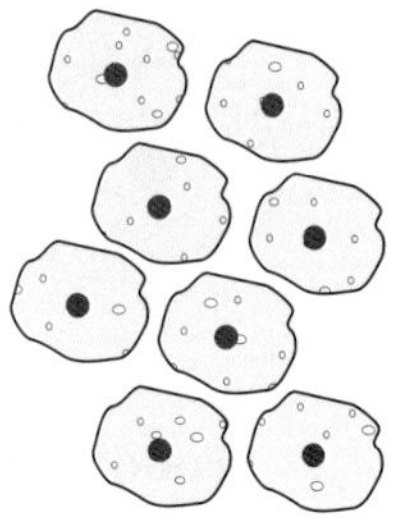

Tissues
(group of similar cells)

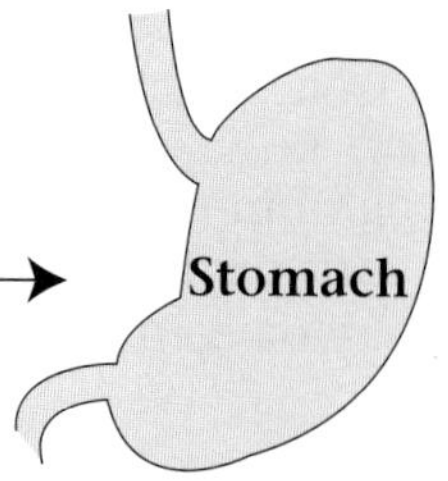

Organ
(group of tissues working together)

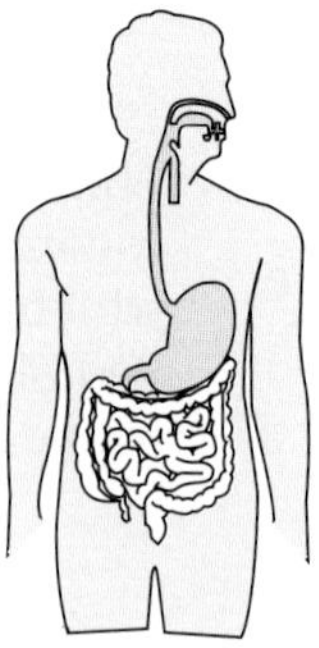

System
(e.g. digestive system)

Digestive system includes the stomach, oesophagus and intestines, i.e. different organs working together.

↓

Systems work together in an organism.

There are nine major systems in the human body		
System		**Function**
Digestive	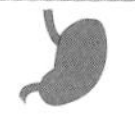	To digest and absorb food.
Breathing	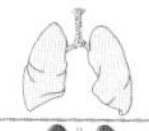	To take oxygen into the body and remove carbon dioxide.
Excretory		To remove waste materials from the body.
Circulatory	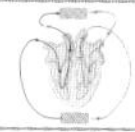	To carry blood round the body.
Nervous	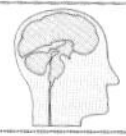	To carry messages round the body.
Sensory	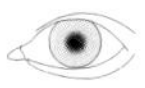	To receive information.
Muscle		To bring about movement.
Skeletal		To provide support, protection and movement.
Reproductive	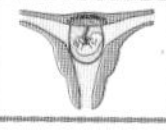	To produce young.

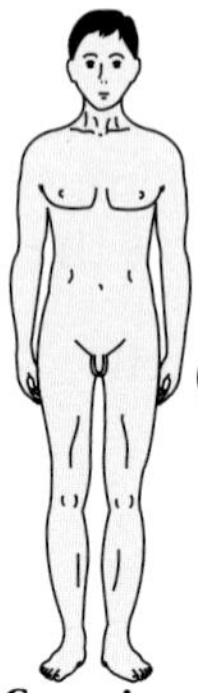

Organism
(member of a species)

Species

↓

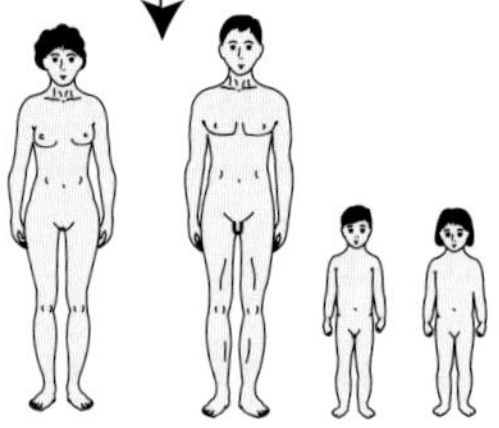

Population
(group of one species)

↓

(non-living part of habitat)

Community
(different populations in a habitat)

↓

Ecosystem
(e.g. woodland, has living and non-living parts)

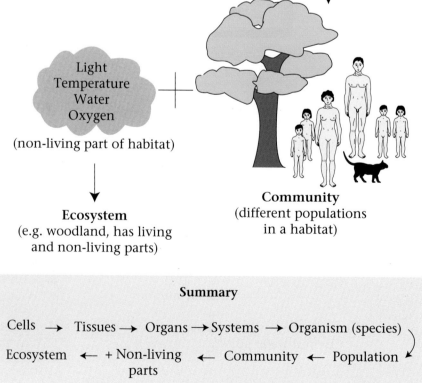

From Bowes: *A Colour Atlas of Plant Propagation*

Woodland ecosystem.

Summary

Cells → Tissues → Organs → Systems → Organism (species)

Ecosystem ← + Non-living parts ← Community ← Population

HOW SUBSTANCES ENTER A CELL

- Diffusion.
- Active transport.
- Osmosis.

1. Diffusion This is the movement of molecules from a region where they are in **high** concentration to a region where they are in **lower** concentration. Diffusion continues until the molecules are evenly mixed and there is no difference in concentration.

Low concentration of perfume

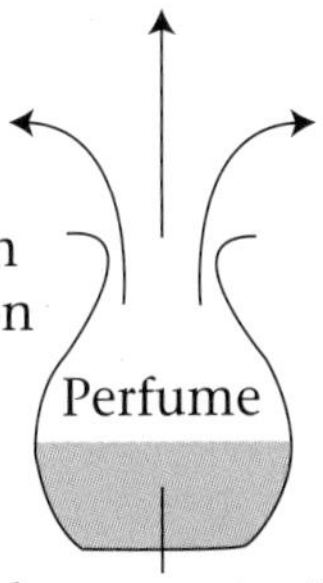

Perfume molecules move out by diffusion to lower concentration

High concentration of perfume

Diffusion occurs due to the random movement of individual particles

Only gases and liquids can diffuse

High concentration — Diffusion → Low concentration

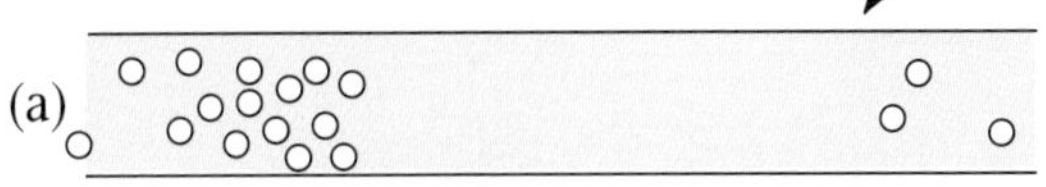

The greater the difference in concentration, the faster the rate of diffusion.

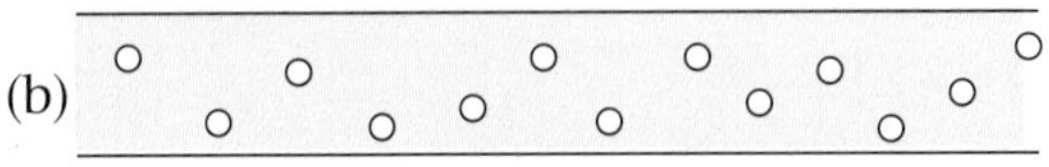

Molecules are evenly mixed – diffusion stops.

---

2. Active transport Molecules move from an area of *low* concentration to an area of *high* concentration (opposite to diffusion).

Low concentration — High concentration

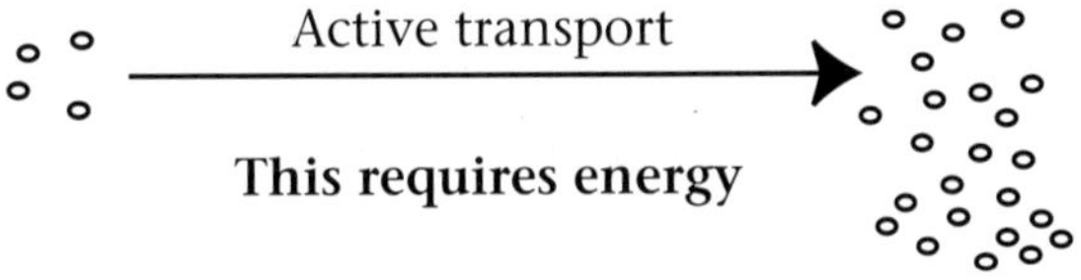

In kidney tubules, glucose passes into blood by diffusion and active transport.

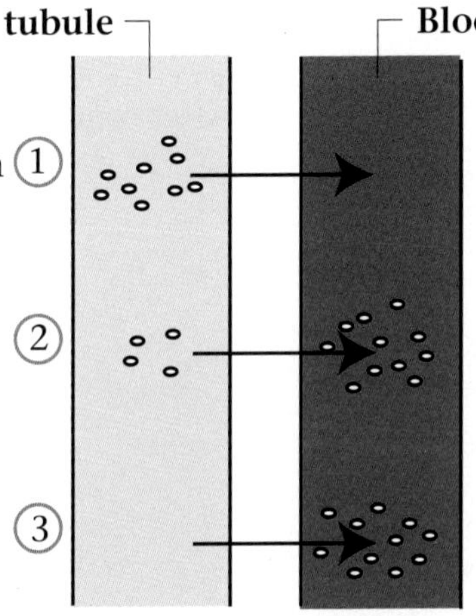

High concentration glucose (1) — **Low** concentration glucose = **diffusion**

Low concentration glucose (2) — **High** concentration glucose = **active transport**

No glucose left in kidney tubule (3) — All glucose has passed into the blood

---

Example of active transport in plants

Root hair cells are able to absorb mineral ions from the soil by active transport.

Questions:
1. A drop of ink in water will spread until all the liquid is blue. What is this process called?
2. How is diffusion involved in attracting insects for pollination?

Uptake of minerals

Minerals can enter by **diffusion** and **active transport** (low to high concentration).

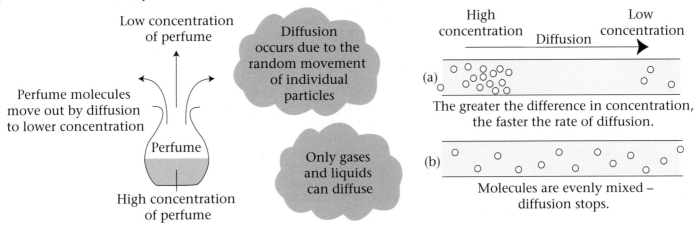

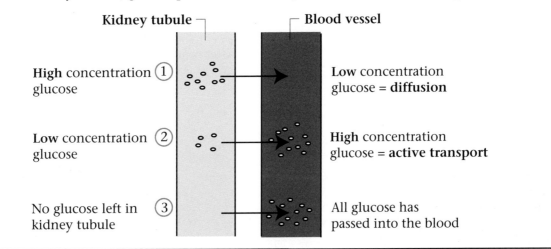

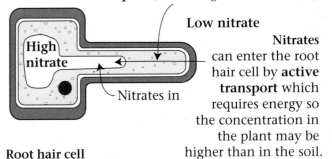

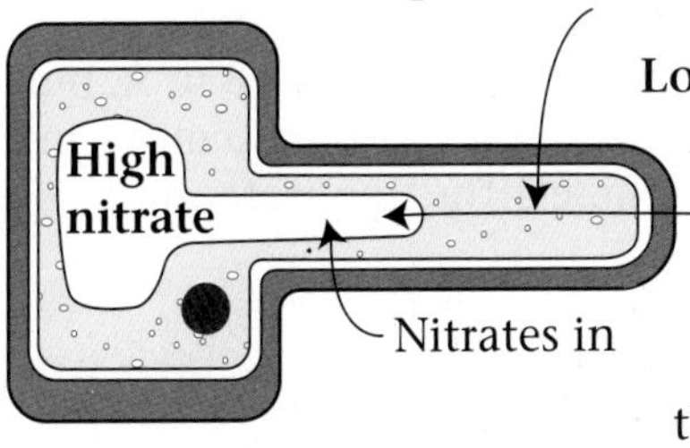

Nitrates can enter the root hair cell by **active transport** which requires energy so the concentration in the plant may be higher than in the soil.

Root hair cell

OSMOSIS The movement of water through a membrane

Osmosis is the movement of *water* from an area of **high water concentration** to an area of **lower water concentration** through a **selectively permeable membrane.**

High water concentration —— WATER ——▶ Lower water concentration

Water molecules are constantly moving due to kinetic energy. Solutes, like sugar, attract water molecules making them less free to move. Therefore solutes affect the ability of water to move. The more solute molecules present, the less free water molecules are to move.

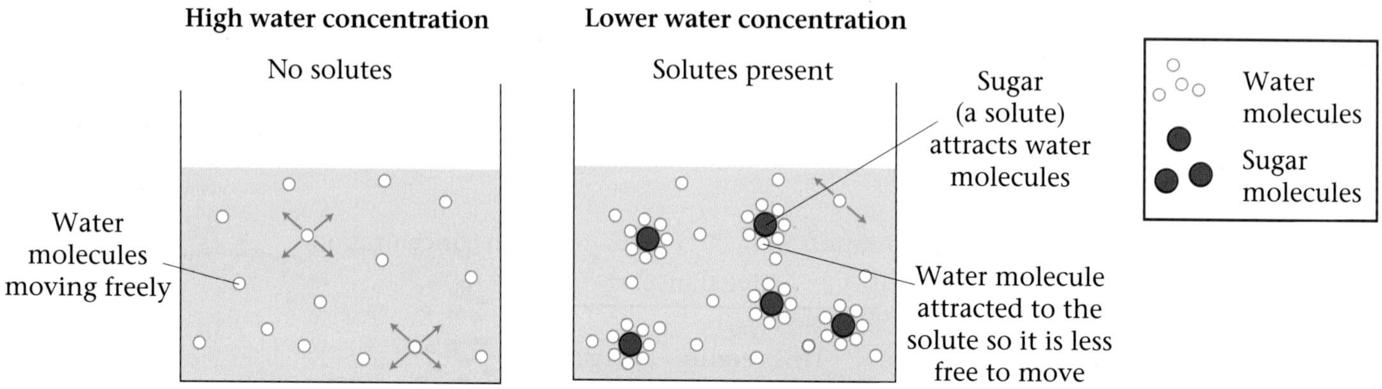

Water able to move freely has a **high water potential.**

Water less free to move has a **low water potential.**

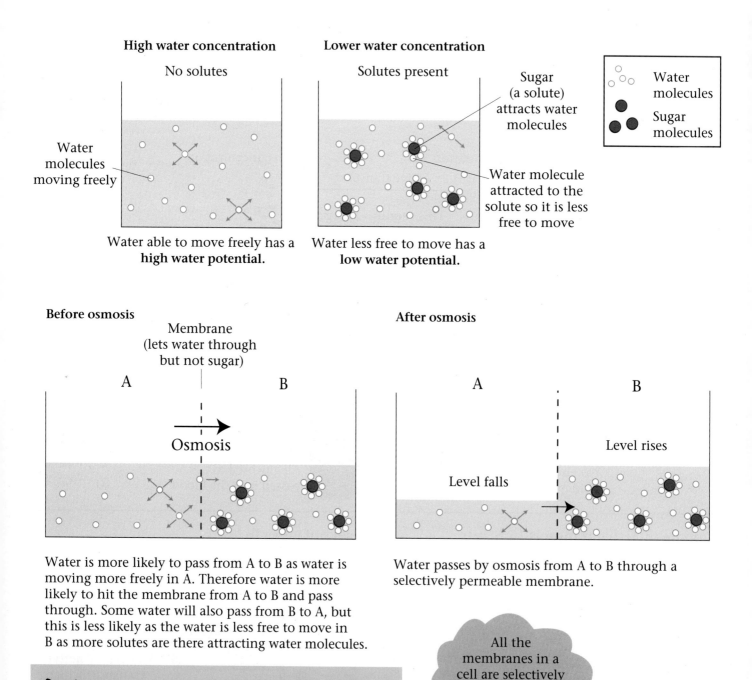

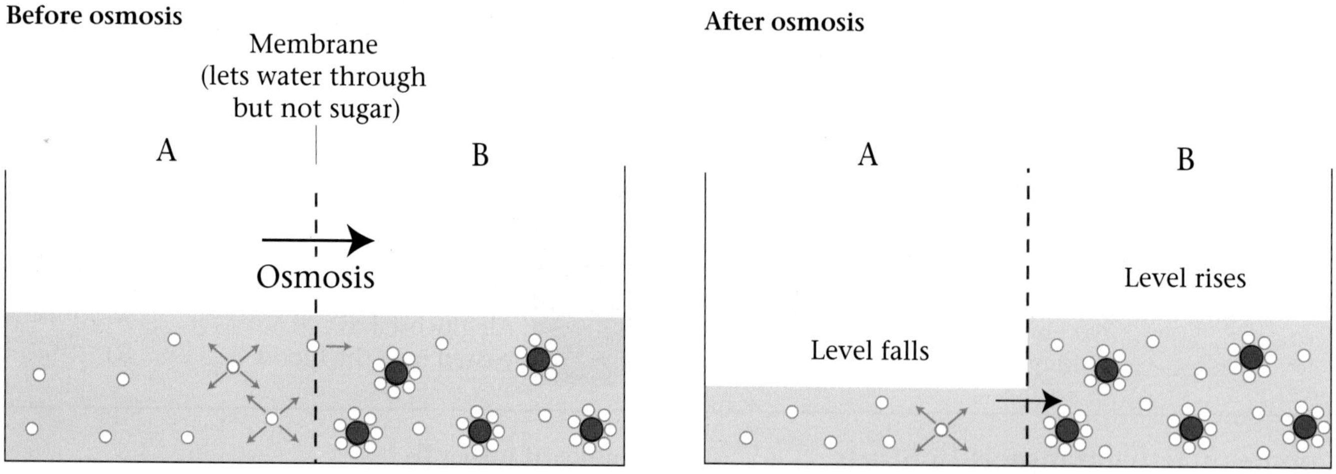

Water is more likely to pass from A to B as water is moving more freely in A. Therefore water is more likely to hit the membrane from A to B and pass through. Some water will also pass from B to A, but this is less likely as the water is less free to move in B as more solutes are there attracting water molecules.

Water passes by osmosis from A to B through a selectively permeable membrane.

All the membranes in a cell are selectively permeable.

Question:

1. A girl watered her pot plants with sea-water instead of fresh water, thus adding solutes to the soil. The plants wilted and died. Using osmosis, can you explain why?

OSMOSIS IN ACTION

1. ***Amoeba***, a single-celled organism that lives in freshwater ponds.

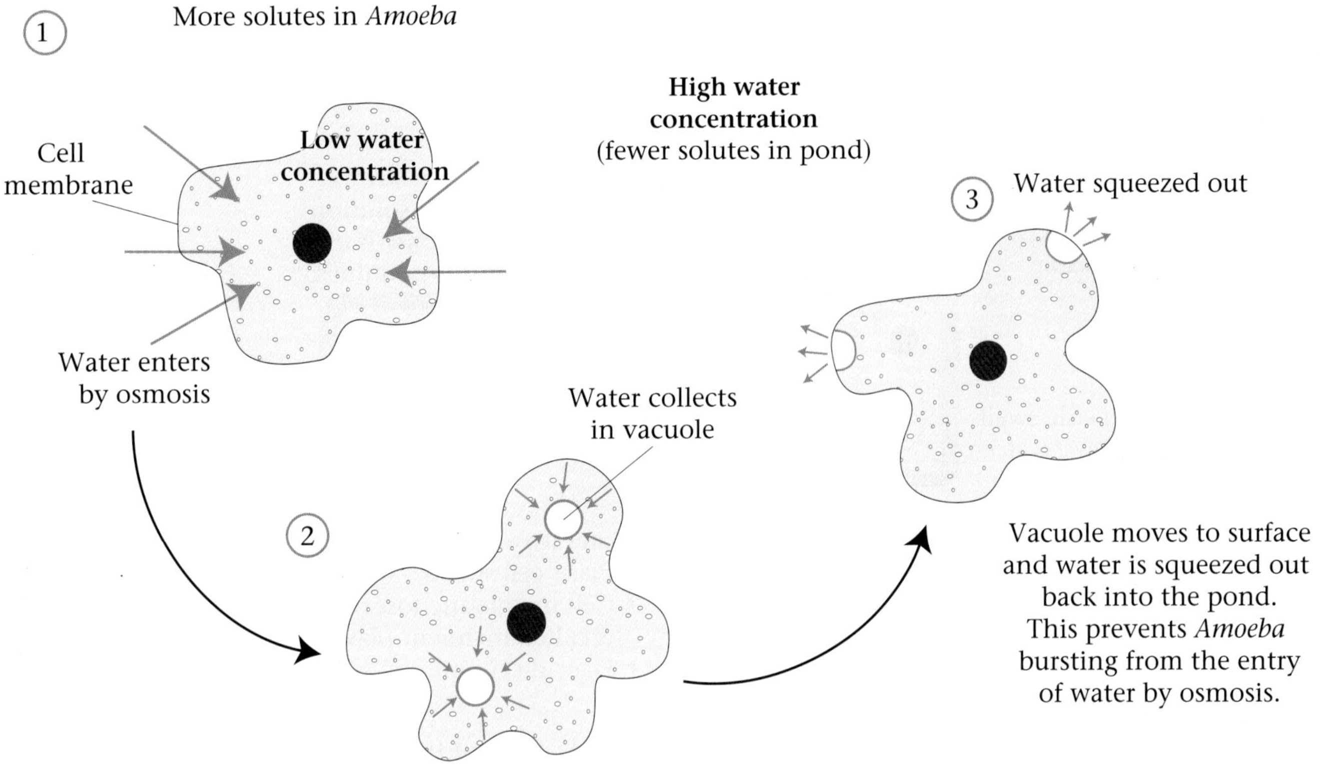

1
More solutes in *Amoeba*
Cell membrane
Low water concentration
High water concentration
(fewer solutes in pond)
Water enters by osmosis
3
Water squeezed out
Water collects in vacuole
2
Vacuole moves to surface and water is squeezed out back into the pond. This prevents *Amoeba* bursting from the entry of water by osmosis.

2. Red blood cells

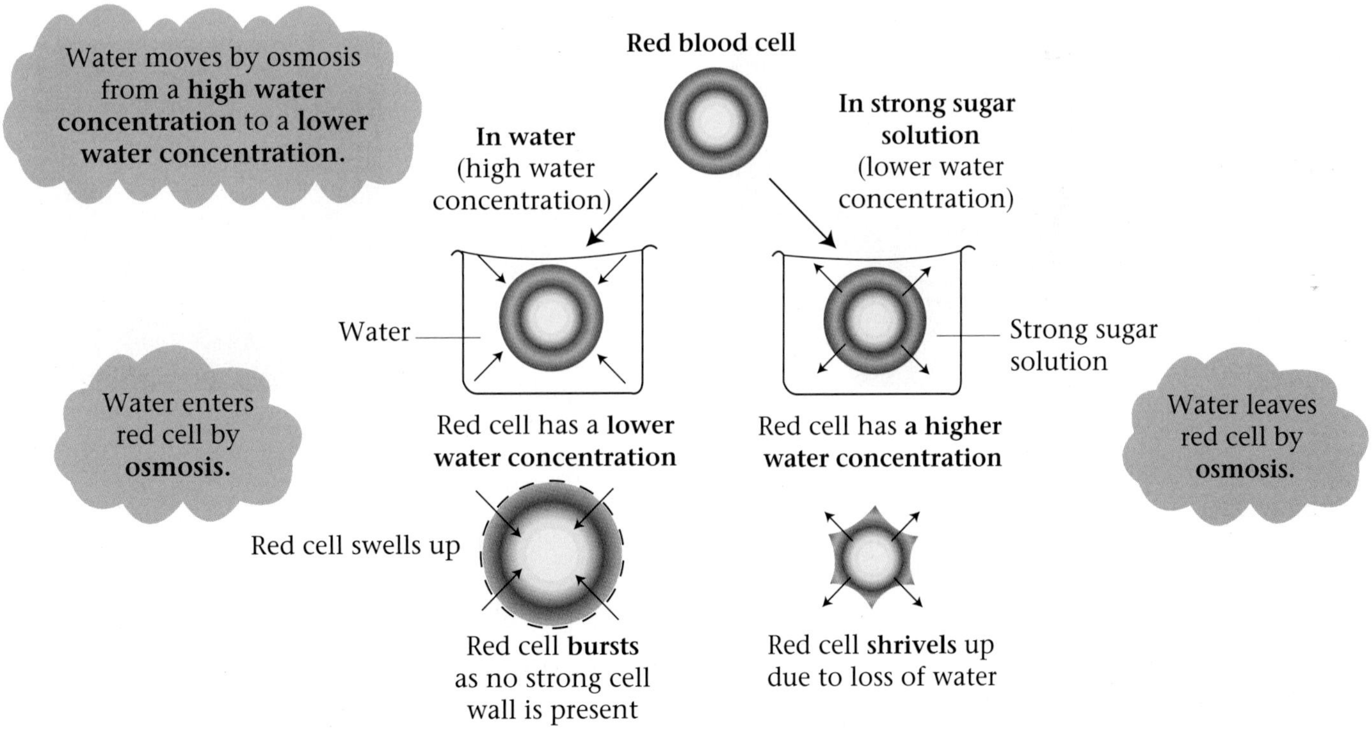

Water moves by osmosis from a **high water concentration** to a **lower water concentration.**
Red blood cell
In water
(high water concentration)
In strong sugar solution
(lower water concentration)
Water
Strong sugar solution
Water enters red cell by **osmosis.**
Water leaves red cell by **osmosis.**
Red cell has a **lower water concentration**
Red cell has **a higher water concentration**
Red cell swells up
Red cell **bursts** as no strong cell wall is present
Red cell **shrivels** up due to loss of water

Questions:

1. Why must plasma (the liquid in which red blood cells are found) have the same water concentration as red and white blood cells?
2. What problems will organisms face if they live in the sea which has a lower water concentration than many organisms?

OSMOSIS AND PLANT CELLS

Plant cells do not burst when water enters by osmosis due to their strong cellulose cell wall. However, the vacuole in plant cells may lose or gain water by osmosis.

Remember
Osmosis is the movement of water from an area of **high water** concentration to an area of **lower water concentration** through a selectively permeable membrane.

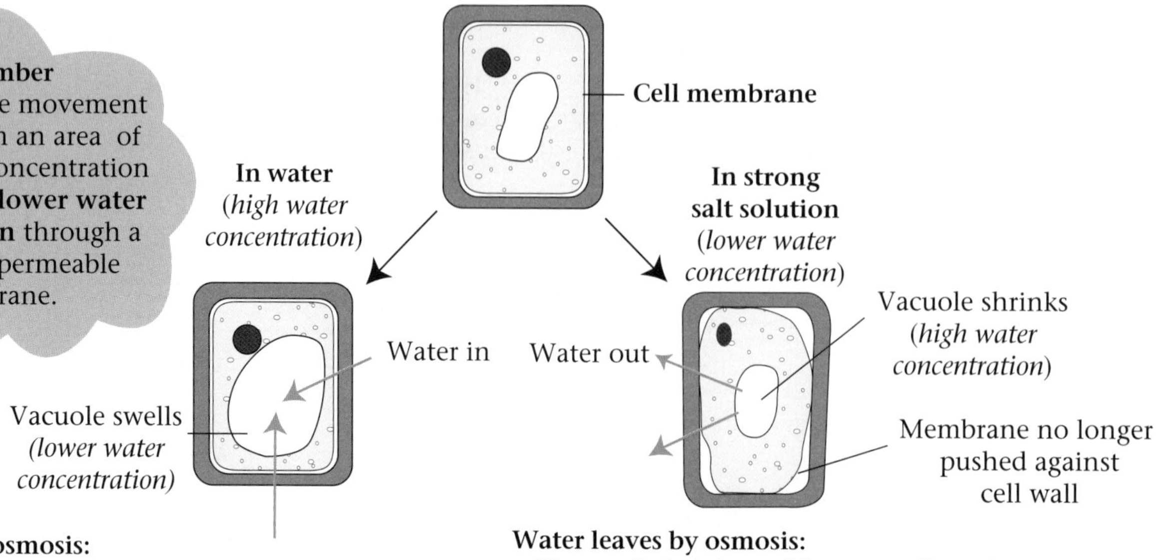

Water enters by osmosis:
- Plant cell swells due to enlarged vacuole.
- This results in turgid plant cells, a healthy condition.
- Plant cells do not burst due to their strong cellulose cell wall.

Water leaves by osmosis:
- Plant cell shrinks and becomes flaccid.
- Cell membrane loses contact with the cell wall. This is called incipient plasmolysis.
- Flaccid cells are unhealthy and the plant could die.

A leaf full of turgid cells stands out firmly and a large surface area is exposed to the sun for photosynthesis.

A leaf full of flaccid cells curls up and less surface area is exposed to the sun for photosynthesis.

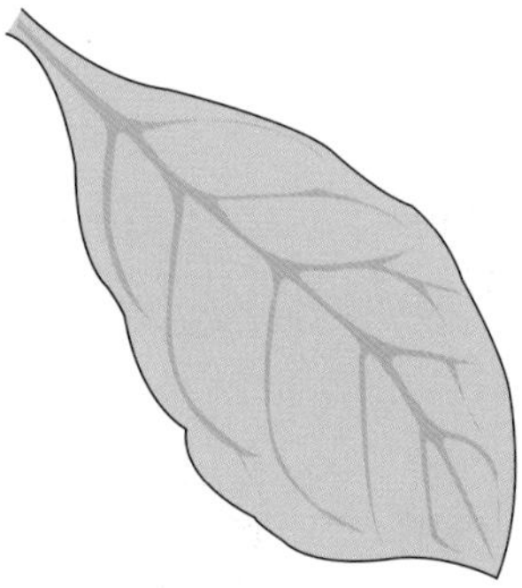

Turgid leaf

From Bowes: *A Colour Atlas of Plant Structure*

Turgid leaf.

Flaccid leaf

From Alford: *A Colour Atlas of Pests of Ornamental Trees, Shrubs and Flowers*

Flaccid leaf.

Making a grape

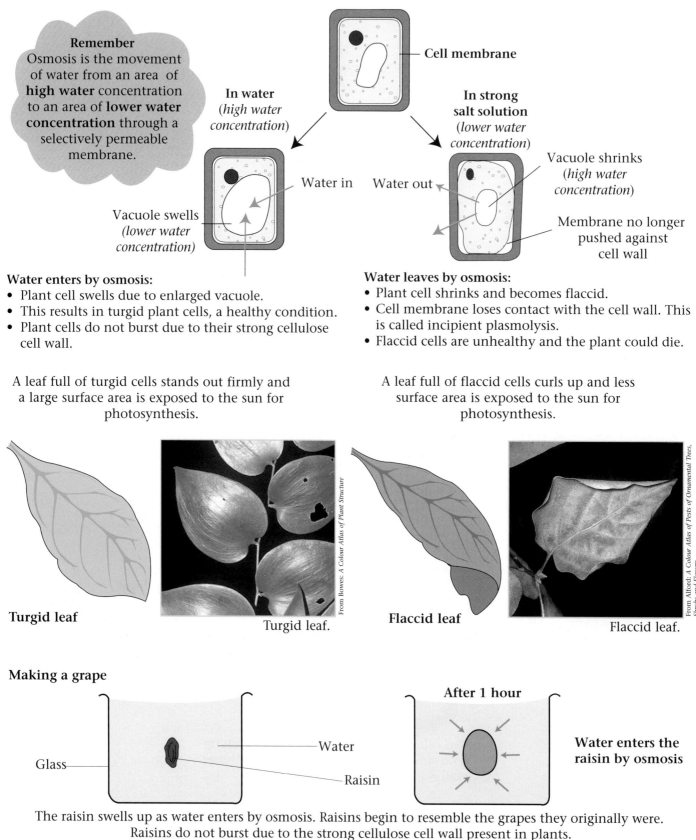

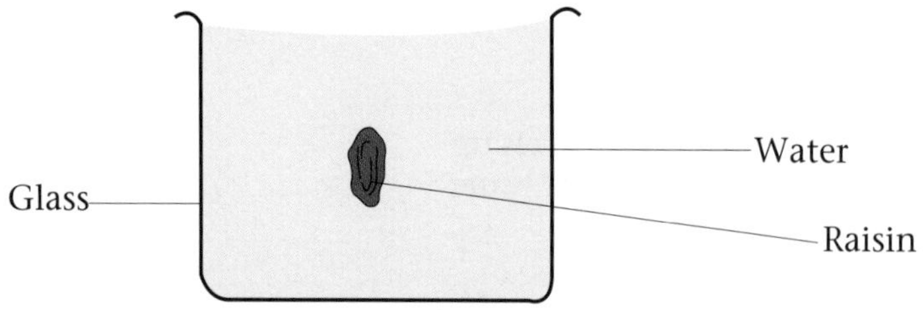

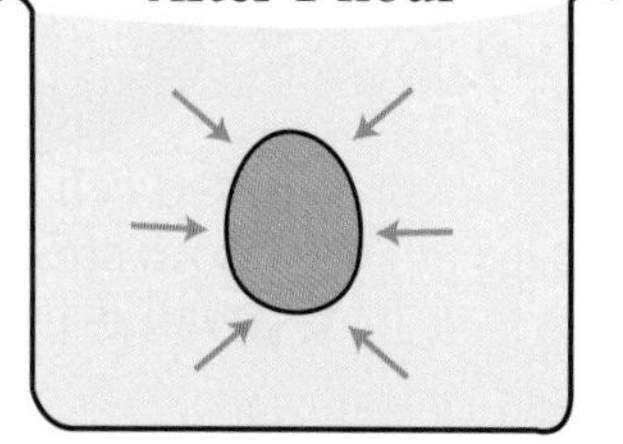

Water enters the raisin by osmosis

The raisin swells up as water enters by osmosis. Raisins begin to resemble the grapes they originally were. Raisins do not burst due to the strong cellulose cell wall present in plants.

Questions:
1. If a piece of raw potato is placed in a strong salt solution, what do you think will happen and why?
2. Pot plants were watered with a salt solution by mistake. What do you think will happen to the plants?

THE IMPORTANCE OF VOLUME AND SURFACE AREA

Surface area is the amount of surface an organism has. If we removed our skin, flattened and measured it, this would be our surface area.

Volume is the space taken up by an organism. Large organisms take up more space, so have larger volumes.

One-celled organisms like *Amoeba,* are able to get all the oxygen they need by simple **diffusion**, i.e. oxygen moves from a higher concentration outside the cell to a lower concentration inside.

Oxygen(O_2) diffuses through the surface area of *Amoeba* and can reach into its small volume so every part of the cell gets oxygen. This is possible as *Amoeba* has a **large surface area** and a **small volume**.

Similarly, carbon dioxide (CO_2) diffuses out to the lower concentration outside the cell

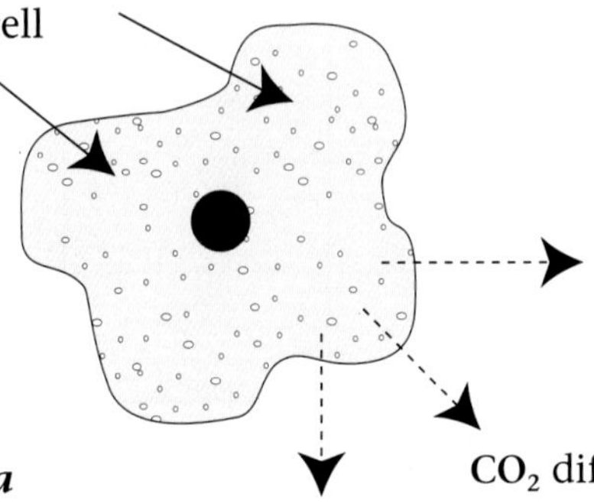

Large organisms have two major problems with gaining oxygen:

1. Their surface area is small compared to a large volume, so insufficient oxygen enters.

The surface area for gas exchange surface is increased by the development of a folded gas exchange surface, e.g. alveoli in humans, gills in fish.

Human thorax

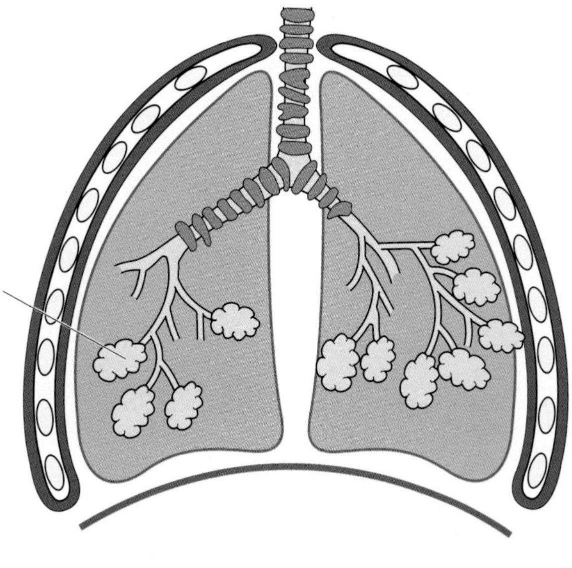

2. With a large volume, the distance from the gas exchange surface to every cell is too far for diffusion to be efficient.

Therefore a **transport system**, blood, is required to carry **oxygen** efficiently to all cells and to remove **carbon dioxide**. The development of a **heart** enabled blood to be pumped all round the body.

Transport system

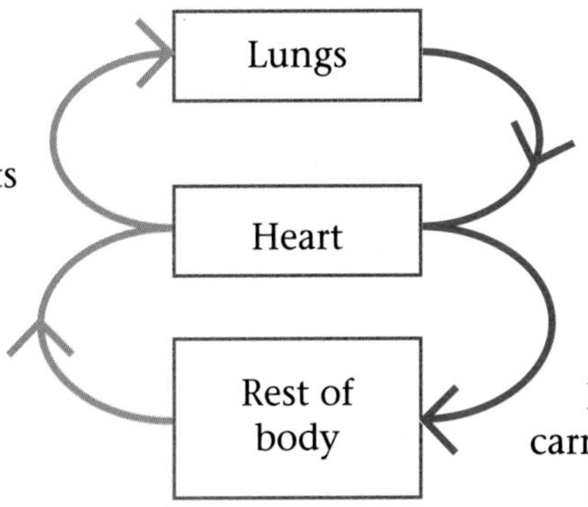

Questions:

1. Why is it possible for one-celled organisms to get all their oxygen by diffusion?
2. What problems do large animals face when getting oxygen and removing carbon dioxide?
3. Why must gas exchange organs be well supplied with blood vessels?
4. What other feature of gas exchange surfaces increase the uptake of oxygen?
5. Our alveoli are moist. Why, in terms of water, is it necessary for alveoli to be deep inside the body?

SURFACE AREA TO VOLUME RATIO

1 cm cube (1 cm^3)

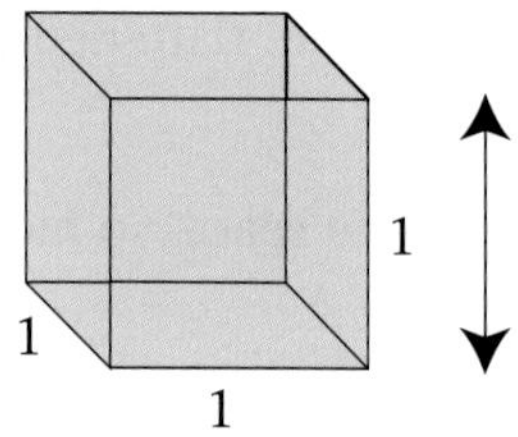

Volume = l × w × d = 1 × 1 × 1 = 1 cm^3
Surface area = l × w × 6 = 1 × 1 × 6 = 6 cm^2

2 cm cube (2 cm^3)

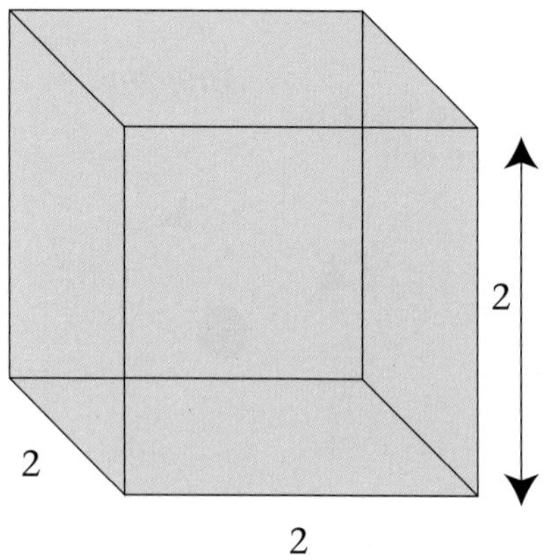

Volume = l × w × d = 2 × 2 × 2 = 8 cm^3
Surface area = l × w × 6 = 2 × 2 × 6 = 24 cm^2

Volume – space taken up = length × width × depth
Surface area – outer surface = length × width × 6 (6 sides of a cube)

Cube size (cm)	l × w × d **Volume (cm^3)**	l × w × 6 **Surface area (cm^2)**	
1	1	6	**Small animals** have a *large* surface area compared to volume.
2	8	24	
3	27	54	
4	64	96	
5	125	150	
6	216	216	
7	343	294	**Large animals** have a *small* surface area compared to volume.

Large animal
(small surface area compared to volume)

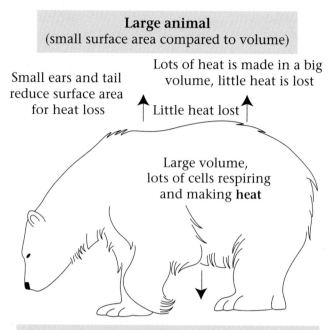

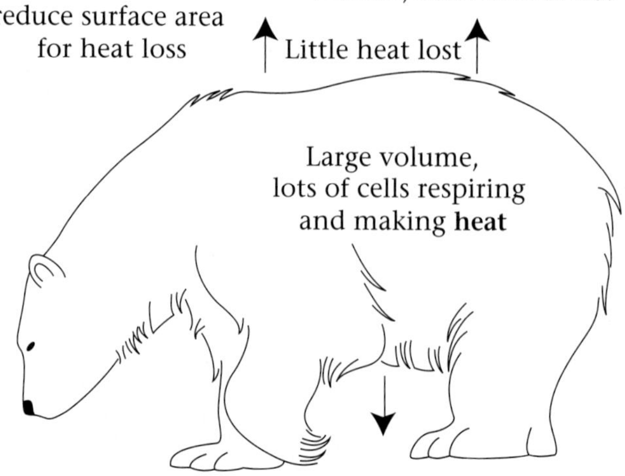

This has a small surface area to volume ratio.

Lives in a **cold** climate.

Small animal
(large surface area compared to volume)

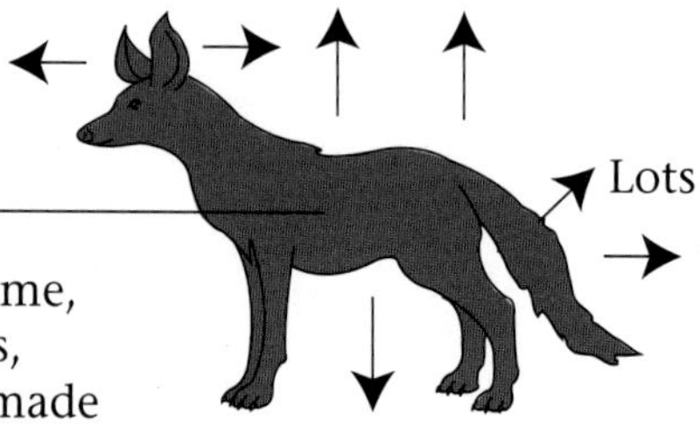

This has a large surface area to volume ratio.

Lives in a **hot** climate.

Questions:

1. Where is heat made in an organism and in what process?
2. How is heat lost from an animal?
3. What features of an animal will increase heat loss?
4. How would you recognise an animal living in a cold climate?
5. How would you describe the surface area to volume ratio of a) a very small animal, b) a large animal?
6. What problems do large animals face if living in hot climate and why?

Volume – heat is made
Heat is *made* in our cells which make up our volume, in *respiration*. Big volume = lots of heat made.
Surface area – heat is lost
Heat is *lost* through our skin or *surface area*. Big surface area = lots of heat lost.

BIOLOGICAL MOLECULES

There are three important biological molecules: 1. Proteins. 2. Lipids. 3. Carbohydrates.

PROTEINS These contain the elements carbon, hydrogen, oxygen, nitrogen.

Protein is needed for **growth**.

Sources of protein – for humans
- Meat (beef, lamb, chicken, pork).
- Fish.
- Egg-white.
- Beans.

Plants have to make their protein in order to grow. They need to combine the four elements carbon (C), hydrogen (H), oxygen (O), and nitrogen (N). Carbon, hydrogen, and oxygen are available from H_2O and CO_2. **Nitrogen** is acquired from the soil in the form of **nitrates**. **Protein** is built up from amino acids linked together by **peptide** bonds.

These chains may fold up to form a specific globular shape.

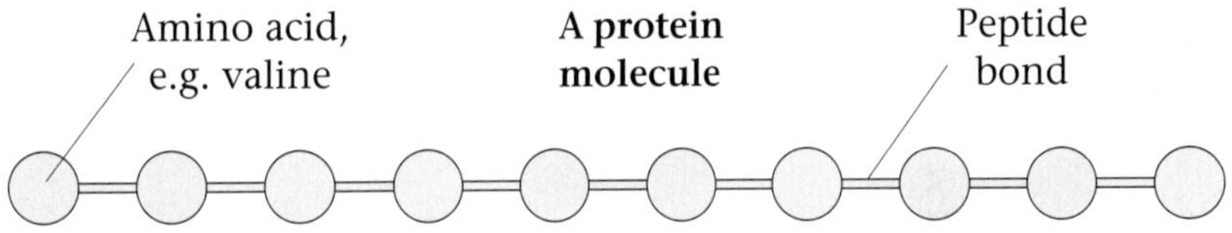

There are 20 different amino acids.

Enzymes are proteins which speed up reactions in living systems. Amylase (the enzyme in saliva) speeds up the breakdown of starch to maltose sugar. Amylase is also produced by the pancreas, to break down any starch remaining into maltose.

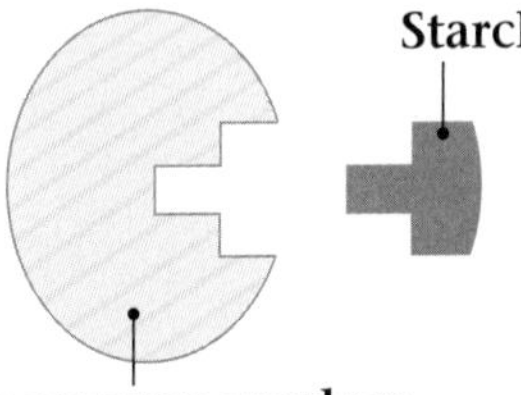

They fit together like a lock and key. Therefore amylase only reacts with starch.

Kwashiorkor

Children who do not have enough protein in their diet fail to grow properly. In parts of Africa, children may suffer from Kwashiorkor (protein deficiency). They are recognised by stick-like arms and legs and swollen abdomen, due to the build-up of tissue fluid, caused by lack of protein in their plasma.

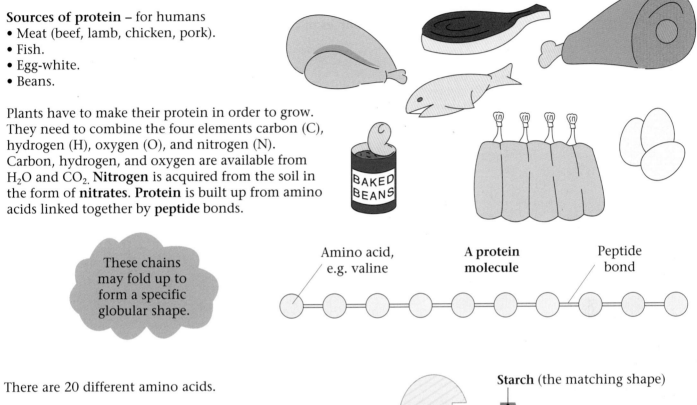

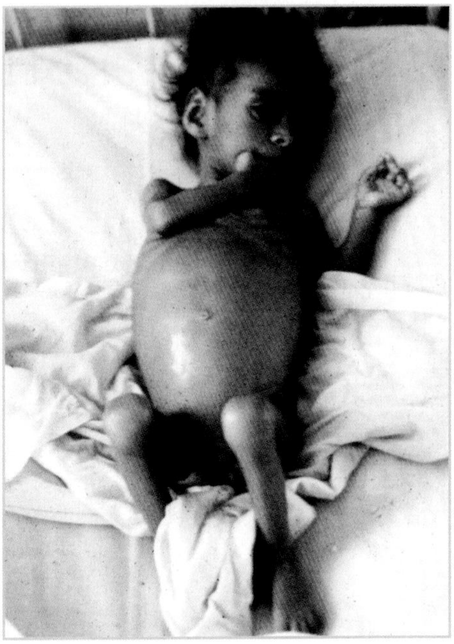

This child exhibits thin limbs and a swollen belly, classic symptoms of kwashiorkor, severe protein deficiency. This child, although looking like an infant, is probably 3–4 years old. Although superficially looking fat, this is a form of malnutrition.

Courtesy of Joseph E Armstrong, Illinois State University, USA.

CARBOHYDRATES These contain the elements carbon, hydrogen, oxygen.

Carbohydrates include:

- An insoluble energy store (starch, glycogen).
- Soluble sugars to transport to cells for respiration.
- The cellulose cell wall in plants.

There are two main sources of carbohydrates:

Sugars

Sugar, jam, honey, treacle
(all sweet, soluble, and form crystals)

Starch

Bread, potatoes, rice, pasta
(all not sweet, insoluble, and do not form crystals)

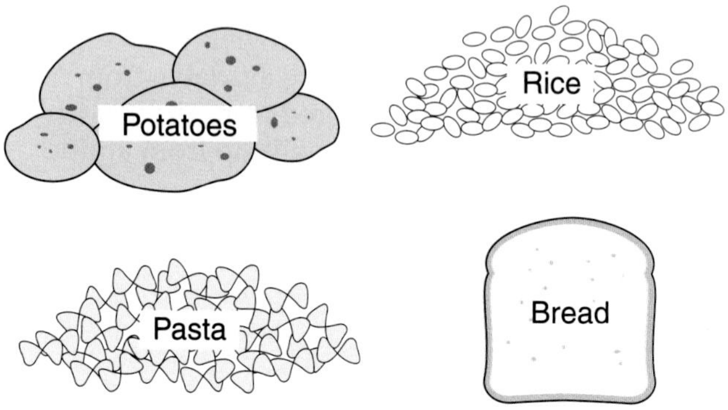

Carbohydrates are made of sugar units, like glucose, joined together by **glycosidic bonds.**

Sugars (one or two glucose units)

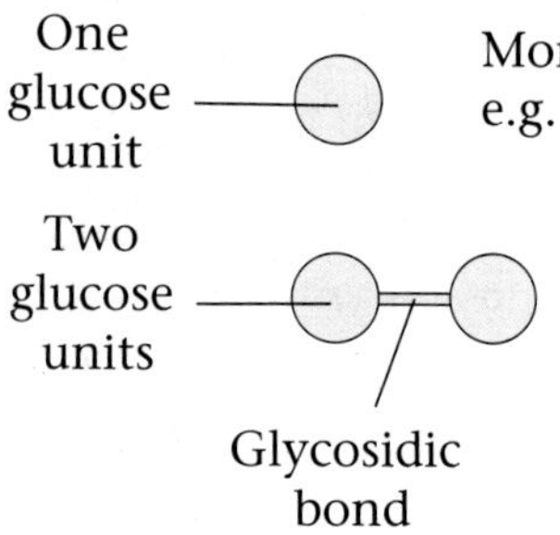

Monosaccharide, e.g. glucose

Disaccharide, e.g. sucrose

Polysaccharides (many glucose units), e.g. starch, glycogen, cellulose

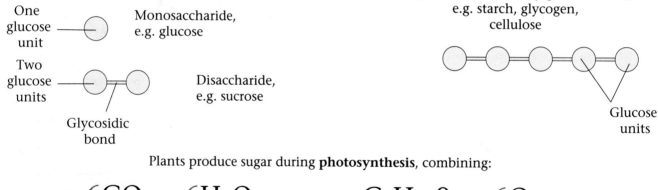

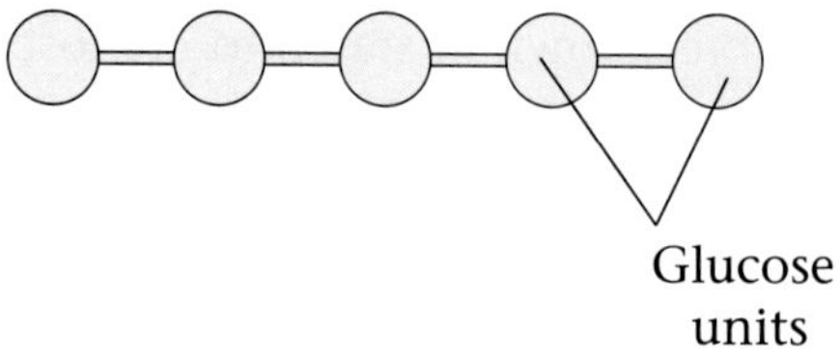

Plants produce sugar during **photosynthesis**, combining:

$$\underset{\text{Carbon dioxide}}{6CO_2} + \underset{\text{Water}}{6H_2O} \longrightarrow \underset{\text{Glucose}}{C_6H_{12}O_6} + \underset{\text{Oxygen}}{6O_2}$$

Glucose is **respired** to release **energy** in both animals and plants.

Animal carbohydrates

- Glucose in blood.
- Glycogen in liver and muscles.

Plant carbohydrates

- Sucrose in phloem.
- Sugar in nectar.
- Starch in leaves.
- Cellulose cell wall.

Fructose is a monosaccharide used as a sweetener in the food industry. It is very sweet, so only small quantities are needed.

LIPIDS These contain the elements carbon, hydrogen, oxygen.

Lipids include fats and oils – fats are **solid** at room temperature, oils are **liquid** at room temperature.

Lipids are needed for:

- Insulation (to stop heat loss).
- Energy store.
- Waterproofing, e.g. waxy cuticle on leaves.
- Buoyancy, e.g. in large aquatic mammals.

Sources of lipids:

- Butter, margarine, vegetable oil, egg-yolk.
- Nuts are also high in lipids.
- Seeds high in lipids, e.g. sunflower seeds, corn, and soya beans, all provide us with a source of oil.

Lipids are made up of glycerol and fatty acids joined by ester bonds.

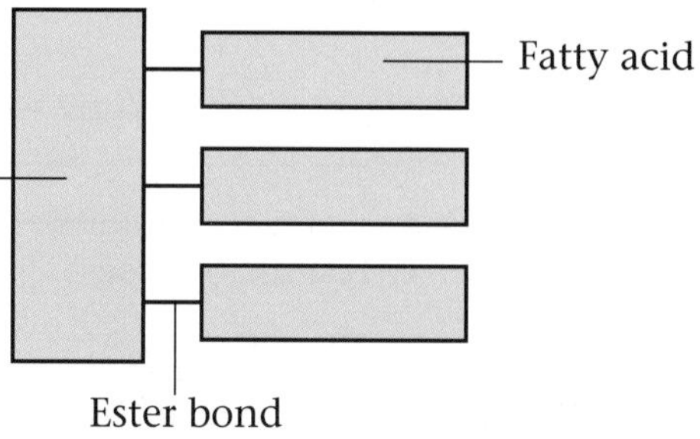

Insulation
A layer of fat under the skin acts as an **insulator** reducing heat loss. Whales and dolphins (mammals) have lots of fat called blubber to reduce heat loss in cold seas because fur cannot insulate underwater.

Waterproofing
Lipid forms the waxy cuticle covering leaves and the waxy outer layer of insects and other arthropods. In both cases, the lipid waterproofs the surface and prevents loss of water, essential for life on land.

Buoyancy
The fat stored also provides buoyancy, helping these large, heavy mammals to float in the water.

Fin whale

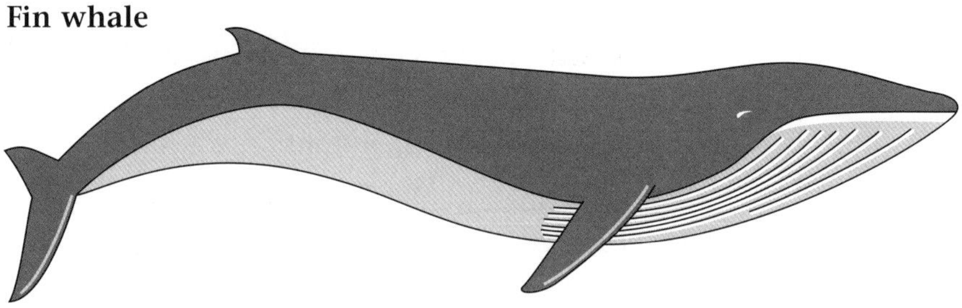

From: Summerhayes and Thorpe: *Oceanography, An Illustrated Guide*, courtesy of Tony Martin.

Hunted fin whale brought to whaling station.

Sea birds are covered in a natural oil that prevents their feathers becoming waterlogged. When there is an oil spillage from a ship, detergents are often used to disperse the oil, but it also removes the natural oil from the sea birds. As a result their feathers absorb water making the birds heavy and they may drown.

From: Summerhayes and Thorpe: *Oceanography, An Illustrated Guide*, courtesy of Professor Olof Linden/ICCE.

Large blowout oil spill from oil rig in the Gulf of Mexico.

Saturated fats increase blood cholesterol levels. Mono-unsaturated and polyunsaturated fats may help to reduce the blood cholesterol levels.

Find out which fats are saturated and which are polyunsaturated.

Questions:
1. What are the three biological molecules?
2. Which elements are found in all three molecules?
3. Which biological molecule contains nitrogen?
4. Why is protein needed by organisms? Give a good source of protein for humans.
5. What is the function of a layer of fat under the skin?
6. What are the two main groups of carbohydrates?
7. In what form is sugar transported in a) plants, b) animals?

FOOD TESTS

Test for	Chemical reagent	Colour change
Glucose	**Benedicts** solution	Blue → orange/brown
Starch	**Iodine** solution	Red → blue/black
Protein	**Biuret** reagent	Blue → purple/mauve
Lipid	**Ethanol** and water	Mixture → white

A **balanced diet** should contain a mixture of proteins, lipids, carbohydrates, vitamins, and minerals.

1. Test for glucose

Benedicts

Blue → Orange/brown

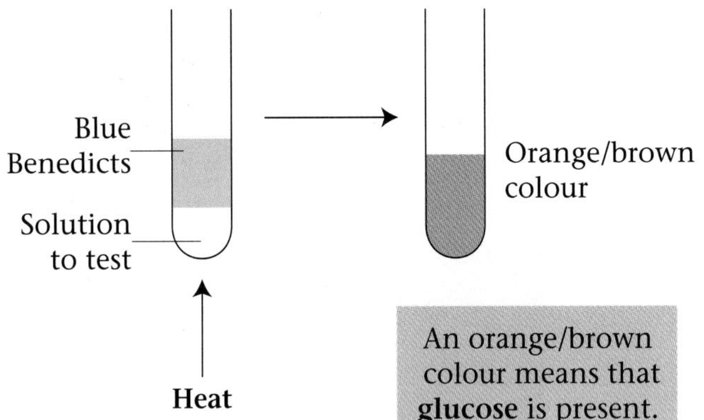

An orange/brown colour means that **glucose** is present.

2. Test for starch

Iodine

Red/brown → Blue/black

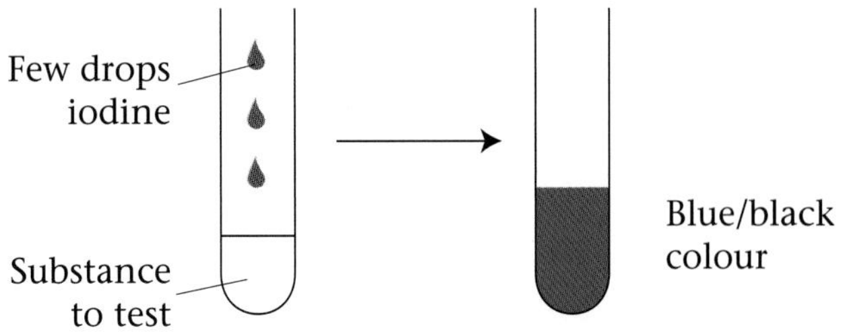

A blue/black colour means **starch** is present.

3. Test for protein

Biuret

Blue → Purple/mauve

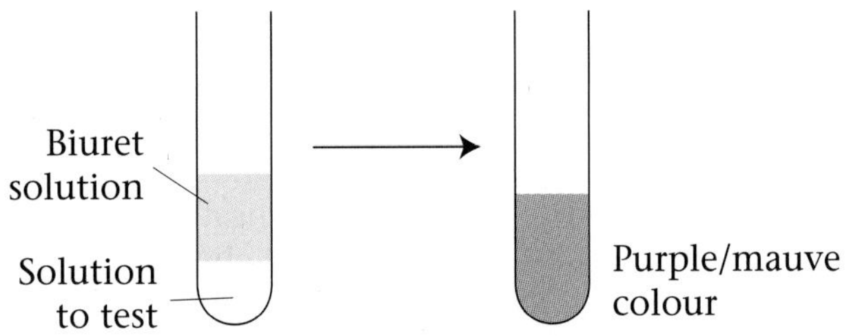

A purple colour means that **protein** is present.

4. Test for lipid (fats + oils)

Ethanol + Test substance —Water→ White

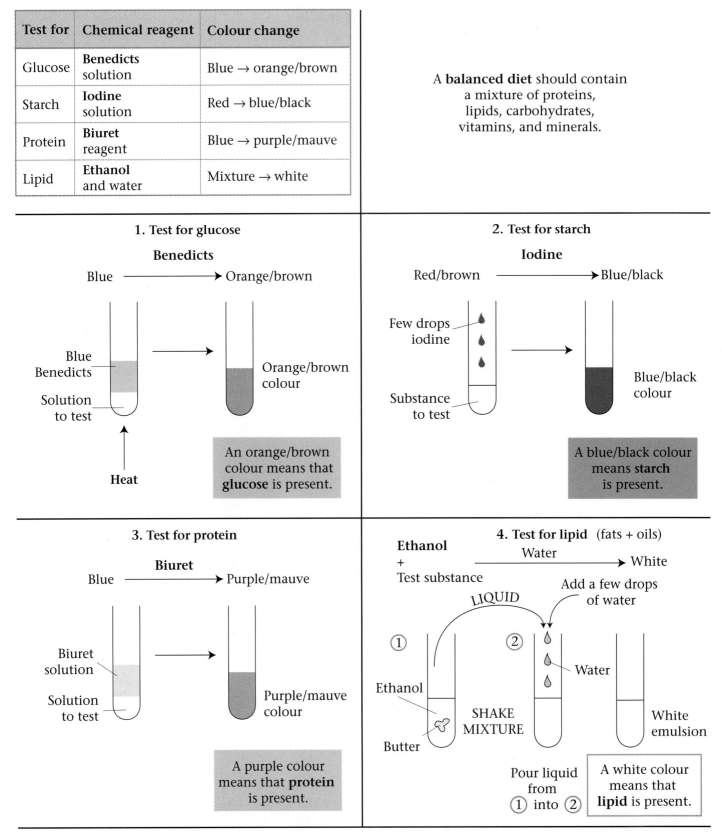

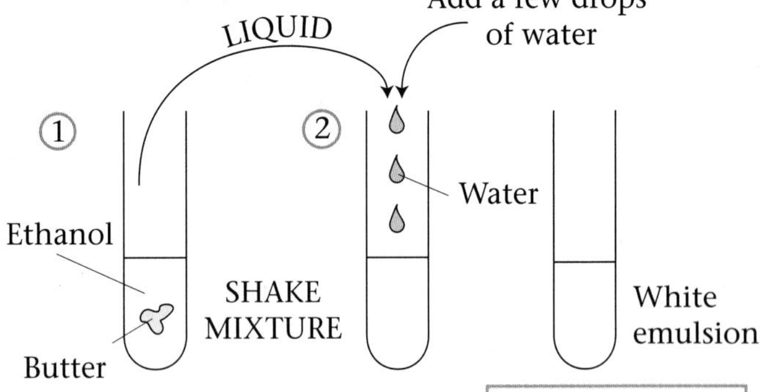

Pour liquid from (1) into (2)

A white colour means that **lipid** is present.

Food	Benedicts	Ethanol	Biuret	Iodine
Peanuts	Blue	White	Purple	Red
Bread	Blue	Clear	Blue	Blue/black
Egg	Blue	White	Purple	Red
Apple	Orange	Clear	Blue	Blue/black
Meat	Blue	White	Purple	Red

Questions:

1. Which foods (left) contain glucose?
2. What food types does egg contain, and can you explain why?
3. Protein is found in which foods?
4. Eating which foods would give you a mixture of glucose, fat, and protein?

ENZYMES

Enzymes control the **rate** of a reaction. They are **biological catalysts**, speeding up reactions in living organisms.

Enzyme action

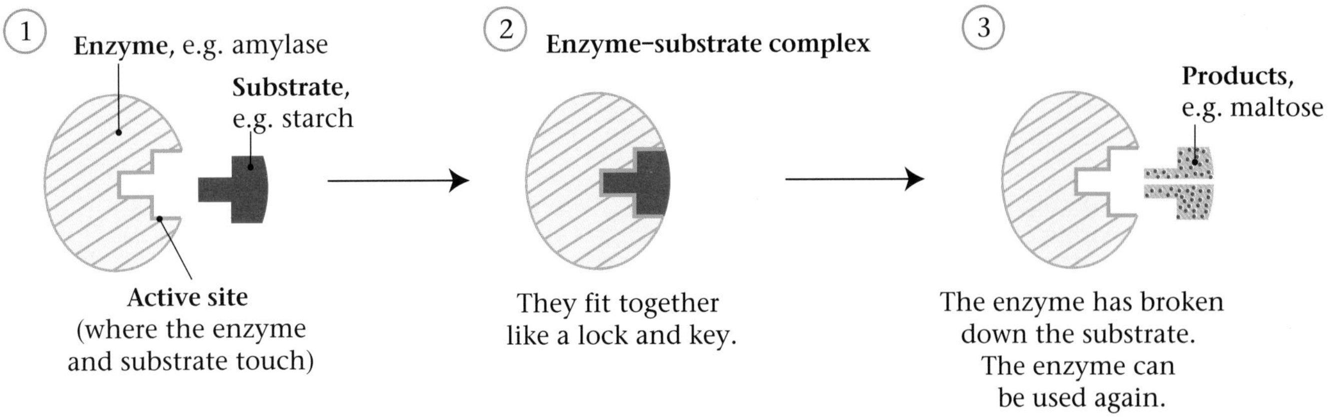

Enzymes are **specific.** They only speed up one reaction by joining with the matching substrate.

Enzymes and temperature

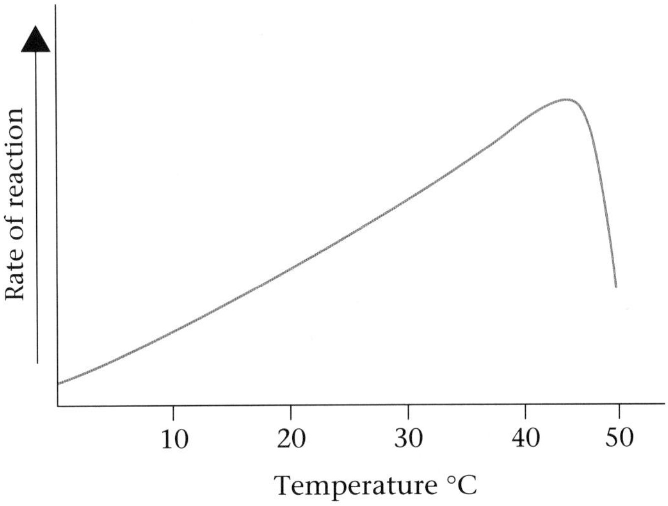

Enzymes and pH

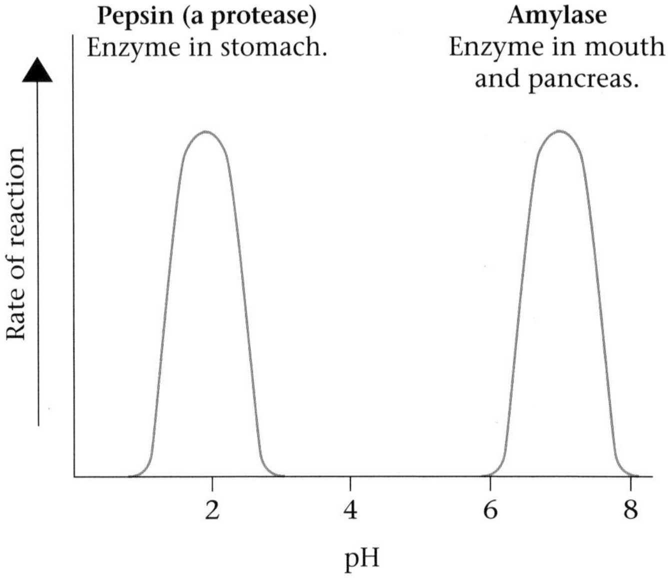

As the temperature rises, the enzyme and substrate molecules gain kinetic energy and move faster. This increases the chance of collision between them and so increase the rate of reaction.

Most enzymes work best at 40°C. Above 40°C, their shape changes and they no longer fit with their substrate. They are denatured or destroyed. Denaturing is irreversible.

Enzymes inside living cells speed up the processes of respiration, photosynthesis and protein synthesis.

Questions:

1. What are enzymes made of?
2. At what temperature do enzymes work best in animals?
3. Why do enzymes stop working at high temperatures?
4. How do enzymes affect the rate of reactions?
5. Why is the shape of enzymes important?
6. What is the name of the substance to which the enzyme attaches?

Enzymes work best at their particular pH range.

Enzymes:
- Speed up reactions – they are known as biological catalysts.
- Are specific.
- Work best at 40°C.
- Are made of protein.
- Work at a particular pH.
- Are not used up in a reaction.
- Most names end in -ase.

Enzymes speed up the digestion of food.
There are three main types of digestive enzymes:
1. **Amylases** break down starch (into maltose).
2. **Lipases** break down lipids (into fatty acids and glycerol).
3. **Proteases** break down proteins (into amino acids).

COMMERCIAL USES OF ENZYMES

Enzymes can be mass produced in factories and are used to produce:

- **Biological washing powders.**
- **Fructose** – a sweetener in the food industry.
- **Clinistix** – to detect diabetes.

Enzymes are both **specific** and **sensitive**

Specific ↓ Their particular shape only allows reaction with a matching shaped substrate, it is **specific** to that substrate.

Sensitive ↓ Enzymes can react with tiny amounts of substrate, i.e. they detect small quantities. They are **sensitive.**

---

1. Biological washing powders

Clothes are stained by proteins (blood, meat, egg-white) and fat (oil, grease, egg-yolk). A protein-digesting enzyme, a protease, and a fat-digesting enzyme, a lipase, are needed. These enzymes are present in biological washing powders to clean our clothes effectively.

---

2. Production of fructose – a sweetener

Extracting sugar from sugar cane is expensive. To produce large quantities of sugar cheaply, enzymes are used.

Cheap form of starch e.g. corn waste in fields → (Enzyme action) → Maltose → (Enzyme action) → Glucose → (Enzyme action) → Fructose (very sweet)

As fructose is so sweet, little is required and profits are high. Fructose is used in fruit drinks, cake mixes and pie fillings.

Courtesy of Holt Studios international

Corn crop.

---

3. Detection of diabetes (caused by lack of the hormone insulin.)

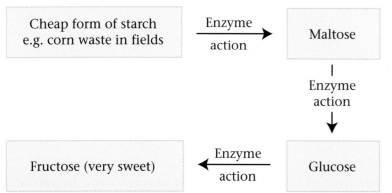

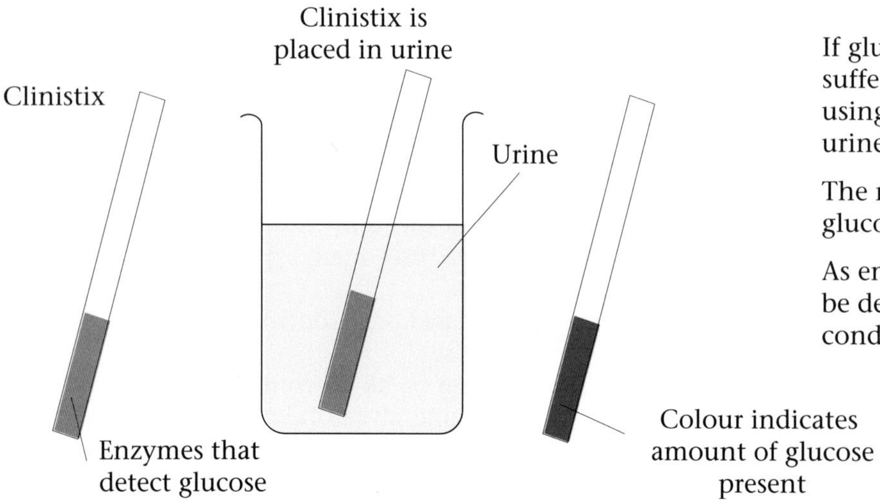

If glucose is present in the urine, a person suffers from **diabetes.** This can be detected using **Clinistix,** which is dipped into the urine.

The resulting colour indicates how much glucose, if any, is present.

As enzymes are sensitive, tiny quantities can be detected allowing early treatment of the condition.

VARIATION AND INHERITANCE

VARIATION

We belong to a species called ***Homo sapiens*** (humans). We do not look the same – there are many differences between us. These differences are caused in two ways.

1. Features caused by the environment (not inherited)

Cause	**Effect**
Sun	Sunburn
Accident	Scar
Weight lifting	Powerful muscles
Over-eating	Lots of fat
Lack of food	Poor growth

These features are caused by our way of life; they are not inherited. You cannot pass these features to your children. They are **acquired**, not inherited.

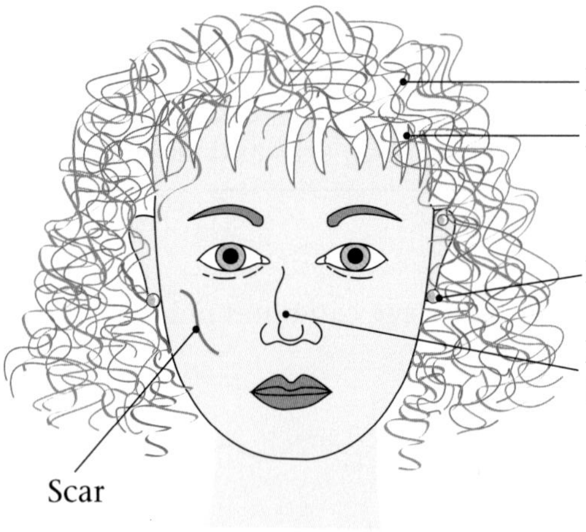

2. Features caused by genes (inherited)

Children **inherit** these features from their parents. This is why members of a **family** look **similar**.

Hair colour

Nose shape

Skin colour

Height

Freckles

Tongue rolling

Eye colour

These features are **inherited** and pass to us from our parents. We will pass these on to our children in our **genes**.

A gene is a section of DNA found on chromosomes.

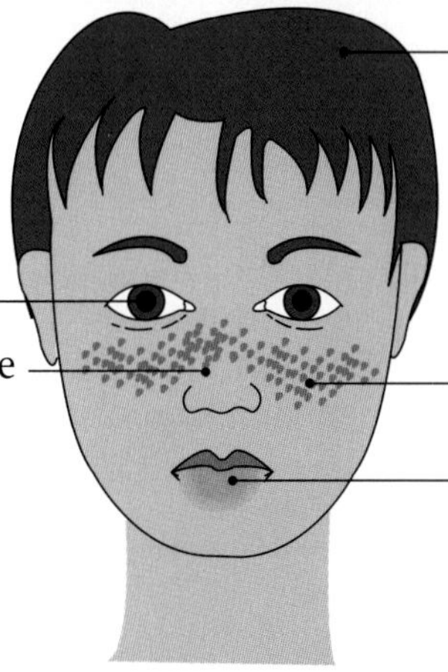

Genes or environment?
Intelligence, sporting ability and health are determined by both genetic and environmental factors.
Which is more important is debatable.

Questions:

1. Give two features that are inherited and not affected by the environment.
2. Name two features which will be affected by the environment.
3. Name one feature that may be affected by both our genes and the environment.
4. Who do we inherit our features from?
5. What is the biological name for humans?

CAUSES OF GENETIC VARIATION

1. Formation of sex cells (meiosis)

One pair of chromosomes — Sex cells are all different

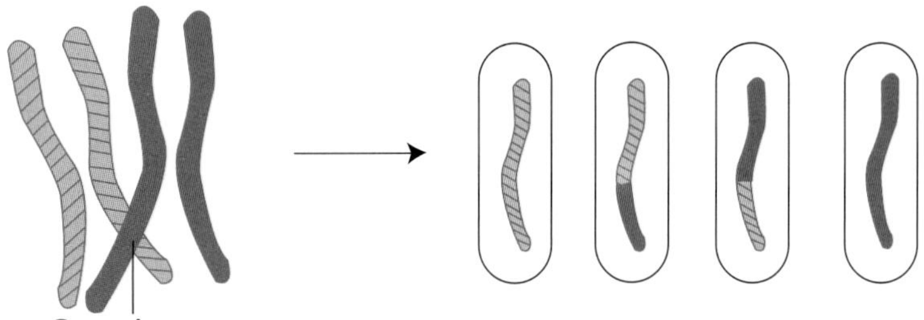

Crossing over

Crossing over of chromosomes during meiosis leads to new combinations of genes. Sex cells are all genetically different. This causes **variation** between sex cells and **variety** in future offspring.

The main causes of variation

- **Meiosis – formation of gametes**, crossing over and independent assortment lead to variation.
- **Fertilisation.**
- **Mutations** lead to new features not present before.
- Meiosis and fertilisation occur during sexual reproduction. Therefore **sexual reproduction** causes variation.

---

2. Fertilisation

Joining of sperm and egg combines unique features.

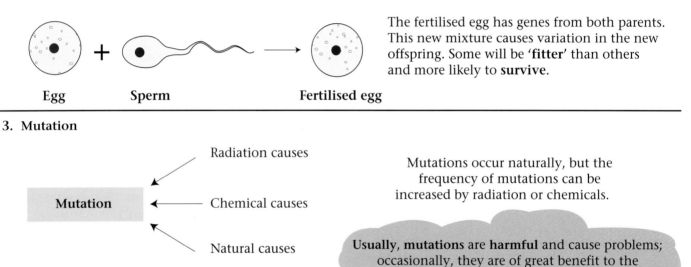

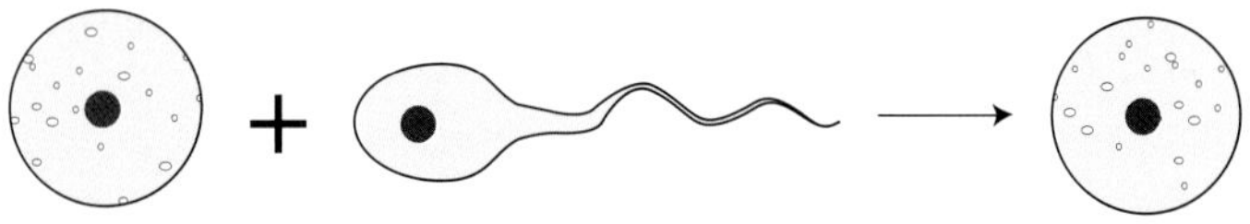

Egg + **Sperm** → **Fertilised egg**

The fertilised egg has genes from both parents. This new mixture causes variation in the new offspring. Some will be **'fitter'** than others and more likely to **survive**.

---

3. Mutation

Mutation ← Radiation causes

Mutation ← Chemical causes

Mutation ← Natural causes

A **mutation** is a sudden change in a gene or chromosome.

Mutations occur naturally, but the frequency of mutations can be increased by radiation or chemicals.

Usually, mutations are **harmful** and cause problems; occasionally, they are of great benefit to the organism.

Mutations cause changes to the DNA making up a gene, so altering the gene. Mutations can also change the number of chromosomes in a cell – both lead to genetic variation.

Mutations can cause the following harmful conditions:	
Sickle cell anaemia	Once the mutation has occurred, the change is passed on to future children. It is inherited.
Cystic fibrosis	
Haemophilia	
Huntington's disease	

The gene causing **haemophilia** appeared by mutation in Queen Victoria, affecting most of the royal families of Europe (see page 28).

Useful mutations help the organism. Mutations have caused some **bacteria** to be **resistant** to **antibiotics**, so increasing their chances of survival and reducing ours.

Mutations always cause variation or **change**.

Questions:

1. What are the two main causes of variation?
2. Which type of variation is passed on to our children?
3. How does the formation of gametes lead to variation?
4. How does the process of fertilisation lead to more variation?
5. What is a mutation?
6. What factors cause mutations?
7. Can mutations be helpful to organisms? Give one example.

DNA (DEOXYRIBONUCLEIC ACID)

DNA is found in the nucleus, making up genes. As genes determine all our features and cell activity, DNA is essential for life.

DNA, or genes, determine what proteins are made by cells. This includes enzymes which control all cell activity. As DNA can copy itself, genes can be inherited.

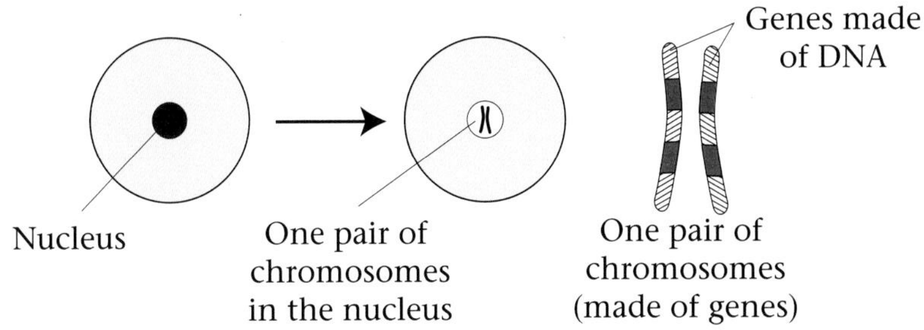

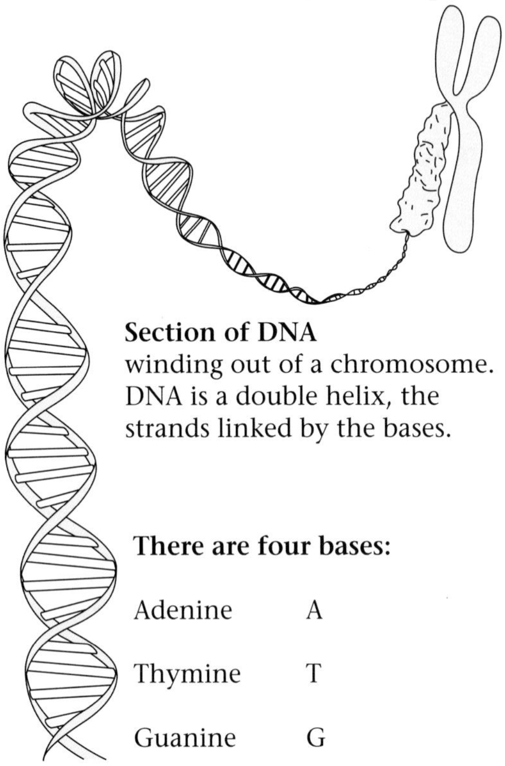

Section of DNA winding out of a chromosome. DNA is a double helix, the strands linked by the bases.

Section of DNA in more detail

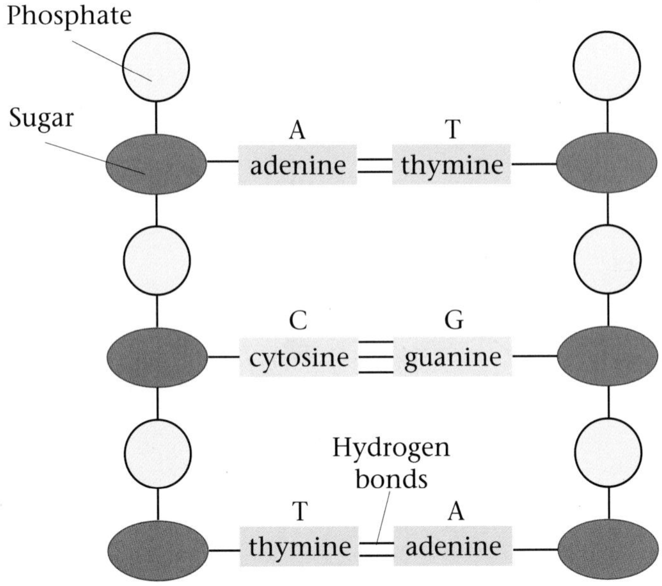

There are four bases:

Adenine	A
Thymine	T
Guanine	G
Cytosine	C

A always joins with T
G always joins with C

DNA can copy itself or replicate

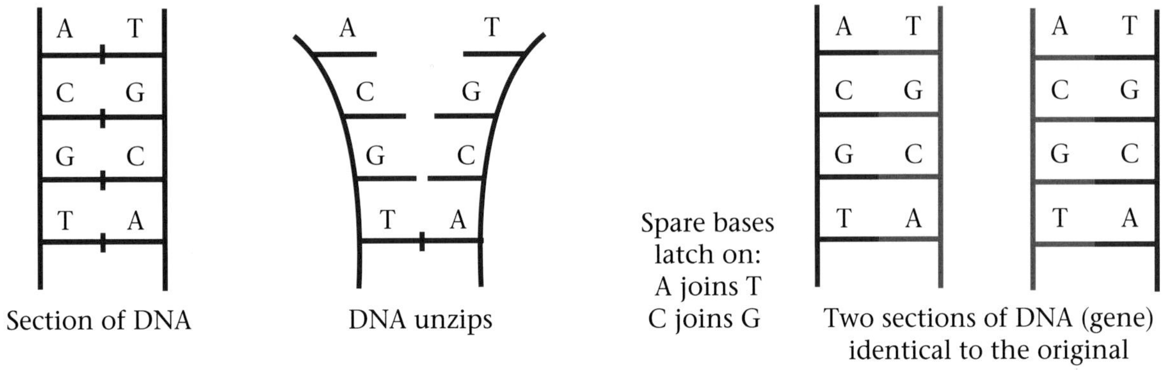

Daughter cells produced by mitosis can therefore have identical DNA to the parent. Replication of DNA also enables genes to pass on to the sex cells, in other words to be inherited.

DNA can be extracted from cells such as onion and kiwi fruit.

Find out about the Human Genome Project.

Questions:
1. Where is DNA found?
2. What is made of DNA?
3. Why is DNA essential for life?
4. DNA controls the formation of proteins. Why is this so important?
5. What four bases make up DNA?
6. Which bases join?
7. Why is the replication of DNA so important?
8. Describe how DNA replicates.

CELLS AND CHROMOSOMES

1. Cells

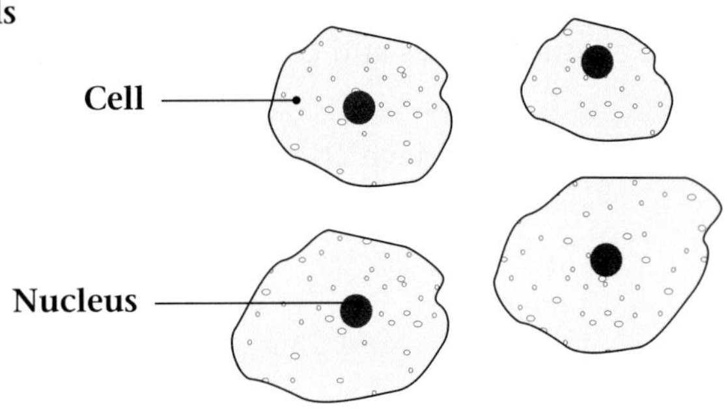

Most cells contain a **nucleus** (not red blood cells).

2. Enlarged nucleus

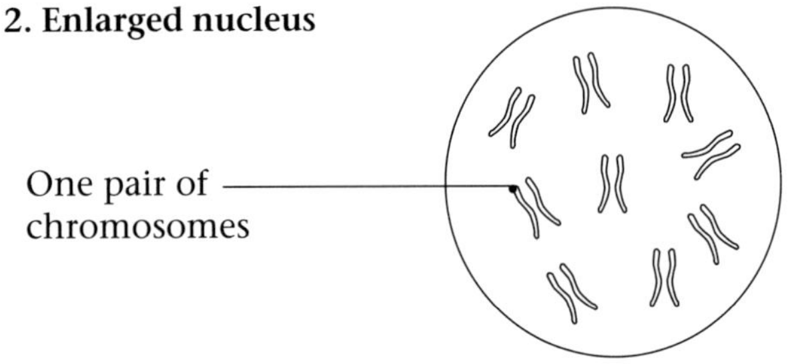

In the nucleus are pairs of chromosomes.

Human body cells have **46 chromosomes** (in 23 pairs).

3. Sex cells

Human egg (ovum) + human sperm → **fertilised egg**

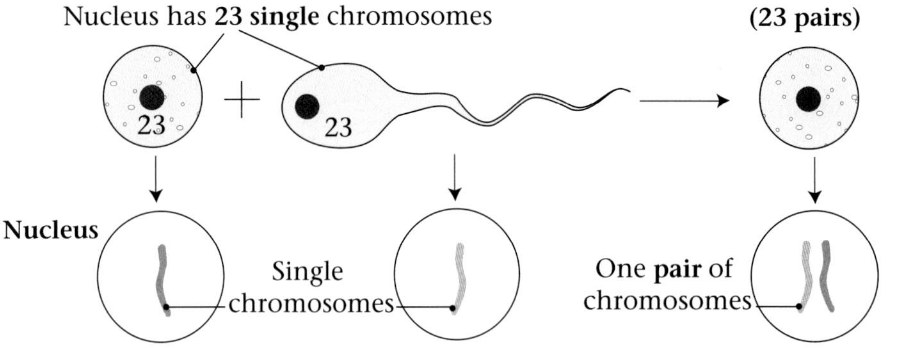

Pairs of chromosomes result from the joining of an egg with a sperm.

Chromosomes total 23 23 46 (23 pairs)

Human sex cells have **23 chromosomes** (no pairs).

4. Nucleus from fertilised egg

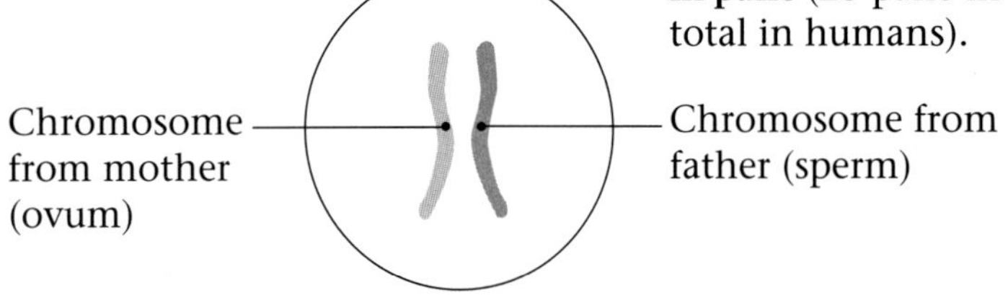

All chromosomes are in pairs (23 pairs in total in humans).

One pair of chromosomes

A **chromosome** is a strand of **genes** which determine our features.

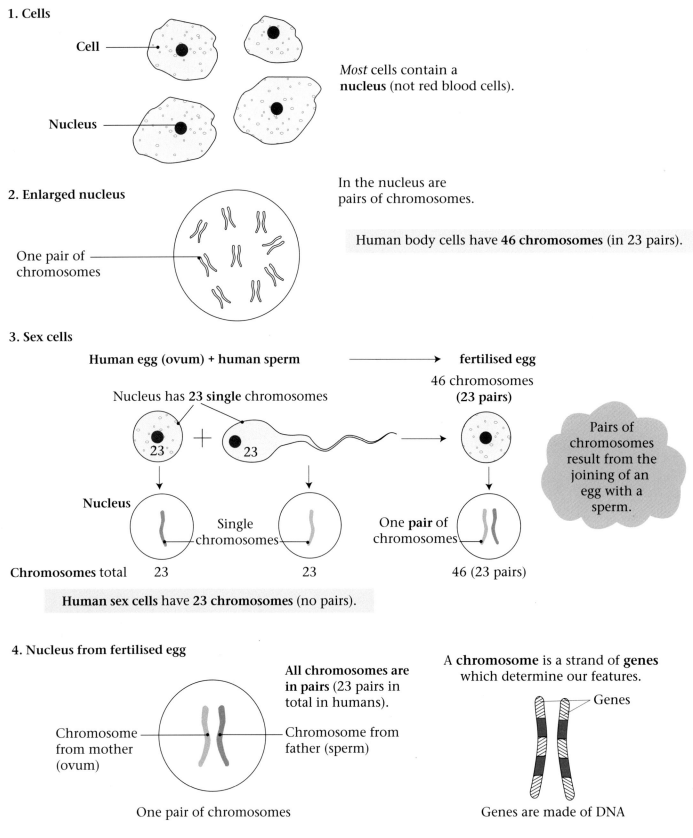

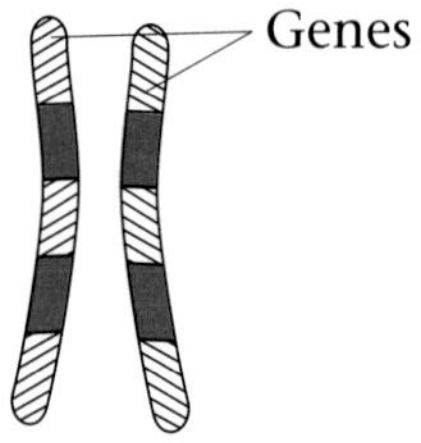

Genes are made of DNA

Questions:
1. How many chromosomes are there in human body cells? How are they arranged?
2. Why do the sex cells have half the usual number of chromosomes?
3. What happens to the chromosome number at fertilisation?

DOMINANT AND RECESSIVE FEATURES

Nucleus from fertilised egg

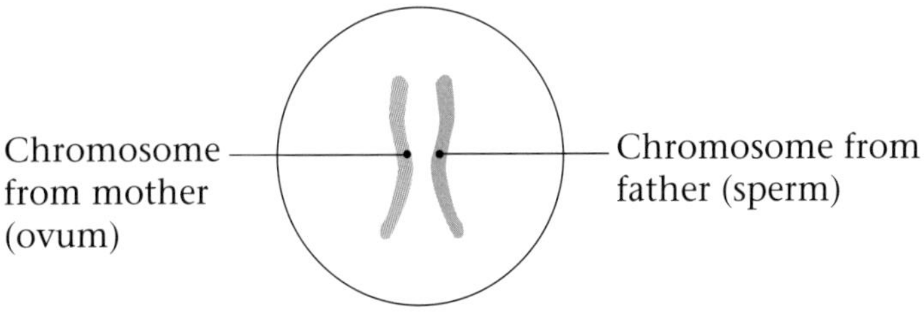

One pair of chromosomes

Every cell, apart from the sex cells, has a full set of genes. Only some of these genes are used in any one cell.

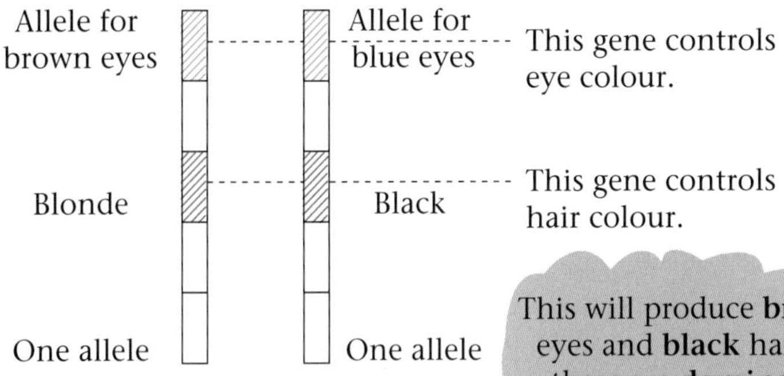

This will produce **brown** eyes and **black** hair as these are **dominant** features.

Alleles are different forms of a **gene**. Eye colour depends on what alleles are present, e.g. brown, blue, green, grey. Normally two alleles (one on each of a pair of chromosomes) determine our features.

There are two different **alleles** here controlling eye colour, brown and blue. This person will have brown eyes, as brown is **dominant** (stronger) to blue. We call blue **recessive**, or weaker. Two alleles determine our features, one from each parent. The stronger, dominant, allele will show.

B represents black hair
b represents blonde hair

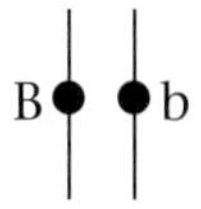

This person will have black hair, it is dominant.

Dominant alleles are shown by a capital letter, e.g. T.
Recessive alleles are shown by a lower case letter, e.g. t.

Dominant features
Black hair
Curly hair
Brown eyes
No freckles

Recessive features
Blonde hair
Straight hair
Blue eyes
Freckles

Courtesy of Dick Williams

Courtesy of D. Colgan

Dominant and recessive alleles have an equal chance of being inherited.

Questions:

1. What does dominant mean?
2. What is an allele?
3. What alleles would produce blue eyes?
4. Why are chromosomes found in pairs?
5. How many alleles usually determine a feature?
6. What does recessive mean?

GENETIC CROSSES (I)

1. Black hair × blonde hair

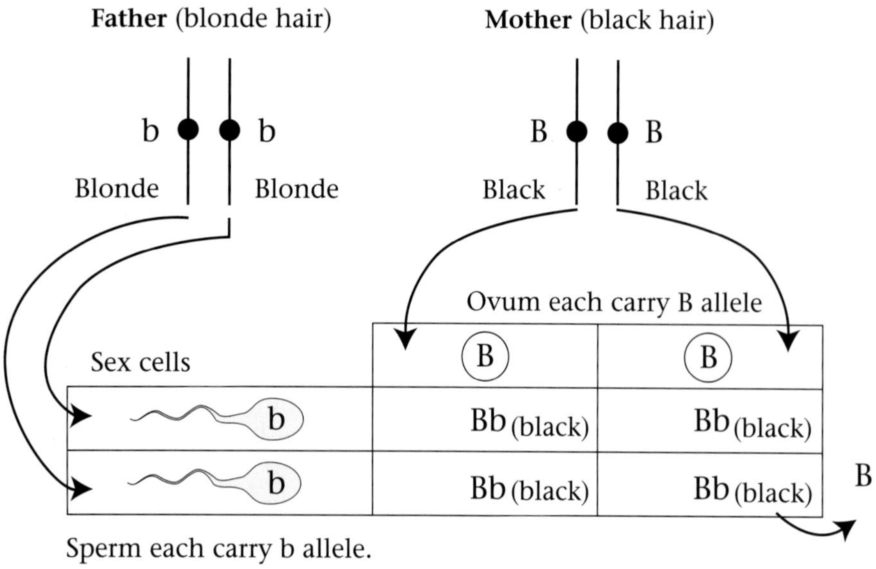

The **sex cells** carry only **one** chromosome of each pair, so only **one** allele.

This child will have **black** hair as B is dominant to b. All the children will have black hair as each is carrying the B allele they inherited from their mother.

2. Black hair × black hair

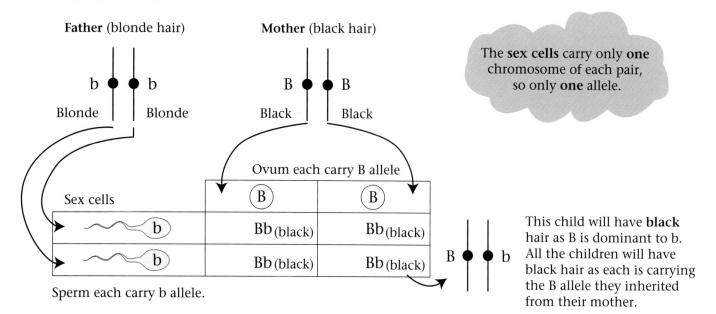

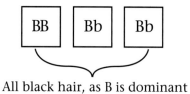

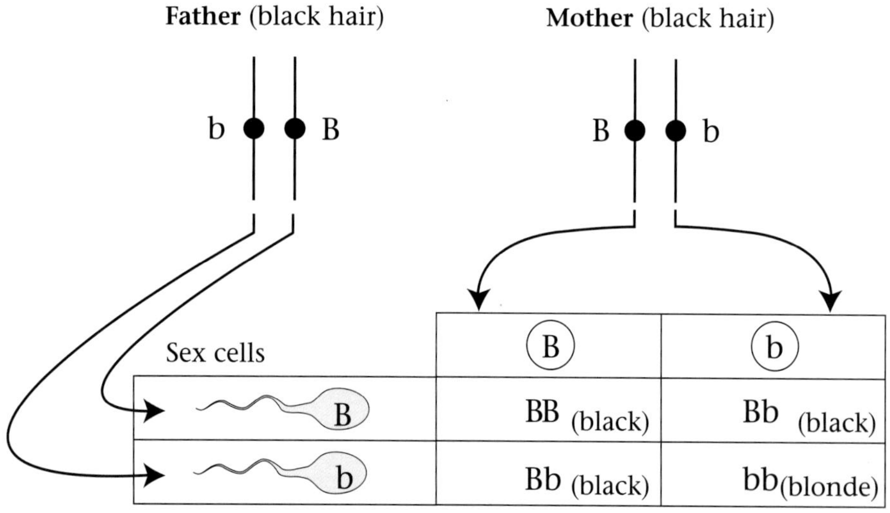

When the sex cells join in fertilisation, pairs of chromosomes are formed and pairs of alleles. The **dominant** allele determines the colour.

BB Bb Bb — All black hair, as B is dominant

bb — Blonde hair

This gives a three black to one blonde ratio or **3:1**, i.e. there is a 1 in 4 chance of these parents having a blonde child

Genetic terms

Chromosome A strand of genes.
Gene A section of DNA controlling a feature.
Allele A different form of a gene.
Dominant Stronger allele (capital letter).
Recessive Weaker allele (small letter).
Genotype The type of alleles present.
Phenotype The appearance of the organism.
Homozygote Two alleles are the same.
Heterozygote Two alleles of a pair are different.

Questions:

1. What type of letters are used for dominant features?
2. What colour hair would the following genotypes produce (B = black hair, b = blonde hair): BB, Bb, bb?
3. What colour hair would the children of two blonde parents have and why?

GENETIC CROSSES (II)

Tongue rolling is caused by a **dominant allele** R. The recessive allele r = non-tongue roller. 84% of the population can tongue roll.

There are three genotypes possible:

RR	Rr	rr
Tongue roll	Tongue roll	Cannot tongue roll
84%		16%

Problem
Two parents, both of whom can tongue roll have a child who cannot. What is the genotype of the parents?

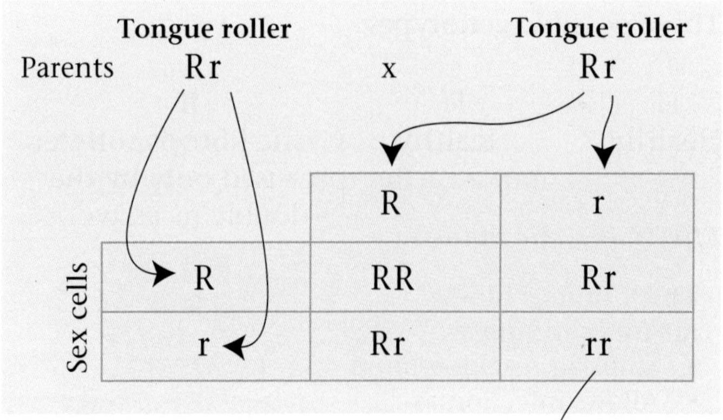

RR Rr Rr (3)

Children who can tongue roll all have a **dominant** allele.

3 out of 4
This is a 3:1 ratio

rr (1)

1 in 4 children cannot tongue roll.

Children who cannot tongue roll inherited a **recessive allele from both parents.**

Gregor Mendel was a monk working in the garden of the monastery in Brno, now in the Czech Republic. He studied the inheritance of features in pea plants. From his studies the basic laws of inheritance were discovered in 1865. However, his ideas were not appreciated until the early 1900's as there was little understanding of genes and chromosomes at that time.

Pea plants
Height is controlled by a pair of alleles: T (Tall) and t (short). Tall pea plants are TT or Tt. Short pea plants are tt.

Two alleles are the **same** (homozygote).

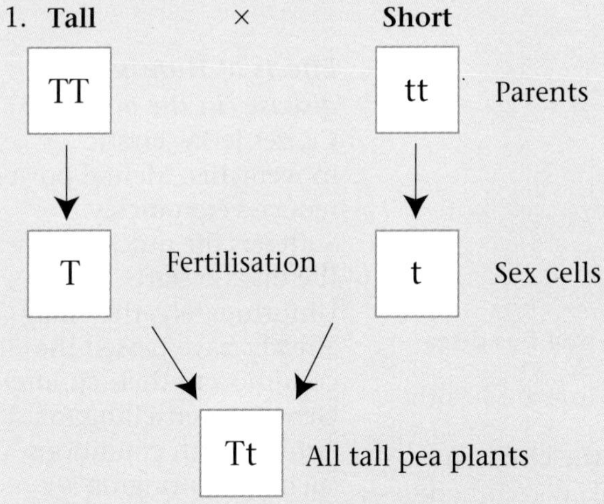

Two alleles are **different** (heterozygote).

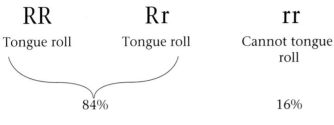

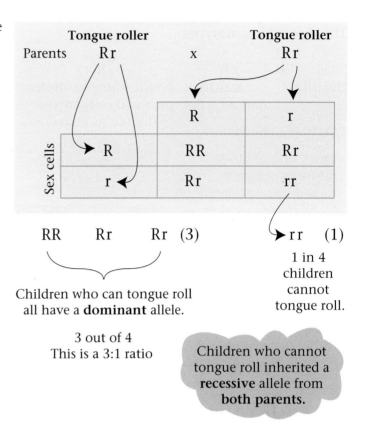

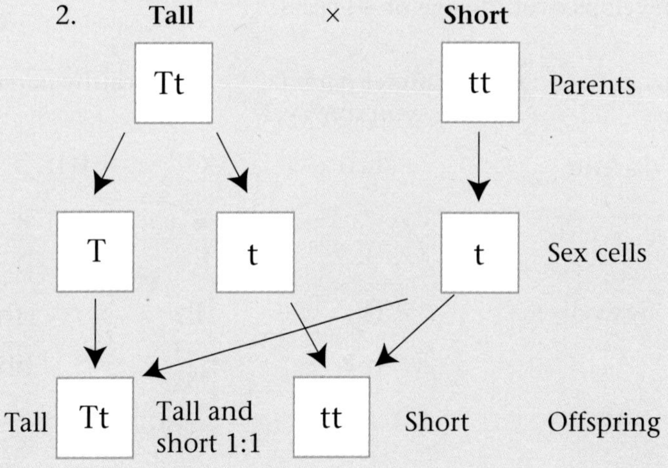

The pea plants were either tall or short, as found in the original parents. No pea plants of intermediate height were produced and therefore no 'blending' of alleles occurred.

Questions:
1. Freckles are caused by a recessive allele. Would the following people have freckles: FF, Ff, ff?
2. What is a homozygote? Give an example.
3. How is it possible to find out if a pea plant is a homozygote or heterozygote?
4. If both parent pea plants are tall but have a short offspring, what must be the genotype of the parents?

CYSTIC FIBROSIS A *recessive* inherited disorder.

Three possible genotypes:

FF	Ff	ff
Healthy	Healthy but a carrier	Cystic fibrosis sufferer (caused only by the double recessive)

Effects of cystic fibrosis

Thick mucus blocks air tubes in lungs:
- Stops air reaching lungs.
- Allows infection as bacteria breed on mucus.
- Causes coughing.

Thick mucus blocks pancreatic duct:
- Stops digestive enzymes leaving the pancreas.
- Food in duodenum is not digested properly.
- Little food is absorbed.

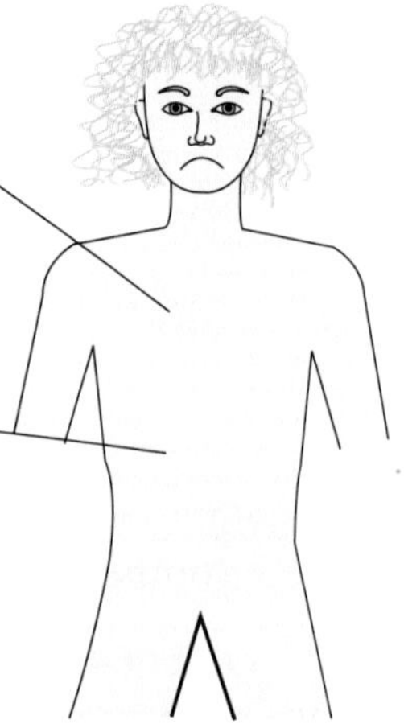

Gene therapy is being used in an attempt to reduce the symptoms of cystic fibrosis. However, targeting the specific cells is difficult and side-effects may be a problem.

Two healthy parents, if carriers, have a 1 in 4 or 25% chance of having a child with cystic fibrosis.

Parents: Ff x Ff

Sex cells	F	f	
F	FF	Ff	3 in 4 are healthy
f	Ff	ff	1 in 4 has disease

Children: FF, Ff, Ff, — Three are healthy; ff — One has cystic fibrosis

1 in 20 people in Europe are carriers and perfectly healthy.
1 in 2,000 babies is born with cystic fibrosis which may lead to an early death.

HUNTINGTON'S DISEASE A *dominant* inherited disorder.

Three possible genotypes:

HH	**Hh**	**hh**
Very rare, probably lethal	Huntington's disease (lethal later)	Healthy

As it is caused by a *dominant* gene, only one dominant allele is needed to cause the condition. The condition develops over the age of 40 years.

Parents: **Sufferer parent** (no symptoms yet) Hh x **Healthy parent** hh

Sex cells	h	h	
H	Hh	Hh	2 in 4 has disease
h	hh	hh	2 in 4 are healthy

Children: Hh, Hh — Two will develop Huntington's disease; hh, hh — Two will be healthy

Half the children are likely to inherit the harmful 'H' gene

Effects of Huntington's disease (in the over-40s)
Causes jerky, erratic movements. Mental powers reduce very quickly. Sufferers die quickly once the disease starts. Unfortunately, they may already have passed the condition to their children. Genetic counselling for families with conditions such as Huntington's disease may be advisable to explain the risks of having affected children.

Polydactyly (having more than 5 fingers) is another dominant, inherited disorder transmitted in the same way as Huntington's disease.

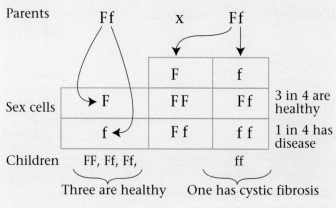

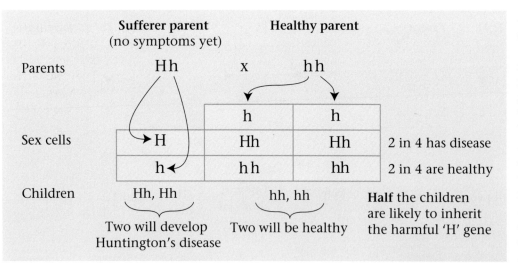

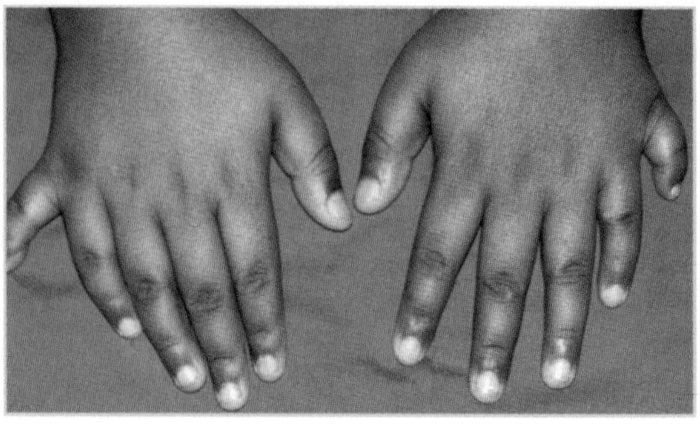

courtesy of Wellcome Photo Library

Polydactyly.

SICKLE CELL ANAEMIA A *recessive* inherited disorder.

Genotype	NN (healthy)	Nn (carrier)	nn (sickle cell anaemia)
Red blood cells	All normal	Most normal few sickled	Many sickled
Oxygen carried	Lots	Lots	Little
Energy available	Lots	Lots	Little
Resistance to malaria	Not resistant	Resistant	Resistant
Anaemia	None	Mild	Severe (can be fatal)

N = normal red blood cell

n = sickled red blood cell

Sickle cell anaemia and malaria

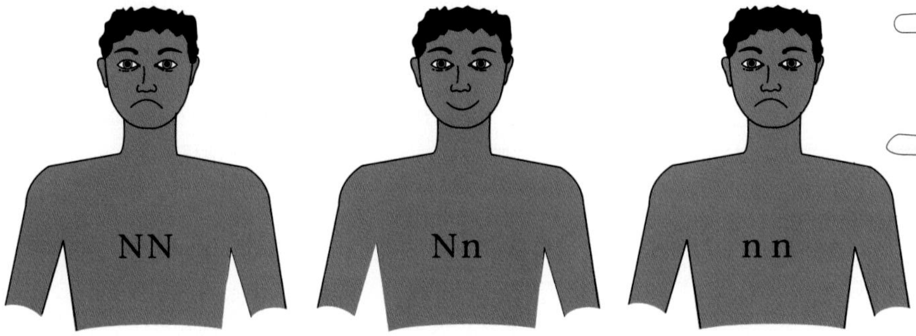

Killed by malaria (healthy in non-malarial regions)

Resistant to malaria (mild anaemia)

Killed by severe anaemia

Blood capillary

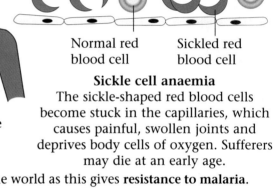

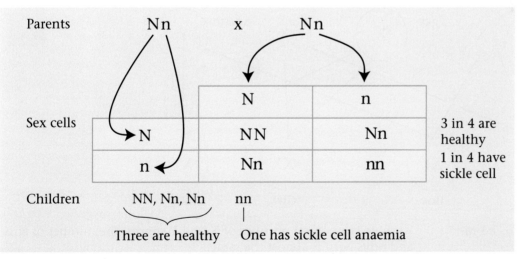

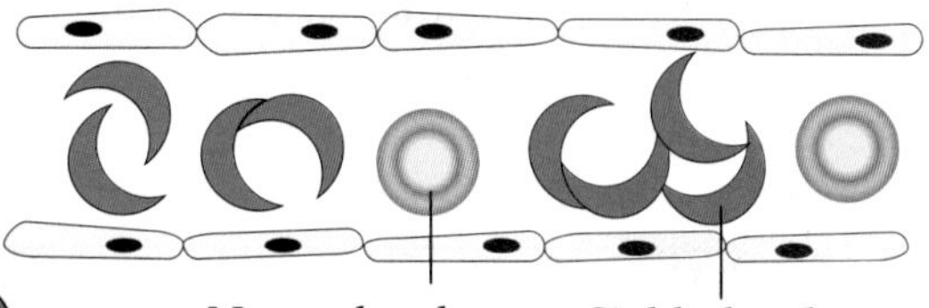

Sickle cell anaemia
The sickle-shaped red blood cells become stuck in the capillaries, which causes painful, swollen joints and deprives body cells of oxygen. Sufferers may die at an early age.

People with the genotype **Nn** are common in malarial regions of the world as this gives **resistance to malaria**. Unfortunately, this increases the number that suffer from sickle cell anaemia, nn.

Parents: Nn x Nn

Sex cells:

	N	n
N	NN	Nn
n	Nn	nn

3 in 4 are healthy

1 in 4 have sickle cell

Children: NN, Nn, Nn — Three are healthy; nn — One has sickle cell anaemia

If both the mother and father are **carriers**, there is a 1 in 4 chance of their having a child with **sickle cell anaemia**.

Questions:

1. What shape can red blood cells be?
2. Draw the shape of red blood cells found in,
   a) healthy people?
   b) people suffering from sickle cell anaemia?
3. How do sickle-shaped red blood cells affect individuals?
4. Describe the symptoms of sickle cell anaemia.
5. What is meant by a carrier for sickle cell anaemia?
6. Why is it an advantage for people to be a carrier for sickle cell anaemia in certain parts of the world?
7. If both parents are carriers, what % of their children may suffer from sickle cell anaemia? Explain this in a genetic diagram.
8. Why do you think there are so many people with sickle cell anaemia in Uganda, Tanzania and Kenya?

SEX CHROMOSOMES

Humans have 23 pairs of chromosomes, of which **one pair** determines whether we are **male** or **female**.

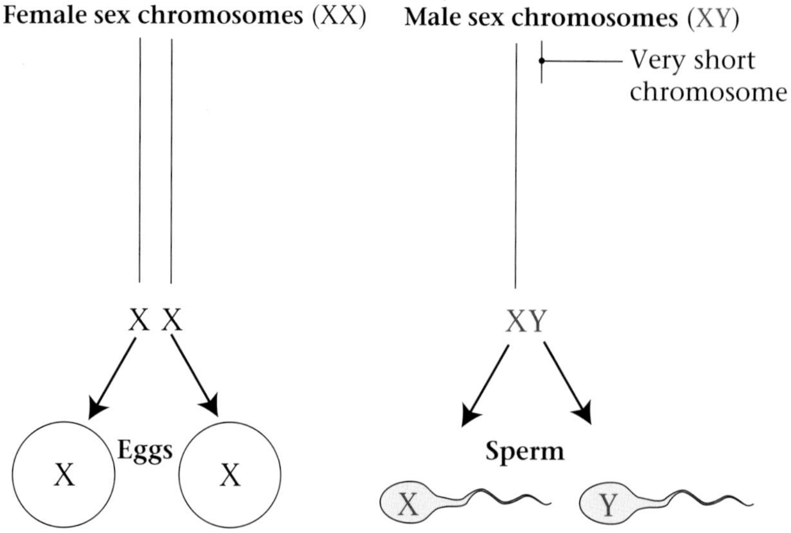

All eggs carry an X chromosome.

Some sperm carry an X chromosome, others carry a Y chromosome.

The type of sperm determines whether a boy or girl is formed.

Sperm

		X sperm	Y sperm
Eggs	X	XX (girl)	XY (boy)
	X	XX (girl)	XY (boy)

If an **X sperm** joins with an egg, it is a **girl**.

If a **Y sperm** joins with an egg, it is a **boy**.

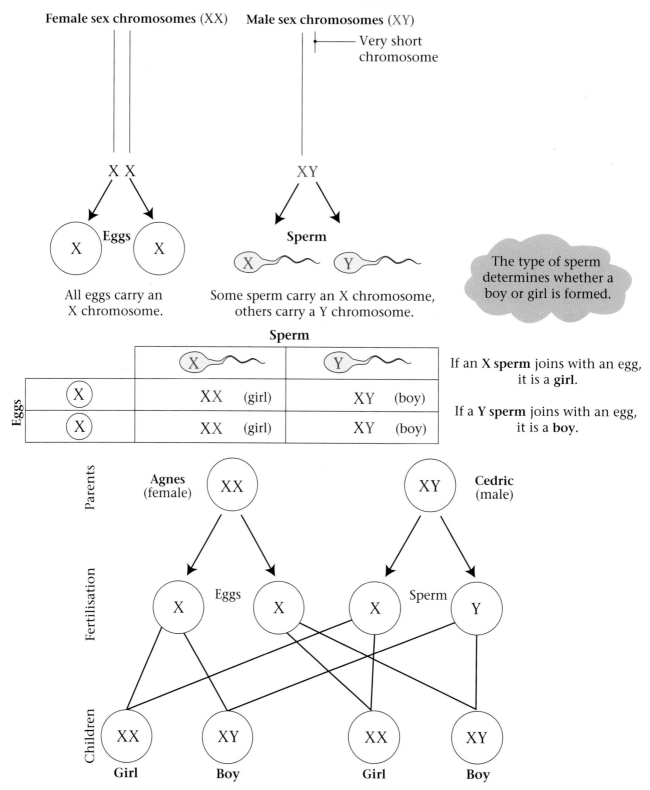

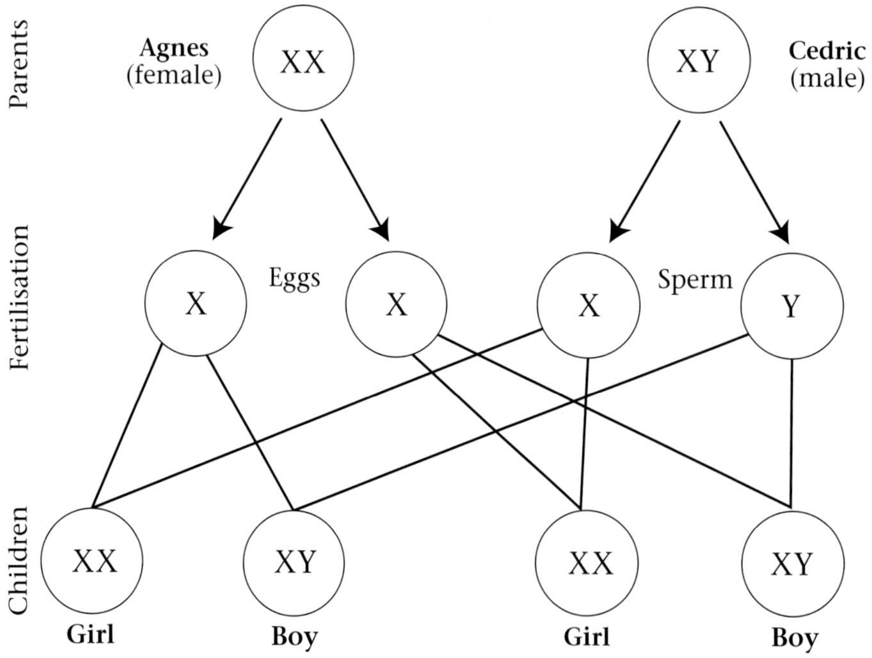

Two girls and two boys (equal), 1:1 ratio.

As there is an **equal number of X** and **Y sperm,** the number of girls and **boys** born is almost the same.

Each baby has a 50% chance of being male and 50% chance of being female.

Questions:

1. Which two chromosomes produce a female?
2. What are the two types of sperm?
3. What sex chromosome is always found in the egg?
4. What are the chances of having a baby boy, compared to a baby girl? Explain.
5. If a family already has five girls, what are the chances that the next baby will be a boy?
6. What determines the sex of the child?

SEX-LINKED CHARACTERISTICS

These are caused by **genes** usually on the **X sex chromosome**, e.g. haemophilia, colour-blindness and baldness. They are recessive genetic disorders.

Sex chromosomes

H = healthy h = haemophilia (a blood disorder)	E = normal vision e = colour blind	B = not bald b = bald

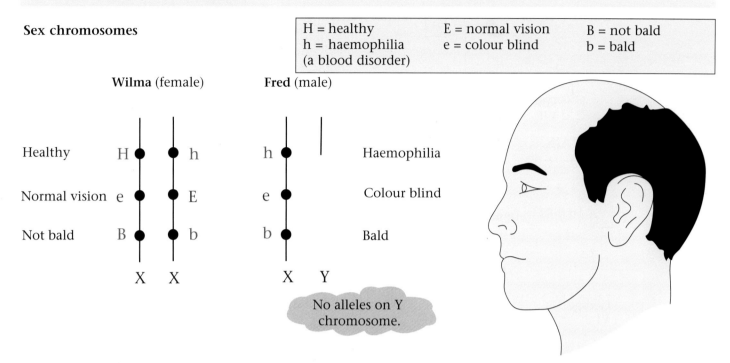

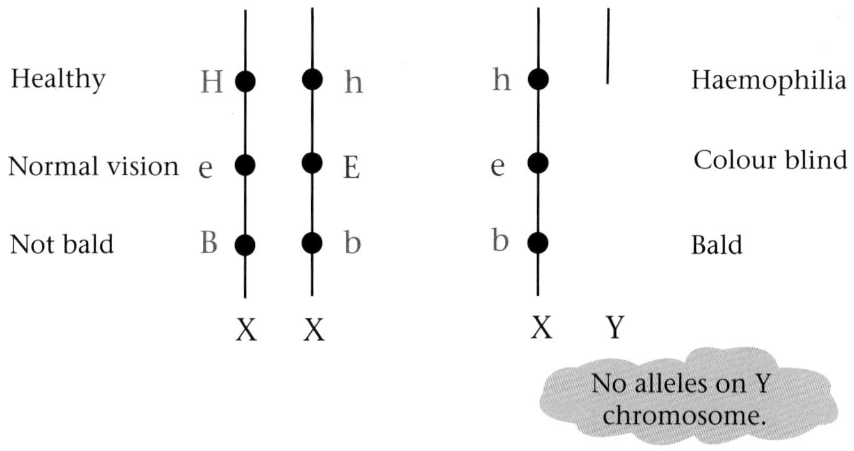

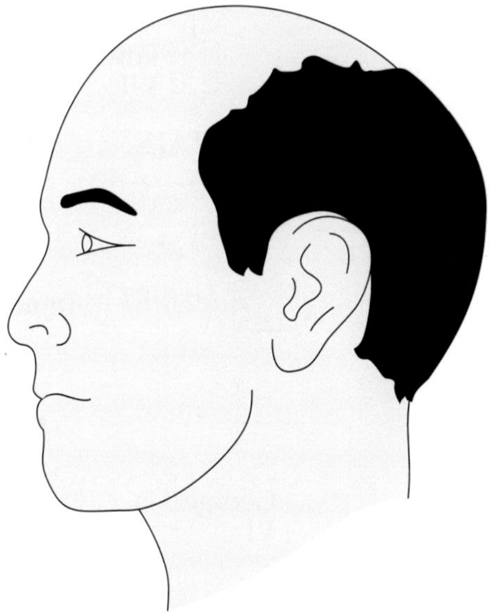

Wilma is healthy, has normal vision and will not go bald as she has dominant alleles present.

Fred has haemophilia, is colour blind and will go bald as only recessive alleles are present on the X chromosome.

Inheritance of haemophilia from Queen Victoria

Queen Victoria
Penny Black Stamp

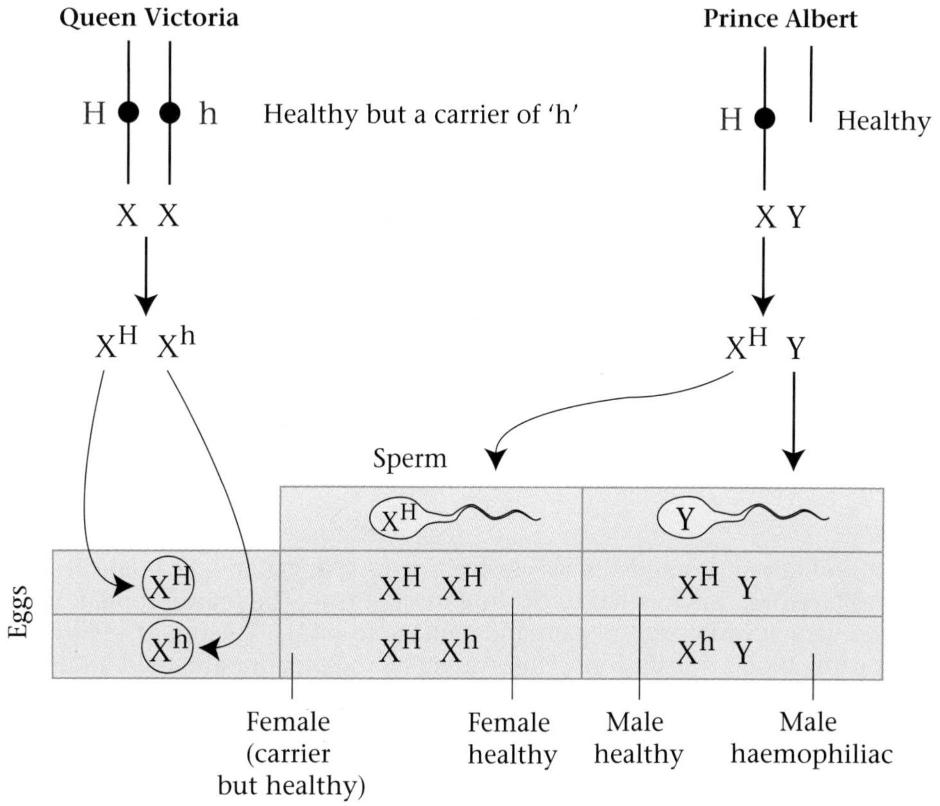

Many descendants of **Queen Victoria** suffered from **haemophilia** and died young.
The present **British Royal Family** do **not** carry the harmful recessive gene for haemophilia.

Men are more likely to suffer from sex-linked conditions as they have no alleles on the Y chromosome to prevent recessive alleles on the X being expressed.

HAEMOPHILIA

Queen Victoria and the inheritance of haemophilia in the royal families of Europe.

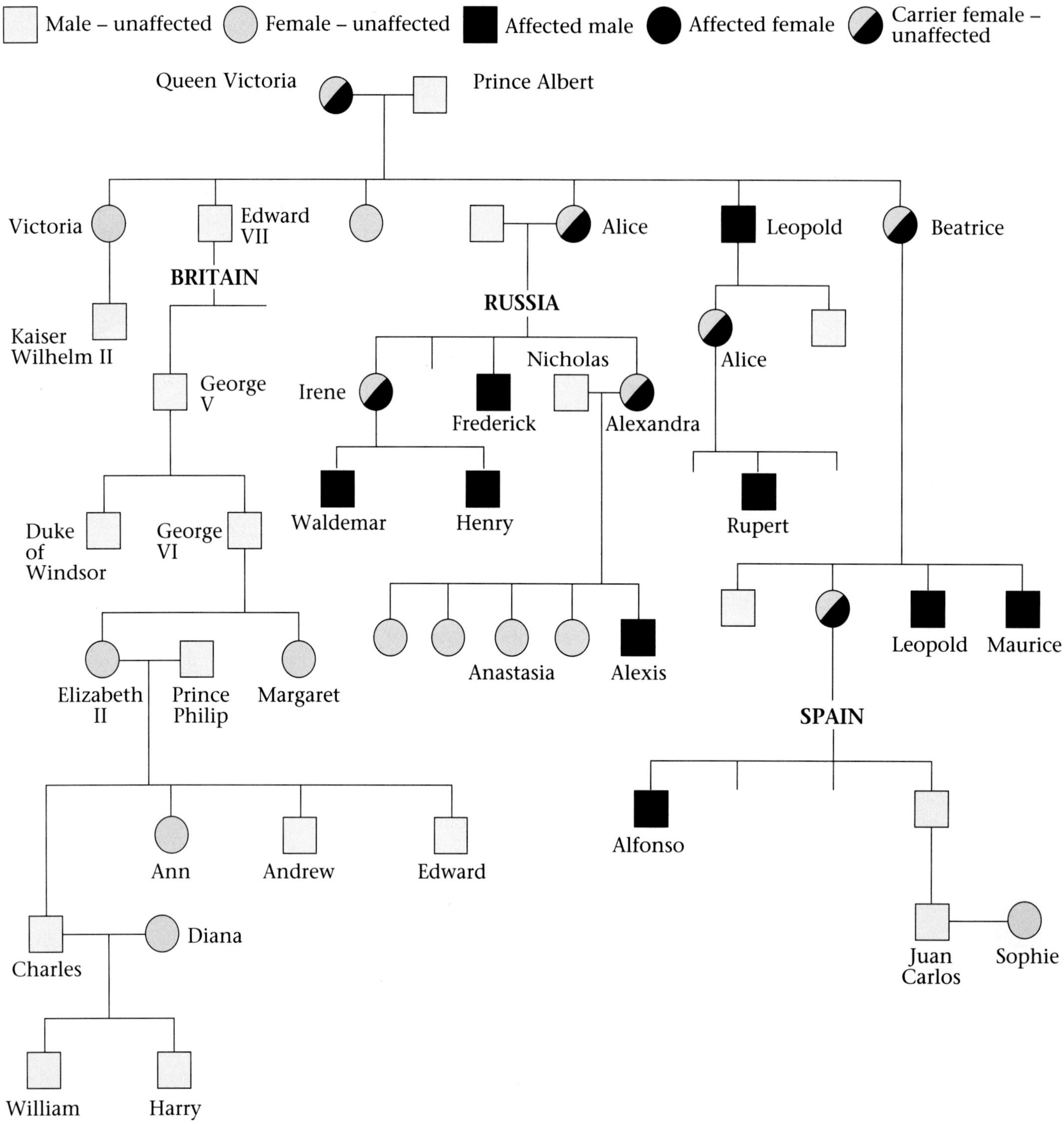

Haemophilia is an example of a recessive, sex-linked, genetic disorder carried on the X sex chromosome. It leads to bleeding due to a faulty clotting mechanism. The main problem is internal bleeding into joints and muscles, which causes severe pain and damage to joints. Treatment involves replacing the missing clotting factor using blood transfusions. Unfortunately, some contaminated blood has been used and this has led to widespread infection of HIV (leading to AIDS), and Hepatitis C amongst some sufferers of haemophilia. Haemophilia is normally carried by females and passed to their sons.

Questions:

1. Why is the British royal family free of haemophilia?
2. How many people suffered from haemophilia in this family tree in total?
   a) How many were males?
   b) How many were females?
3. Can you explain the difference in the number of males and females affected?
4. Why is it not possible for males to be carriers of haemophilia?

CODOMINANCE When two alleles of a pair are equally dominant.

Parents **Red flowers** × **White flowers**
R R × W W

RR = Red
WW = White
RW = Pink

Sex cells	W	W
R	RW (pink)	RW (pink)
R	RW (pink)	RW (pink)

Red × white
↓
Pink

Offspring

All are RW, and have **pink** flowers.

Both red and white
are dominant.

If red alleles were **dominant**, flowers would be **red**.

If white alleles were **dominant**, flowers would be **white**.

A **third colour** indicates **both** are dominant **= codominance.**

Snapdragons

Parents **Pink flowers** × **Pink flowers**
R W × R W

Sex cells	R	W
R	RR (red)	RW (pink)
W	WR (pink)	WW (white)

Pink × pink
↓
Red, pink, white
1:2:1

Offspring

RR	WR, RW	WW
One red	Two pink	One white

1:2:1 ratio

Blood groups

There are four blood groups:
A, B, AB, and O.

- A and B are dominant.
- O is recessive.

(Note that there are three types of alleles: A, B, and O.)

Blood group	Genotype	% in UK
A	AA or AO	40
B	BB or BO	12
AB	AB	3
O	OO	45

Group AB is an example of **codominance**.

Coat colour in cats is sex-linked and involves codominance. It is carried by alleles on the 'X' sex chromosome.

As female cats are XX, they can have 2 alleles controlling coat colour, whereas males only have one X chromosome, so only one allele controlling coat colour.

Only females can have both a 'B' and 'Y' allele resulting in **tortoiseshell** colour.

A tortoiseshell cat

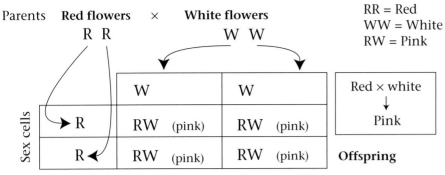

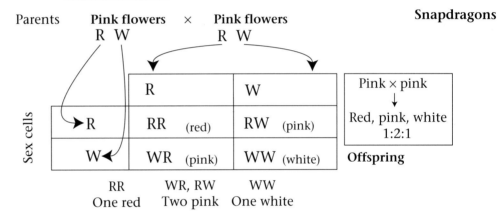

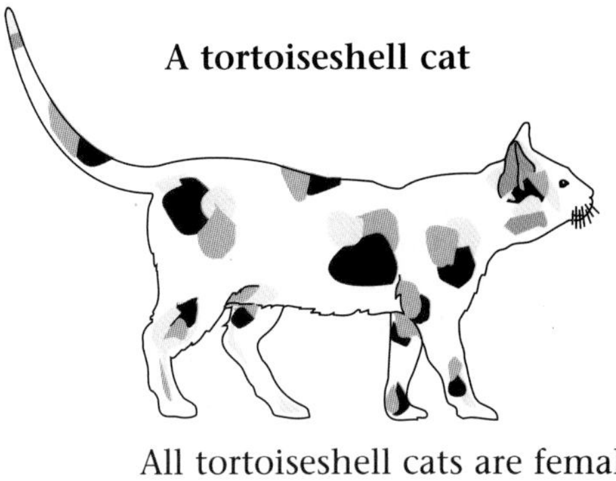

All tortoiseshell cats are female

APPLICATIONS OF GENETICS

GENETIC ENGINEERING Altering genes to make a useful product.

For example, making insulin (needed for diabetics).

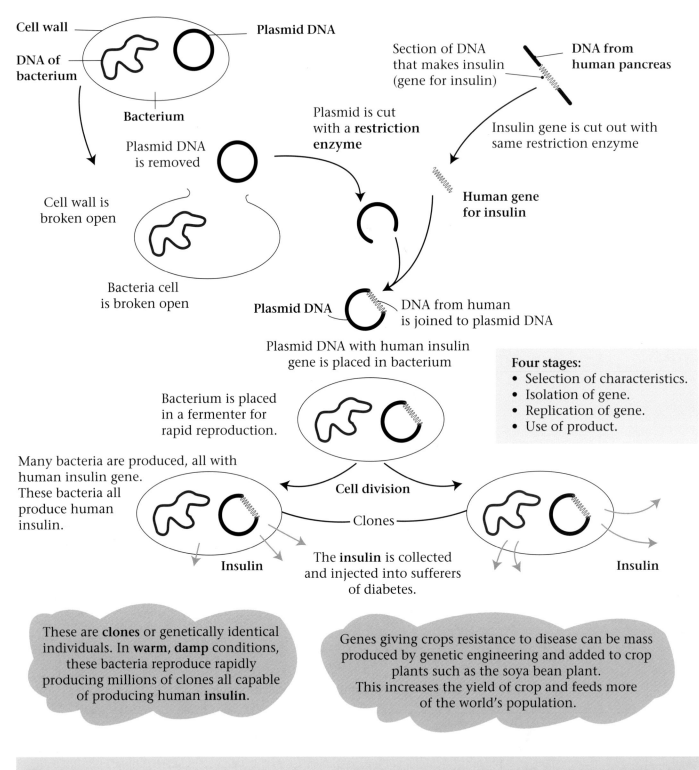

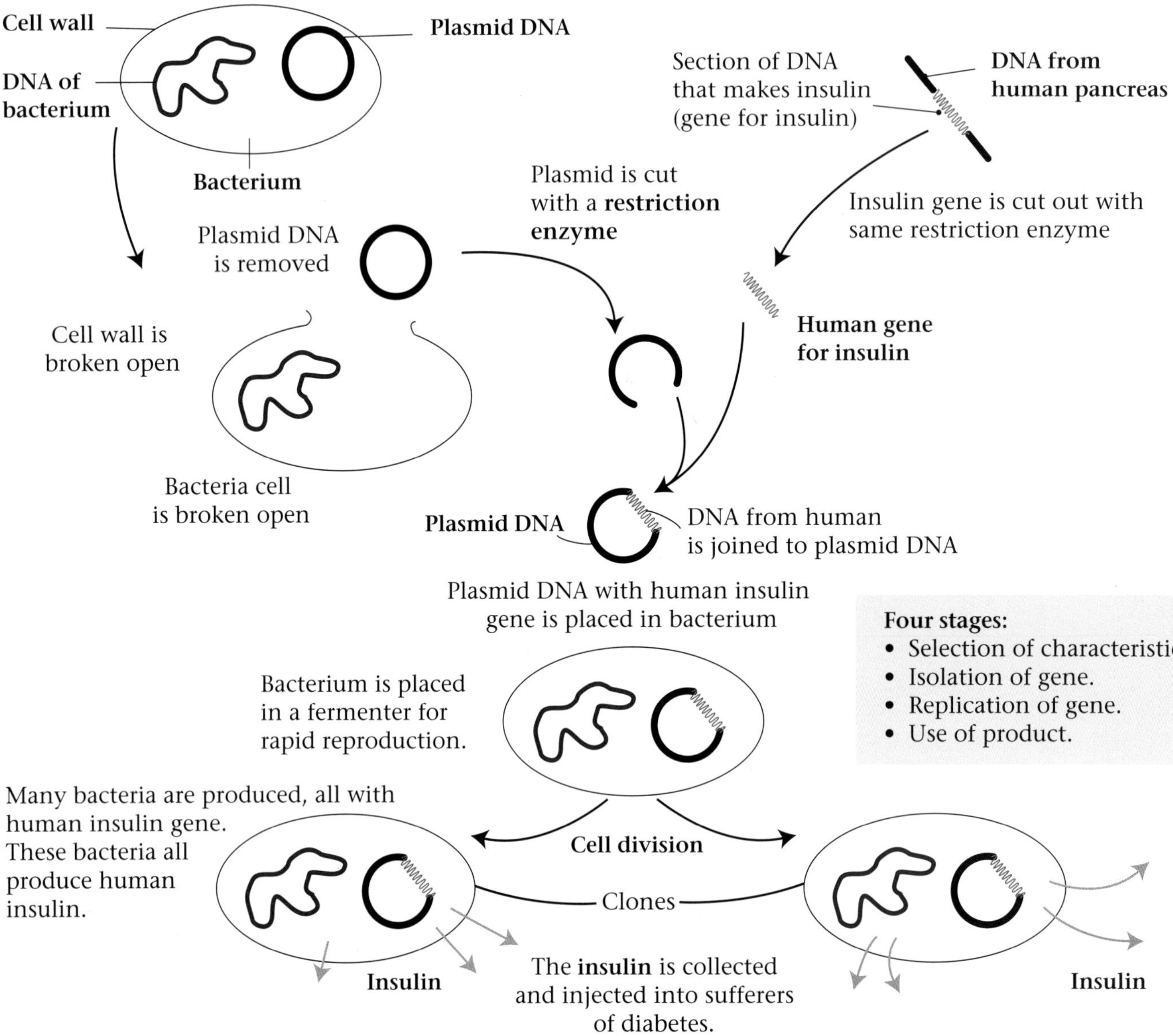

Cell wall
DNA of bacterium
Plasmid DNA
Bacterium
Section of DNA that makes insulin (gene for insulin)
DNA from human pancreas
Plasmid DNA is removed
Plasmid is cut with a restriction enzyme
Insulin gene is cut out with same restriction enzyme
Cell wall is broken open
Human gene for insulin
Bacteria cell is broken open
Plasmid DNA
DNA from human is joined to plasmid DNA
Plasmid DNA with human insulin gene is placed in bacterium
Bacterium is placed in a fermenter for rapid reproduction.
Many bacteria are produced, all with human insulin gene. These bacteria all produce human insulin.
Cell division
Clones
Insulin
The insulin is collected and injected into sufferers of diabetes.
Insulin

Four stages:
- Selection of characteristics.
- Isolation of gene.
- Replication of gene.
- Use of product.

These are **clones** or genetically identical individuals. In **warm, damp** conditions, these bacteria reproduce rapidly producing millions of clones all capable of producing human **insulin**.

Genes giving crops resistance to disease can be mass produced by genetic engineering and added to crop plants such as the soya bean plant. This increases the yield of crop and feeds more of the world's population.

Questions:

1. Which micro-organisms can be used in genetic engineering?
2. What product can be made by genetic engineering?
3. Where is the gene found that makes insulin in humans?
4. How is this human insulin gene cut out?
5. Which part of the bacterium is joined to the human gene?
6. Where are the bacteria with the human gene put to encourage rapid reproduction?

GENETIC FINGERPRINTING

Every individual has a unique pattern of DNA, (except for identical twins), inherited from both parents. This pattern can be seen as a sequence of black lines similar to a 'bar code'.

In the same way that a person's fingerprint is unique, a genetic fingerprint can be used to identify an individual. A genetic fingerprint can be used in forensic work to catch criminals from DNA left at the crime scene. It can also be used to establish paternity, where the true father is uncertain and to establish evolutionary relationships between organisms.

The pattern of DNA is **a genetic fingerprint**.

A genetic fingerprint

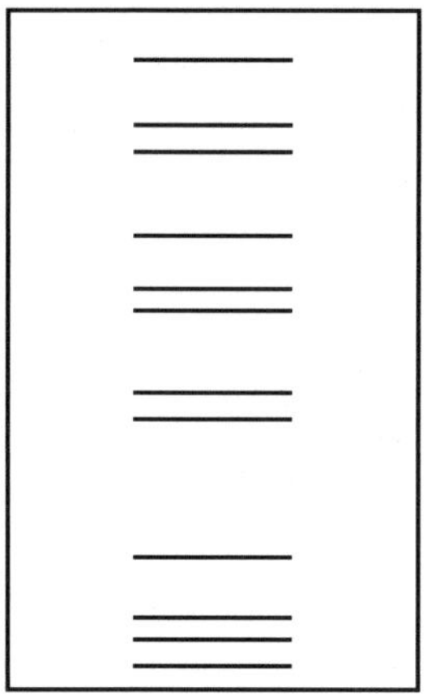

How to produce a genetic fingerprint

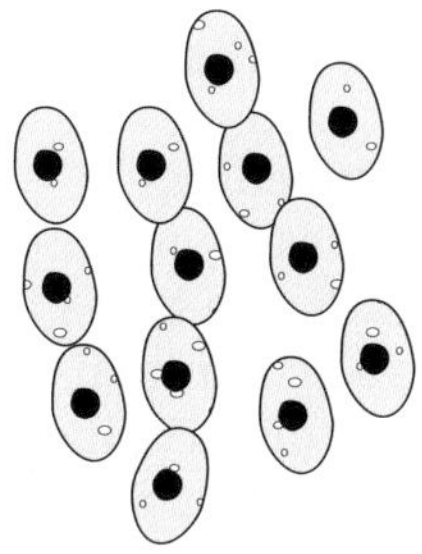

Skin cells

Collect a source of DNA (blood, skin, hair). Increase the quantity of DNA, if necessary, by the **polymerase chain reaction**.

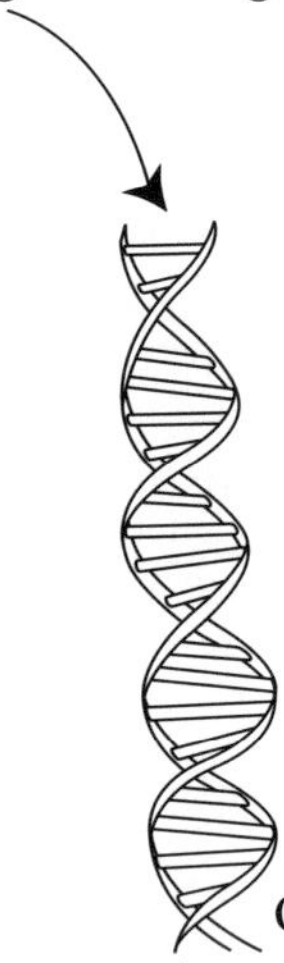

Extract the DNA.

Cut the DNA into smaller fragments using a restriction enzyme.

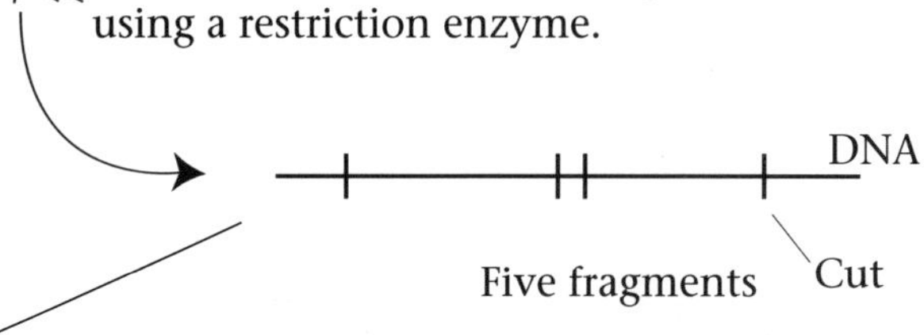

Separate the DNA fragments by gel **electrophoresis**. Small fragments move more easily through the gel and are separated from larger,slower-moving fragments

Gel electrophoresis separates DNA fragments according to size

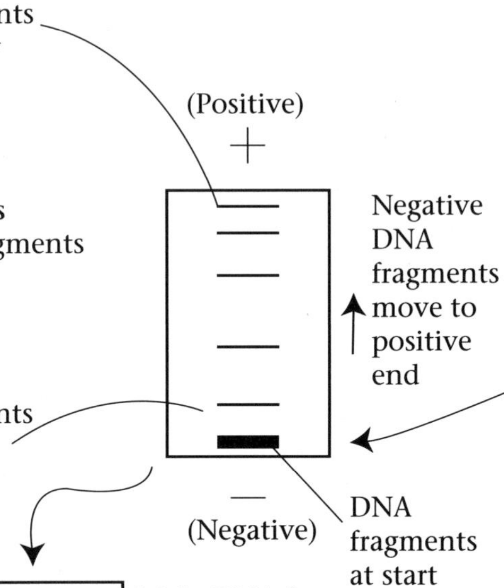

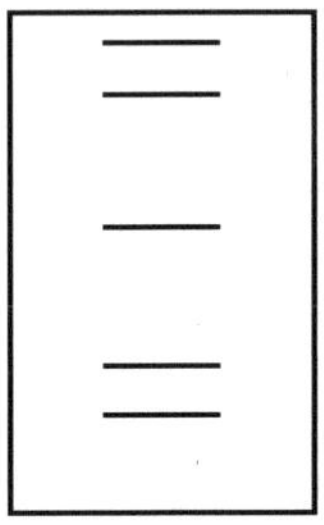

Make DNA fragments radioactive and therefore visible as black lines

A genetic fingerprint

Use of genetic fingerprinting as criminal evidence

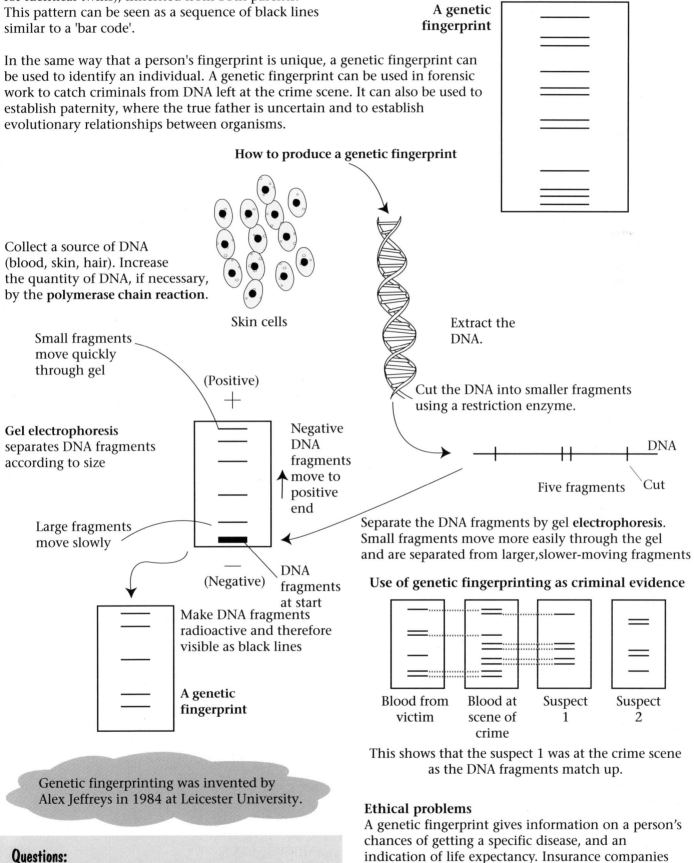

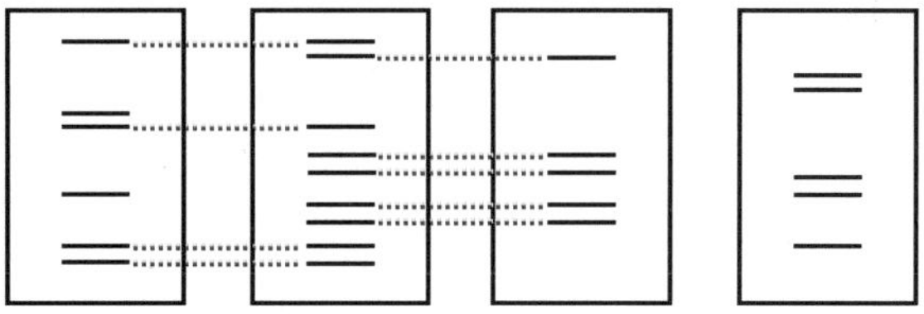

This shows that the suspect 1 was at the crime scene as the DNA fragments match up.

Genetic fingerprinting was invented by Alex Jeffreys in 1984 at Leicester University.

Questions:

1. Which part of a person makes up a genetic fingerprint?
2. What must be extracted from cells to produce the genetic fingerprint?
3. Draw a flow diagram to show the main stages involved in producing a genetic fingerprint.

Ethical problems

A genetic fingerprint gives information on a person's chances of getting a specific disease, and an indication of life expectancy. Insurance companies could use this information before issuing a life policy. Police could use a bank of genetic fingerprints to identify criminals quickly.

Do you think that everyone's genetic fingerprints should be taken at birth and be kept on file?

SELECTIVE BREEDING (artificial selection)

Selective breeding is used by animal and plant breeders to produce a **desired variety or species**. Humans select which organisms should mate.

1. In animals

This is used to produce good guide dogs for the blind.

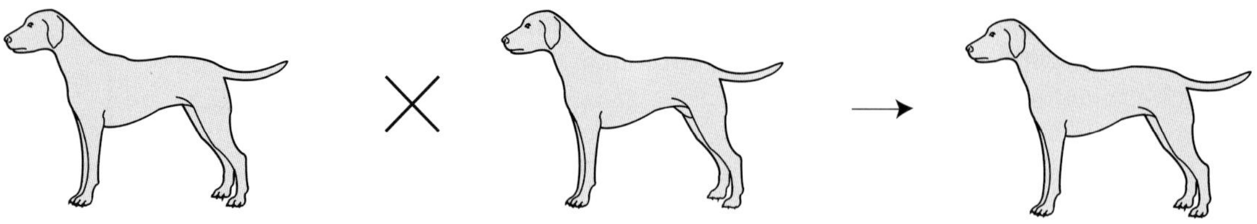

This combines the desirable characteristics of both parents, e.g. intelligence, obedience and gentleness.

Problems with selective breeding

Selected animals and plants are similar; all of them have the same features and therefore similar genes. Breeds which are not selected for mating will die out. This leads to a **loss of genes** which might be required in the future. If a new disease emerges, it could kill all the selected animals, since all of them are similar and none may have the gene for resistance to the disease. The **loss of variation** reduces the long-term survival of a species. They are unable to adapt to changes, as the genes needed may have been lost.

2. In plants

This is used to produce a better crop and more profit for the farmers.

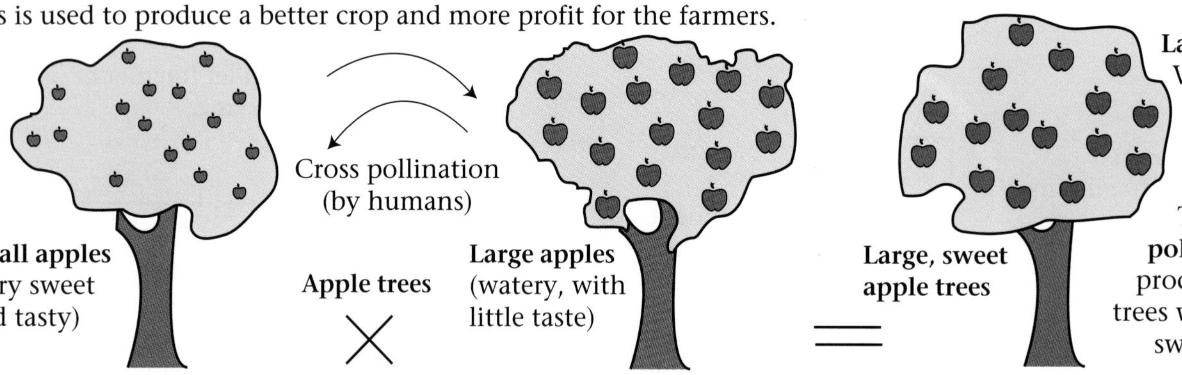

Pollen is transferred from one tree to the other to try to combine the good features of both, i.e. humans select which trees will 'mate' – **artificial selection**.

Any trees with small, watery apples will be **destroyed** and **not** used for **selection.** A new variety of apple tree has been produced.

Selective breeding – the process

- Choose an animal (or plant) with the desired feature, e.g. long legs, resistance to disease or a high milk yield.
- Mate with another desirable organism.
- Identify which offspring have inherited the desired gene.
- Continue to breed with *selected* offspring; do not allow the rejects to mate.
- This increases the number of animals or plants with the desired feature.

The process:

- Select feature.
- Crossbreeding
- Selection of suitable offspring over many generations.

Questions:

1. Why is selective breeding carried out?
2. When breeding cattle, what features might be selected?
   Which offspring would be used for further breeding and why?
   Which offspring would not be selected for further mating and why?
3. How is selective breeding of value to farmers?

CELL DIVISION AND EVOLUTION

CELL DIVISION Mitosis and meiosis.

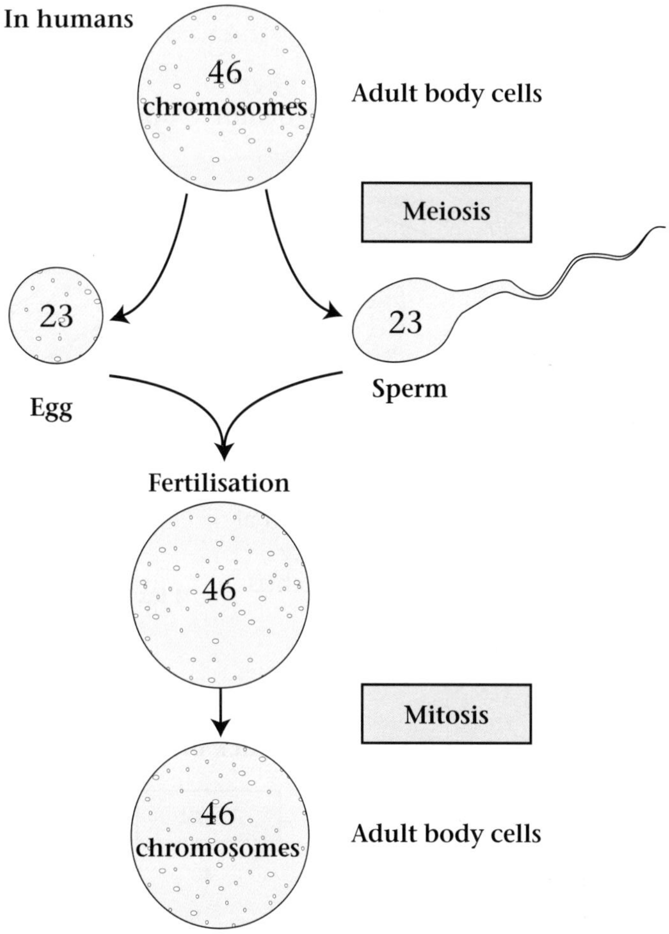
In humans
46 chromosomes
Adult body cells
Meiosis
23
Egg
23
Sperm
Fertilisation
46
Mitosis
46 chromosomes
Adult body cells

Meiosis (sex cells)
This cell division produces the **sex cells** with **half** the usual chromosome number. Meiosis is called ***reduction division,*** as the chromosome number is reduced.

Mitosis (growth)
This cell division causes **growth**. The cells produced all have the full chromosome number.

The **sex cells** must have **half** the usual **chromosome** number. When **fertilisation** takes place, the number of **chromosomes doubles**, so the **fertilised egg** has the full number of chromosomes. The fertilised egg then grows by **mitosis** into an **adult**.

In humans

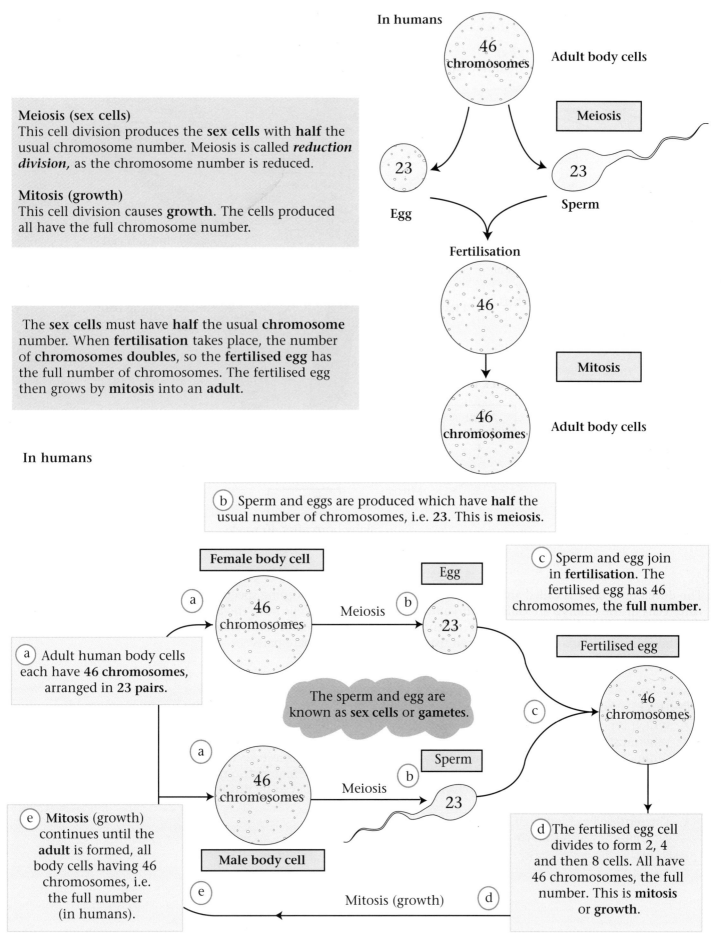

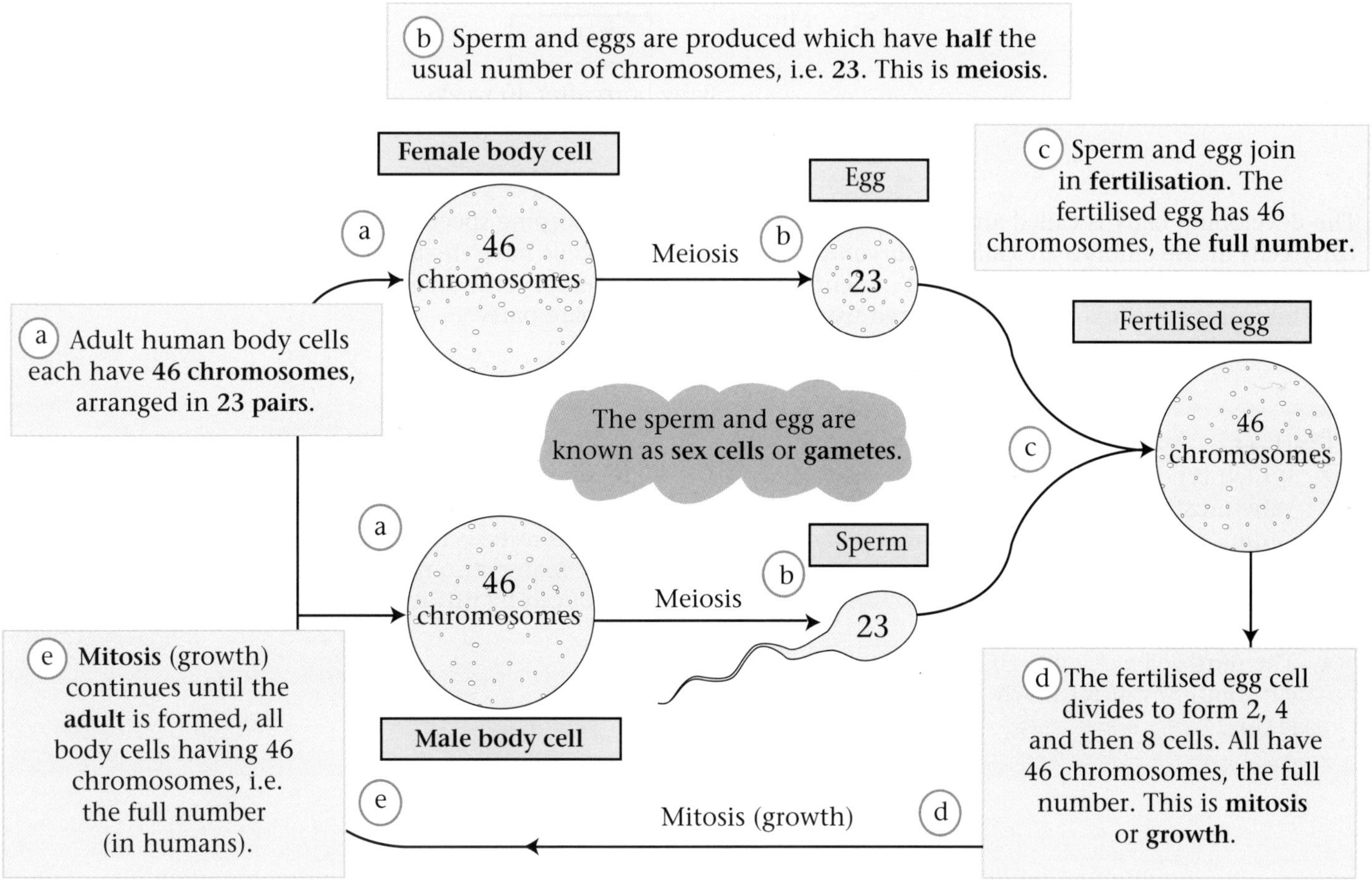
(b) Sperm and eggs are produced which have **half** the usual number of chromosomes, i.e. **23**. This is **meiosis**.
Female body cell
(a)
46 chromosomes
Meiosis
(b)
Egg
23
(c) Sperm and egg join in **fertilisation**. The fertilised egg has 46 chromosomes, the **full number.**
Fertilised egg
(a) Adult human body cells each have **46 chromosomes,** arranged in **23 pairs.**
The sperm and egg are known as **sex cells** or **gametes**.
(c)
46 chromosomes
(a)
46 chromosomes
Sperm
(b)
Meiosis
23
(e) **Mitosis** (growth) continues until the **adult** is formed, all body cells having 46 chromosomes, i.e. the full number (in humans).
Male body cell
(d) The fertilised egg cell divides to form 2, 4 and then 8 cells. All have 46 chromosomes, the full number. This is **mitosis** or **growth.**
(e)
Mitosis (growth)
(d)

CELL DIVISION AND THE HUMAN LIFE CYCLE

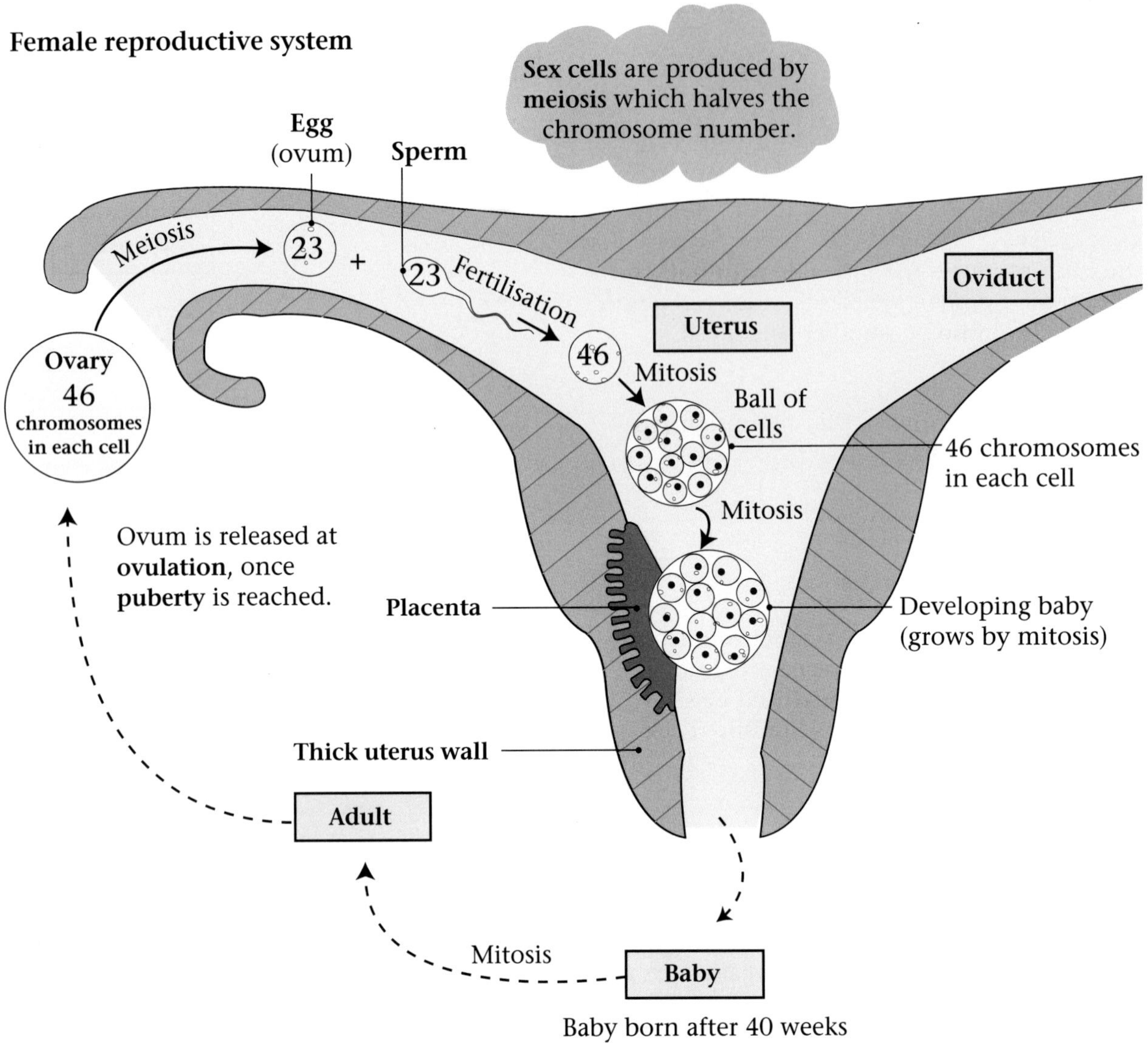

The developing baby is called an embryo. Embryo cells gradually become specialised for a particular function. Early cells in the embryo are called ***embryonic stem cells***. Theses cells have the ability to become many different types of cells. Their future is not 'fixed'. Scientists are able to use embryonic stem cells to replace damaged tissue, but this raises concerns regarding embryo tissue being used for 'spare-parts'

Questions:

1. Which type of cell division produces the sex cells?
2. Why must the sex cells have half the chromosome number of the normal body cells?
3. The human fertilised egg contains how many chromosomes? Where have these chromosomes come from?
4. The fertilised egg, one cell, grows into an adult with millions of cells. Which type of cell division is this?
5. In what process does an ovary release an egg?
6. Where do the egg and sperm join in fertilisation?
7. Why is mitosis necessary in a fertilised egg?
8. How many chromosomes are in each cell produced by mitosis in humans?

MITOSIS AND MEIOSIS

Mitosis (growth)

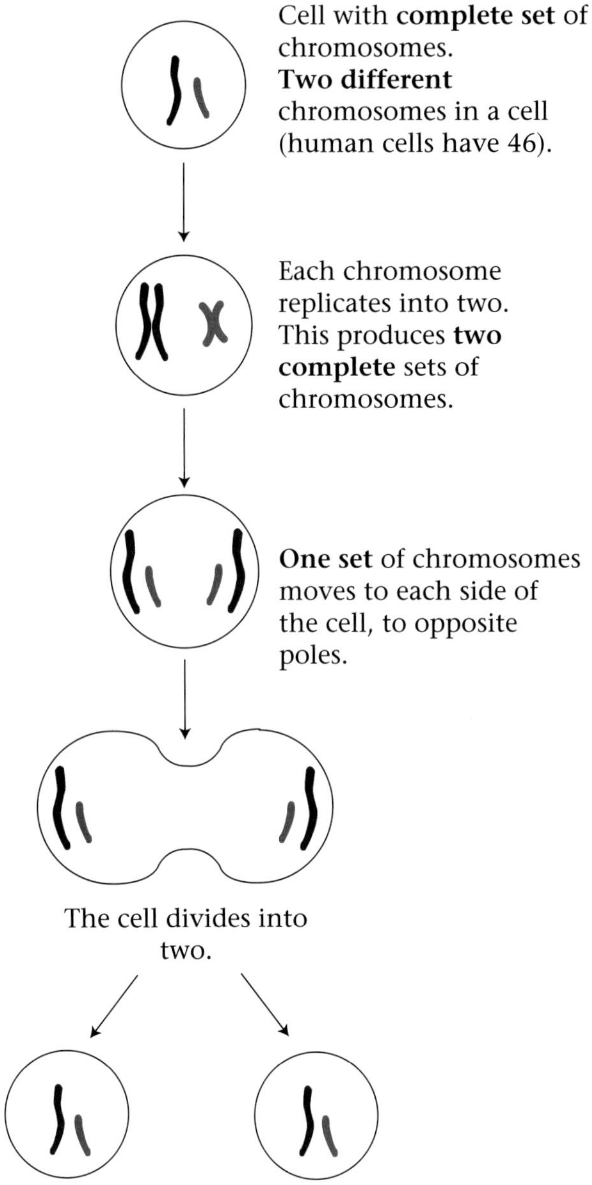
Cell with **complete set** of chromosomes. **Two different** chromosomes in a cell (human cells have 46).
Each chromosome replicates into two. This produces **two complete** sets of chromosomes.
One set of chromosomes moves to each side of the cell, to opposite poles.
The cell divides into two.

The two new cells have the **same chromosomes** as the original cell. These two cells are **identical** to the **parent** cell.

Meiosis (reduction division forming sex cells)

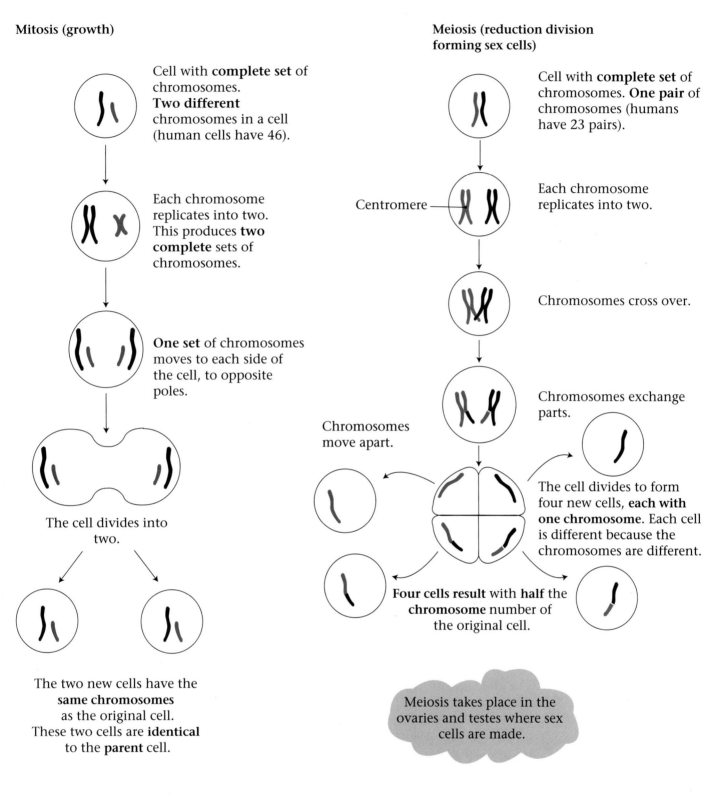

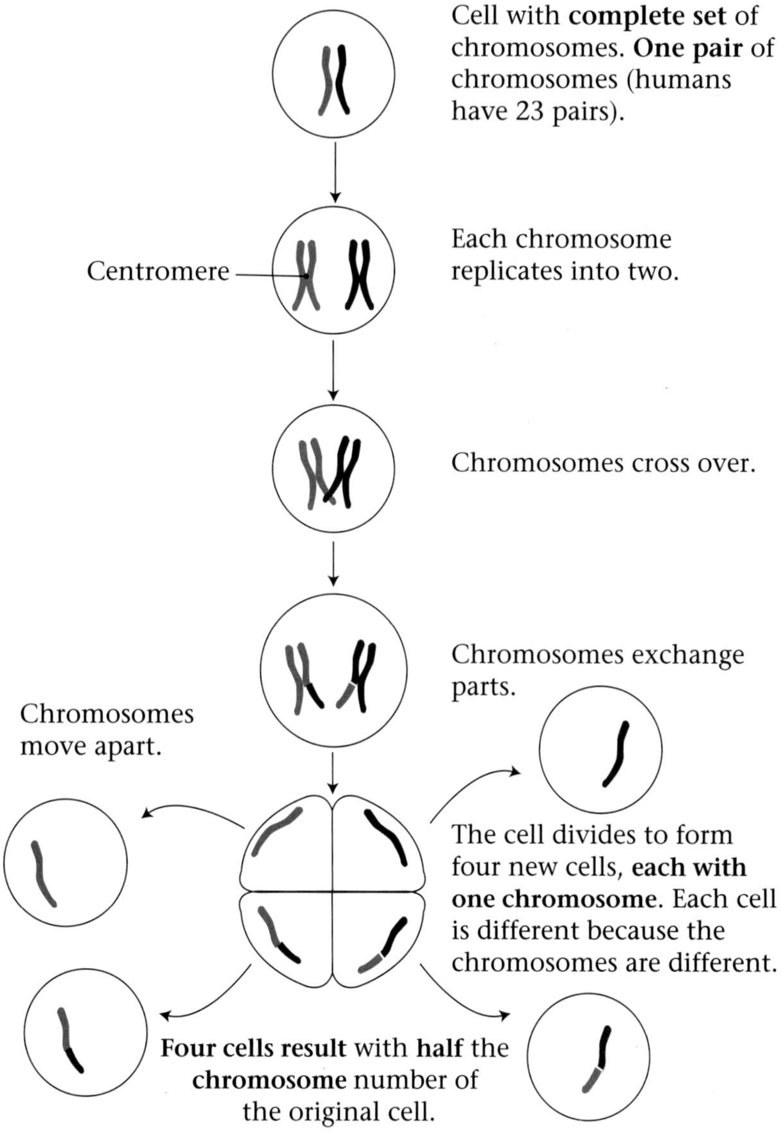
Cell with **complete set** of chromosomes. **One pair** of chromosomes (humans have 23 pairs).
Centromere
Each chromosome replicates into two.
Chromosomes cross over.
Chromosomes exchange parts.
Chromosomes move apart.
The cell divides to form four new cells, **each with one chromosome**. Each cell is different because the chromosomes are different.
Four cells result with **half** the **chromosome** number of the original cell.

Meiosis takes place in the ovaries and testes where sex cells are made.

Questions:

1. Name two places where meiosis occurs in humans.
2. How many sex cells are produced in meiosis from the parent cell?
3. What is the chromosome number in the sex cells compared with the parent cell?
4. What causes variation in the sex cells?
5. Name one place where mitosis occurs in humans.
6. How do the daughter cells in mitosis compare to the parent cell?
7. How many cells result from mitosis?

GROWTH

Growth is the permanent increase in size of an organism.

Three stages of growth

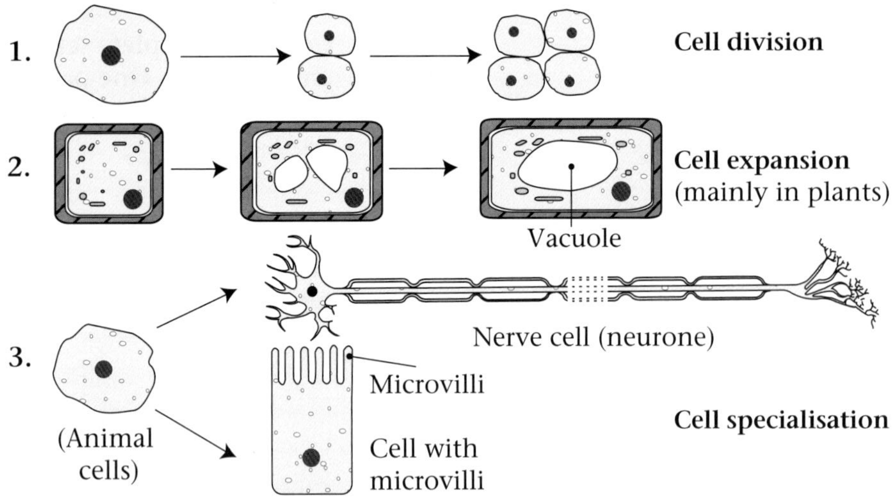

SUMMARY

Three stages of growth:
- Cell division (mitosis).
- Cell expansion.
- Cell specialisation (differentiation).

Types of growth:
- Continuous.
- Discontinuous.

Growth in most animals and plants

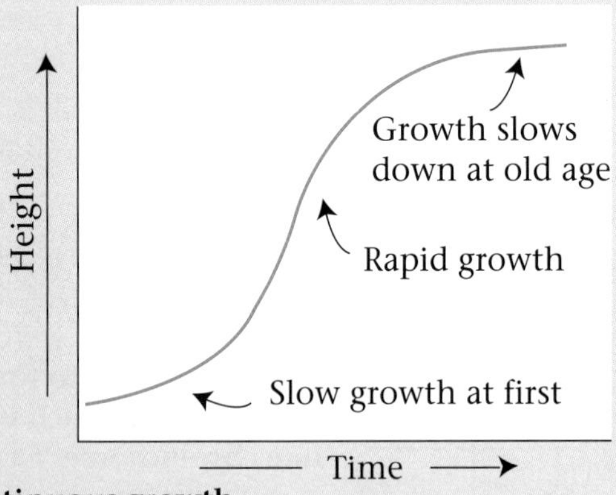

Continuous growth

Growth in arthropods, e.g. locust

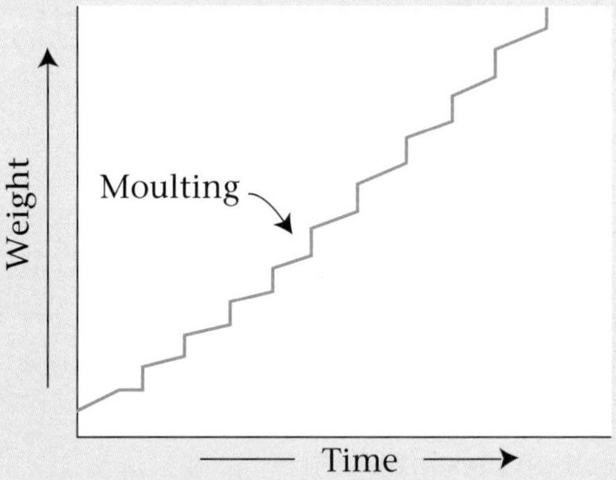

Discontinuous growth
When the outer skin is shed, insects take in water or air to expand their bodies, before new skin hardens

Human growth

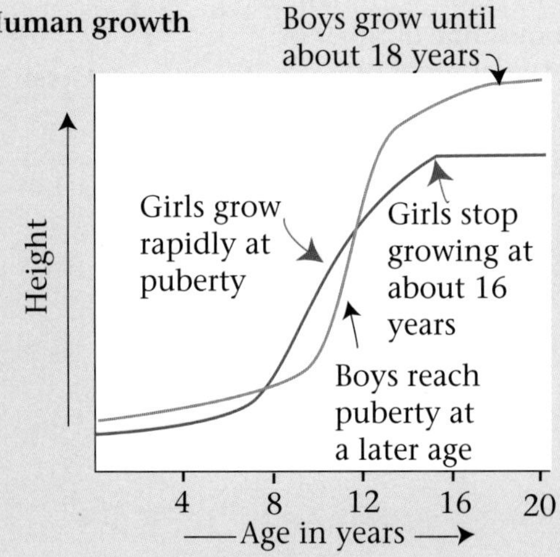

Cancer

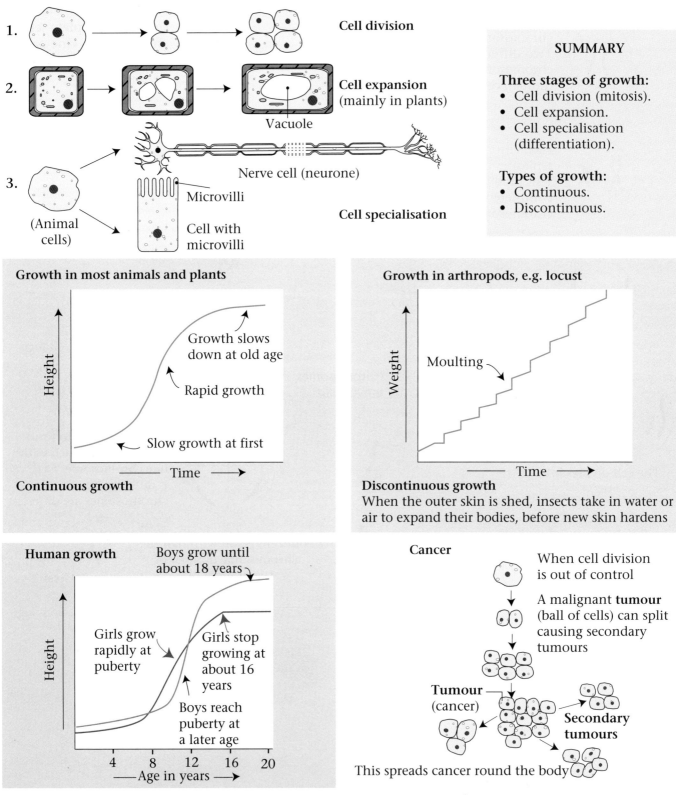

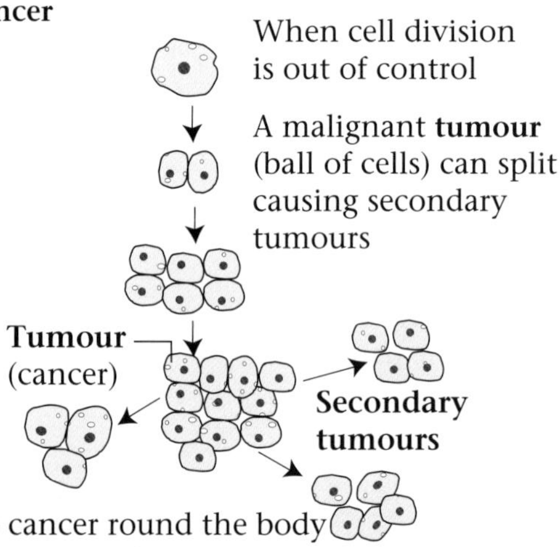

This spreads cancer round the body

Did you know that most animals reach a particular maximum size and then stop growing, whereas plants can continue growing?

Questions:

1. What causes cell expansion in plants?
2. What kind of cell division causes growth?
3. Name four specialised cell in animals.
4. What are the three stages of growth?
5. When are girls taller than boys (refer to the human growth chart)?
6. At what age do boys grow most rapidly?

ASEXUAL REPRODUCTION IN PLANTS (NATURAL) A rapid method of reproduction, but all offspring are genetically identical to the parent (clones).

Natural clones in potato plants

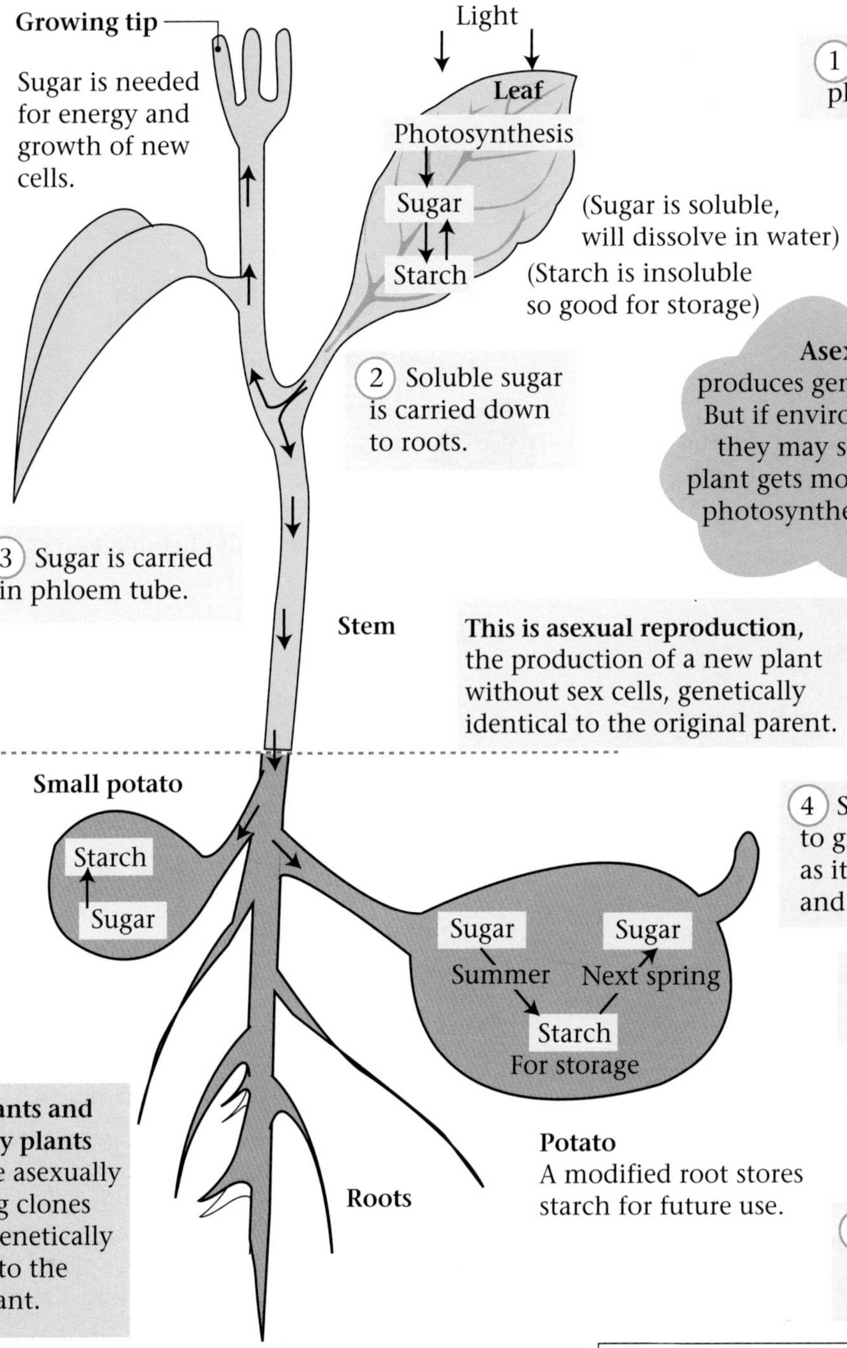

1) Sugar is made during photosynthesis in leaves.

2) Soluble sugar is carried down to roots.

3) Sugar is carried in phloem tube.

Asexual reproduction produces genetically identical offspring. But if environmental conditions differ, they may show variation, e.g. if one plant gets more light than another it will photosynthesise more and grow better.

This is asexual reproduction, the production of a new plant without sex cells, genetically identical to the original parent.

4) Sugar is carried to growth points as it is soluble and transportable.

5) This causes growth of a new plant.

6) Extra sugar made in the summer is stored as starch.

7) Starch stored in the summer is used for growth the following spring.

Spider plants and strawberry plants reproduce asexually producing clones that are genetically identical to the parent plant.

Growth of new plant

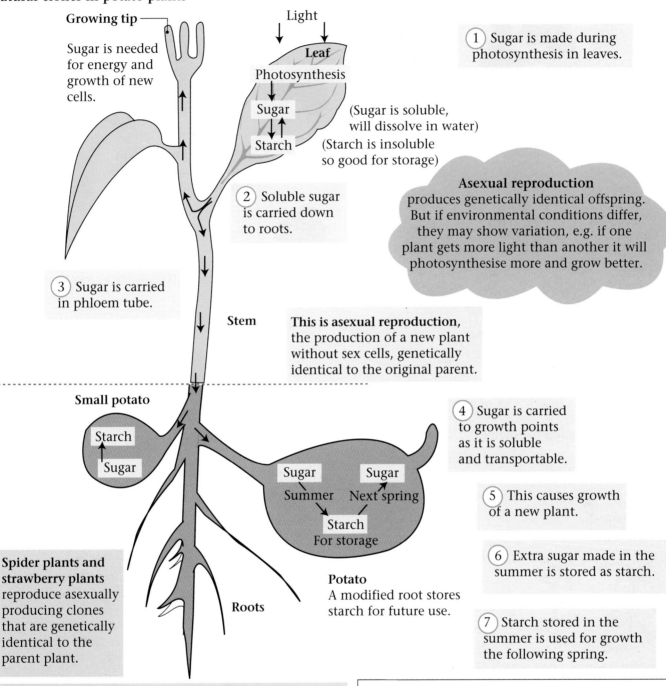

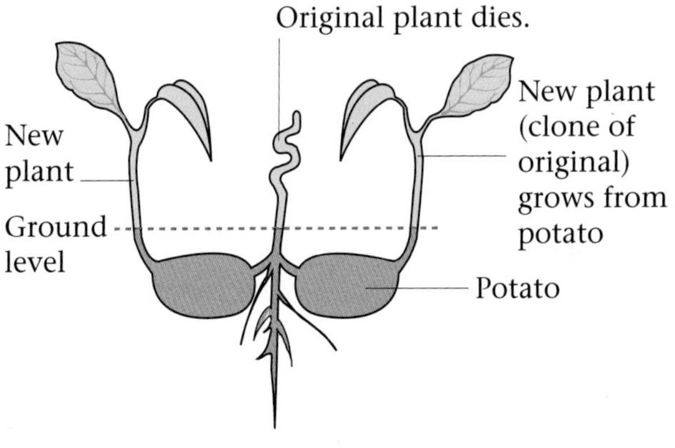

Questions:

1. Asexual reproduction is reproduction without the production of gametes. As a result, there is no variation between the parent and offspring. Why is this, a) a good method of reproduction, b) a poor method of reproduction; compared to sexual reproduction?
2. Where is food made in plants and at what time of year?
3. How is the food transported to a storage area and in what form?
4. How can this food be used for the growth of new plants and when?
5. What evidence for the growth of new plants can be seen in a potato?

ASEXUAL REPRODUCTION IN PLANTS (ARTIFICIAL)

The production of new plants from existing ones (clones). There is no variation.

1. Cuttings from geranium

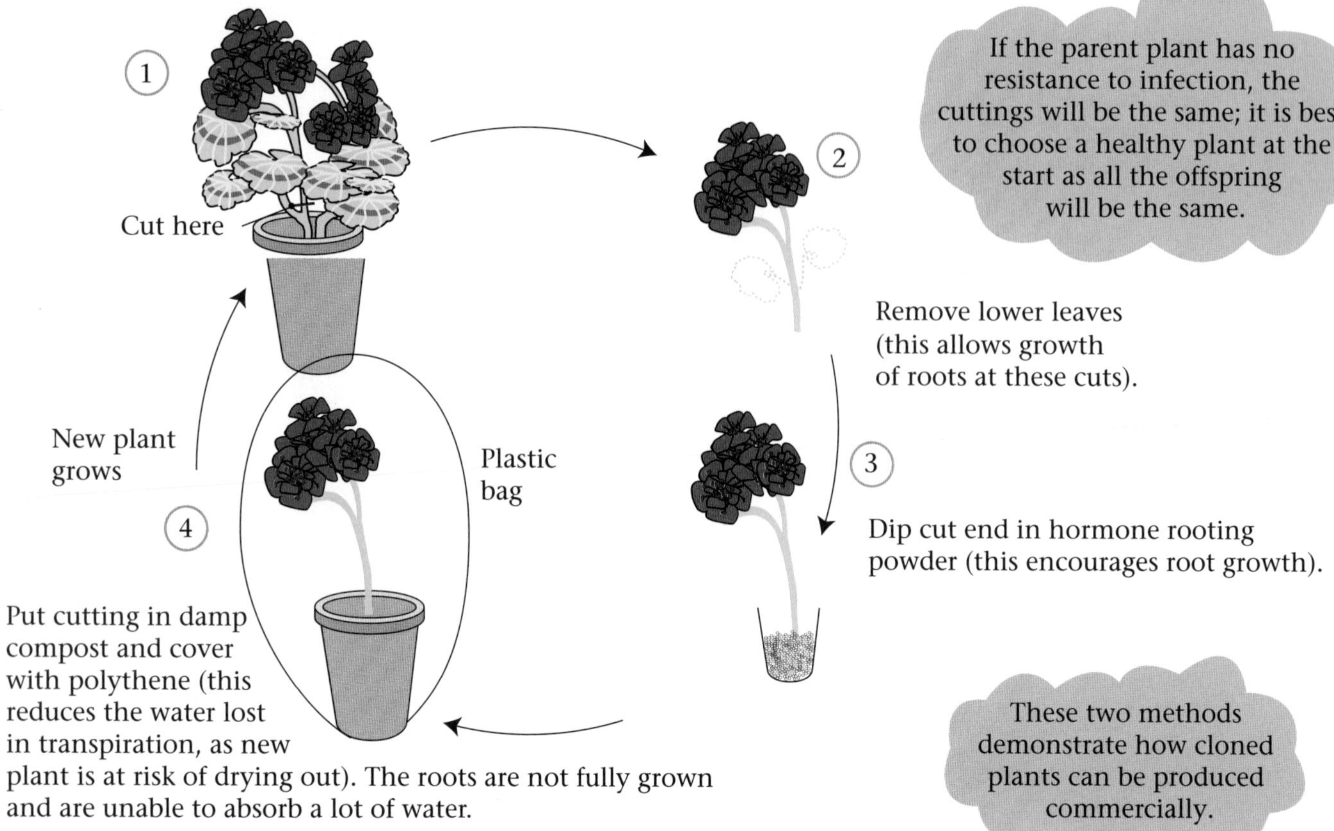
1
Cut here
2
Remove lower leaves (this allows growth of roots at these cuts).
3
Dip cut end in hormone rooting powder (this encourages root growth).
New plant grows
4
Plastic bag
Put cutting in damp compost and cover with polythene (this reduces the water lost in transpiration, as new plant is at risk of drying out). The roots are not fully grown and are unable to absorb a lot of water.

If the parent plant has no resistance to infection, the cuttings will be the same; it is best to choose a healthy plant at the start as all the offspring will be the same.

These two methods demonstrate how cloned plants can be produced commercially.

2. Micropropagation in potatoes

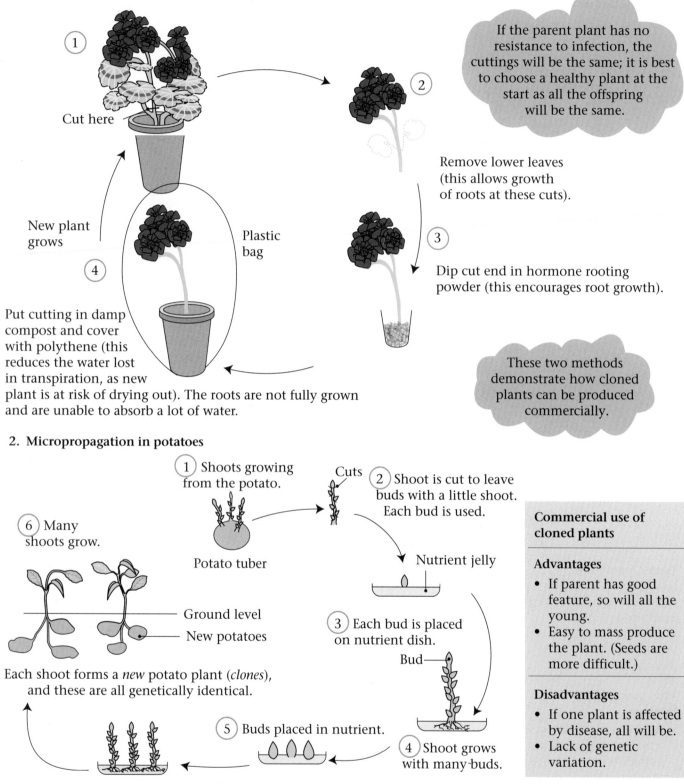

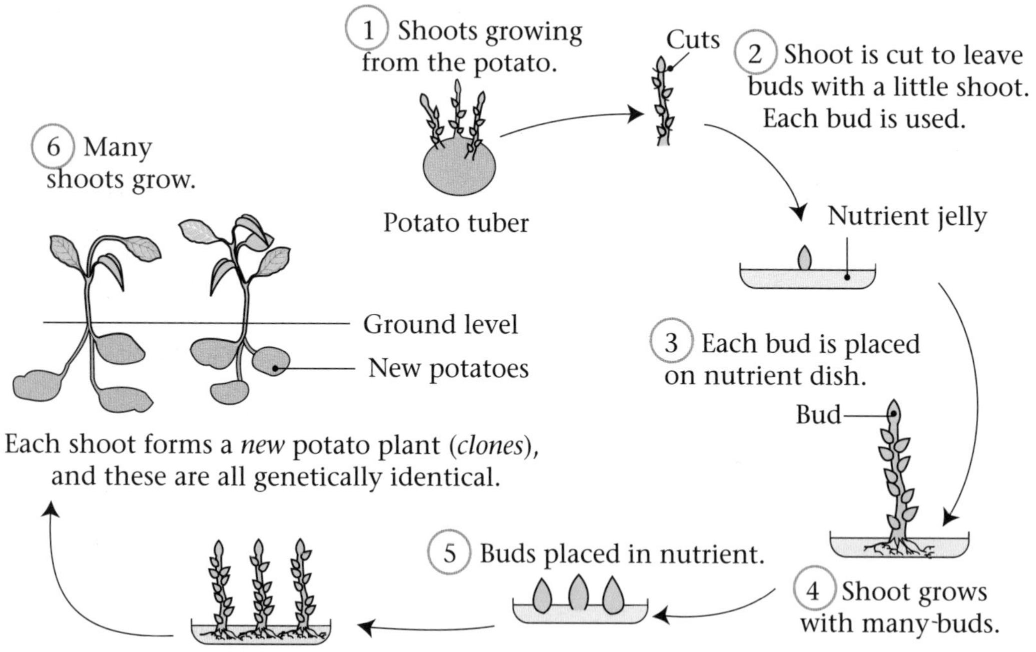
1 Shoots growing from the potato.
Potato tuber
Cuts
2 Shoot is cut to leave buds with a little shoot. Each bud is used.
Nutrient jelly
3 Each bud is placed on nutrient dish.
Bud
4 Shoot grows with many buds.
5 Buds placed in nutrient.
6 Many shoots grow.
Ground level
New potatoes
Each shoot forms a *new* potato plant (*clones*), and these are all genetically identical.

Commercial use of cloned plants

Advantages

- If parent has good feature, so will all the young.
- Easy to mass produce the plant. (Seeds are more difficult.)

Disadvantages

- If one plant is affected by disease, all will be.
- Lack of genetic variation.

Questions:

1. Draw a flow chart to show how new geranium plants can be produced from one plant using cuttings.
2. Why are the lower leaves removed?
3. How does dipping the cut stem in rooting powder help the process?
4. Why is the new plant covered with a plastic bag?
5. What is a clone?
6. Micropropagation involves using small pieces of the plant, such as buds. How can this be used to produce clones?
7. What is the advantage of both cuttings and micropropagation to plant growers?

CLONING BY TISSUE CULTURE

Summary

- Select parent plant.
- Cut off small pieces.
- Sterilise, i.e. ensure aseptic conditions.
- Growth.

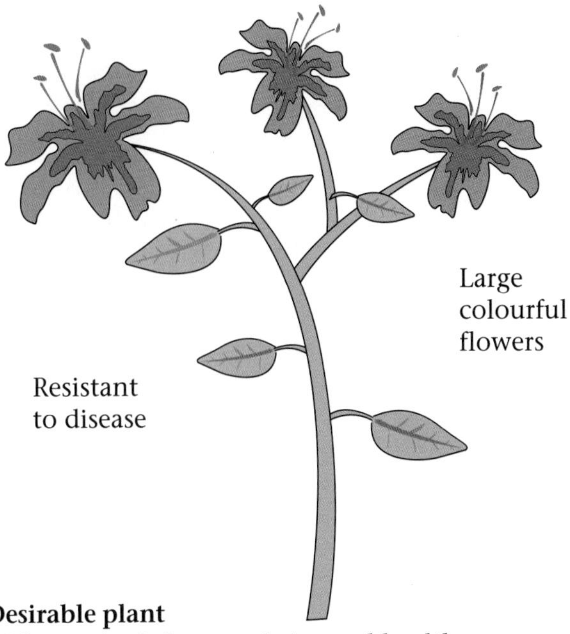

Desirable plant
with required characteristics and healthy

Enormous numbers of identical plants can be produced from small pieces of the selected parent plant.

- A healthy plant with all the desired characteristics is selected.
- The stem, or root, is cut into many small pieces, called **explants.**

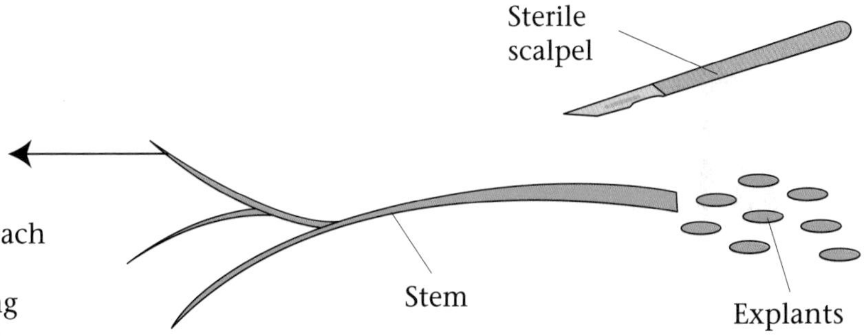

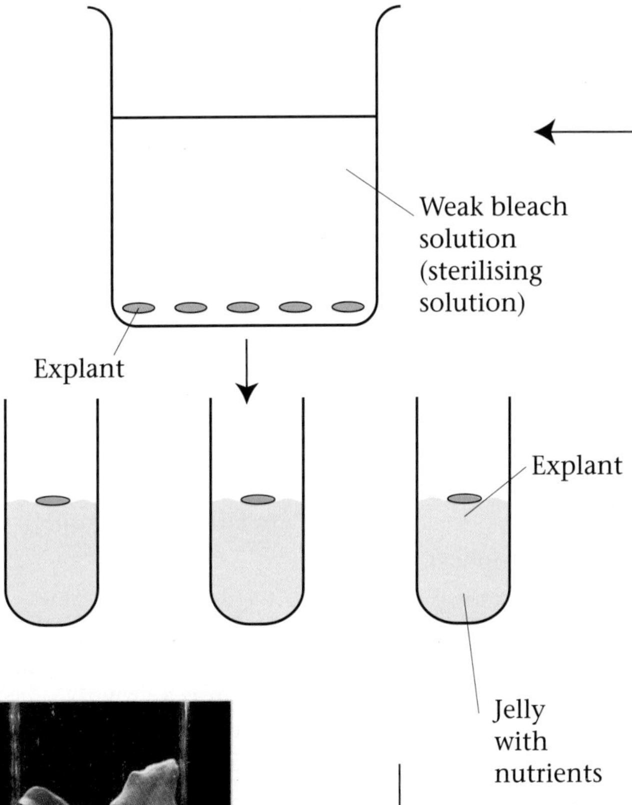

- The small pieces are placed in weak bleach to sterilise them, i.e. to remove all microbes.
- The sterilised explants are placed in jelly containing all the substances required for growth, i.e. sugar, vitamins, minerals, and plant growth hormones. They are kept in a room where the light and temperature are controlled to encourage growth.
- The plants grow into clones of the parent, all genetically identical.

Explant in jelly with full range of nutrients.

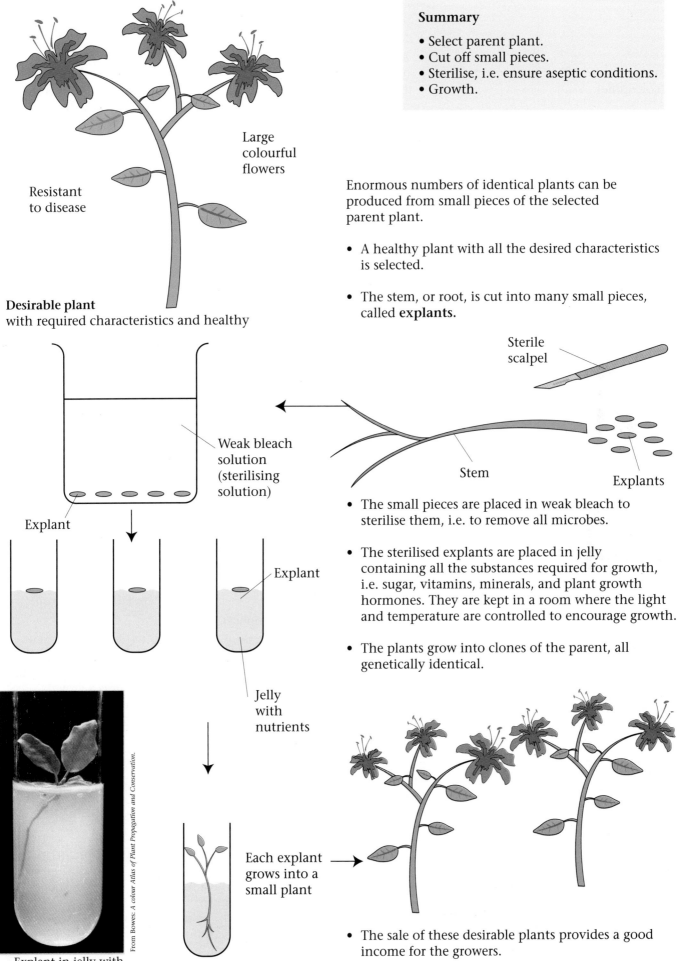

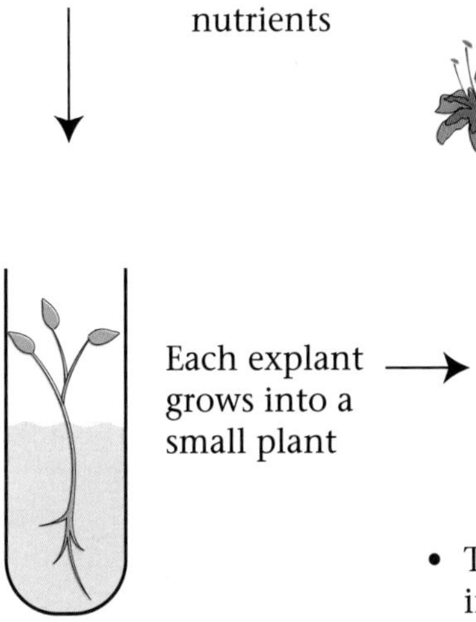

- The sale of these desirable plants provides a good income for the growers.

PLANT HORMONES — AUXINS

Plant growth is controlled by hormones called **auxins**.

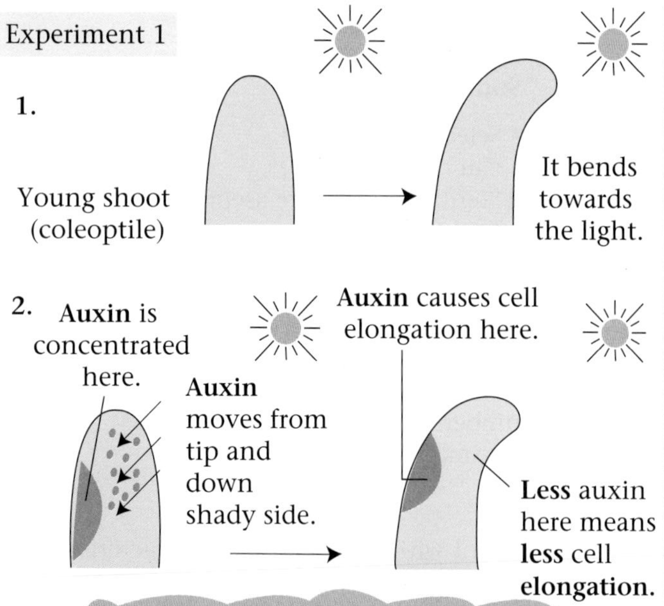

Experiment 1
1.
Young shoot (coleoptile)
It bends towards the light.
2. Auxin is concentrated here.
Auxin moves from tip and down shady side.
Auxin causes cell elongation here.
Less auxin here means less cell elongation.

Bending is caused by unequal distribution of auxin.
More auxin causes more growth on one side of the tip.

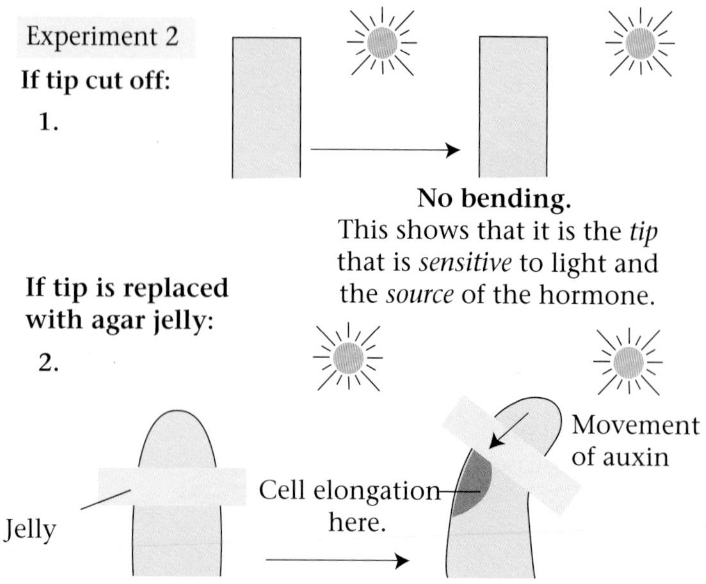

Experiment 2
If tip cut off:
1.
No bending.
If tip is replaced with agar jelly:
2.
Jelly
Cell elongation here.
Movement of auxin

This shows that it is the *tip* that is *sensitive* to light and the *source* of the hormone.

The auxin produced in the tip must **diffuse** through the jelly and concentrate on the shady side. Here *cell* **elongation** causes **bending.** This shows that the **hormone** can move by **diffusion.** Light, gravity, and water cause growth movements in plants called **tropisms.** Tropisms are controlled by auxins.

Growth in young seedlings

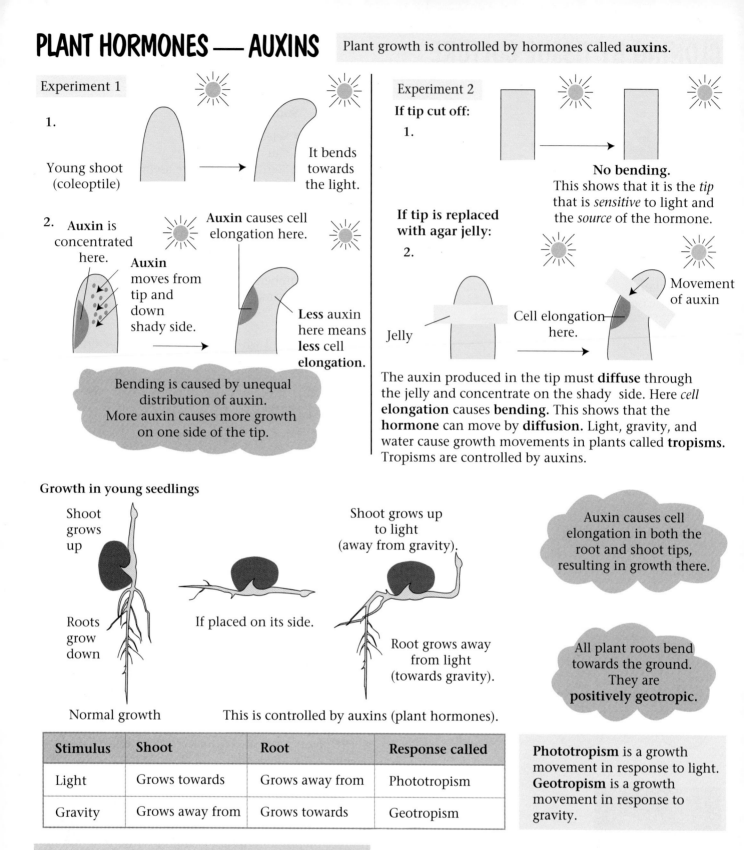

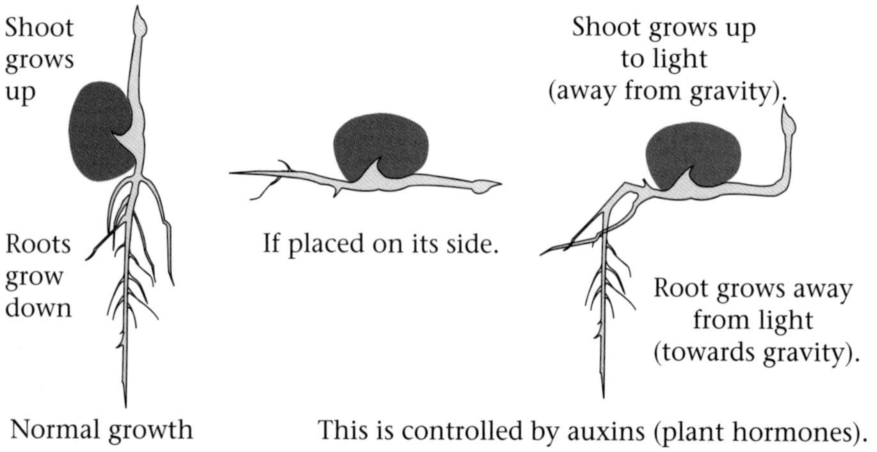

Shoot grows up
Roots grow down
Normal growth
If placed on its side.
Shoot grows up to light (away from gravity).
Root grows away from light (towards gravity).

This is controlled by auxins (plant hormones).

Auxin causes cell elongation in both the root and shoot tips, resulting in growth there.

All plant roots bend towards the ground. They are **positively geotropic.**

Stimulus	Shoot	Root	Response called
Light	Grows towards	Grows away from	Phototropism
Gravity	Grows away from	Grows towards	Geotropism

Phototropism is a growth movement in response to light. **Geotropism** is a growth movement in response to gravity.

Auxins are used in **agriculture** to improve production of fruit. They can cause **fruit formation** in grapes without fertilisation. This results in seedless grapes. Auxins are used to encourage **root growth** in cuttings (rooting powder) and to prevent the growth of side branches from the stem. They are also used as **weed killers.** Some auxins are absorbed through leaves, so broad leaved plants absorb more causing their death. Lawns are kept free of broad leaved plants by this method. The auxins cause plants to grow rapidly,disrupting their normal growth pattern. These plants develop long weak stems and die.

Questions:

1. What is a tropism?
2. Which part of a shoot is sensitive to light? How do you know?
3. Where does the auxin concentrate in a shoot exposed to light?
4. What effect does the auxin have on the shoot?
5. What is phototropism?
6. How does phototropism help plant growth?

EVOLUTION The changes that take place in living organisms over a long period of time.

The theory of evolution by natural selection was first proposed by Charles Darwin in 1859. It explains why animals and plants seem to fit in with their surroundings. Modern genetics confirms this theory, e.g. **Giraffes**.

1. There are too many offspring.
2. There is competition for food, space, and other things.
3. Within a species there is variation, due to differing genes.
4. Some organisms have features which help them live long enough to mate and have more young like themselves.
5. Those less fit, do not survive to breed, so their unhelpful features are not passed on.
6. The survivors are different from their ancestors because they only have the successful genes. This slight change is **evolution**.

Originally giraffes' neck length varied

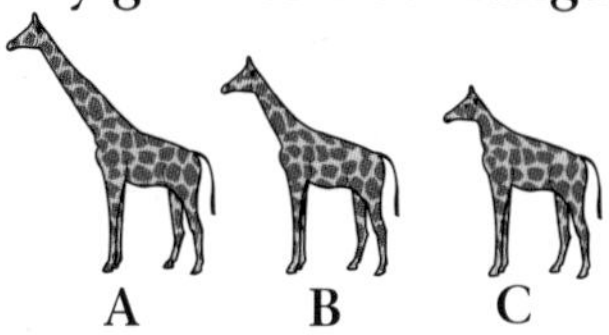

With plenty of leaves to eat, all giraffes survived to breed.

Variety of neck lengths

In harsh times, leaves were left only at top of trees. Only those with long necks survived to mate with other long-necked survivors.

Only tall giraffes can feed

All giraffes now have long necks

The gene for a long neck is passed on to the young. The gene for shorter necks disappears with the giraffes. Eventually only giraffes with long necks remain. This is **natural selection** or **survival of the fittest** (the long-necked giraffe is the most fit and survives).

Originally fish colouring varied

With plenty of food and no predators all the fish types survive to breed and pass on their colouring.

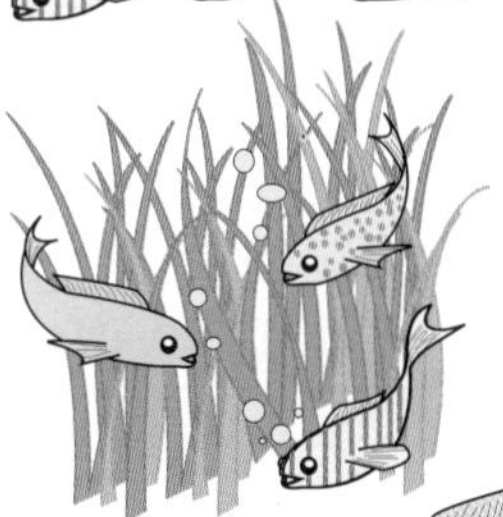

Predators appear and start to eat the visible fish. Those that blend in survive better. They are camouflaged.

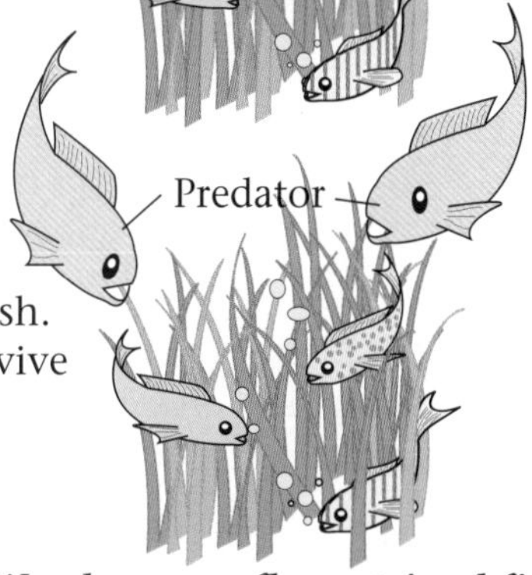

Weeds camouflage striped fish

Only the striped fish survive to breed and pass on the striped gene. Now all these fish are striped. The striped fish is the most fit and will survive in the weed.

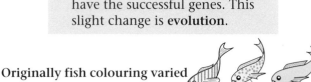

Survival of the fittest

Other ideas on evolution
In 1809 Jean Baptiste Lamarck proposed that organisms acquire new features which can then be passed to their young. e.g. if giraffes constantly stretch their necks to reach leaves their longer neck will pass to their offspring. A knowledge of genetics completely disproves this idea. (Stretching does not alter genes.)

Questions:

1. What do rabbits compete for in the wild?
2. How might a longer neck help a giraffe to survive?
3. Why might only long neck giraffes remain to mate with?
4. What will their young be like and why?
5. How is this the survival of the fittest? Explain.
6. Why are the survivors different from their ancestors?
7. What happened to the gene for short necks in giraffes?

EVIDENCE FOR EVOLUTION

1. Observed natural selection

Natural selection – the process	• Variety within one species. • Struggle for existence. • Survival of the fittest. • Most successful type emerges.

Warfarin-resistance in rats

Warfarin is used both as a rat poison and to 'thin' the blood in human's at risk from strokes.

As a rat poison, warfarin causes rats that eat it to bleed excessively and die. Warfarin is widely used and originally was very successful at reducing rat numbers. However, the few rats with resistance survived to reproduce and pass on their alleles for resistance to their young, so giving them resistance too. Rats with no resistance died out leaving only resistant ones to reproduce. Now populations of warfarin-resistant rats mean that warfarin is of no use as a rat poison. The change in the rat population over time is an example of evolution in action and has been well documented.

The resistant rats were the 'fittest', if warfarin was used. This is an example of survival of the fittest.

Alford: *A Colour Atlas of Pests of Ornamental Trees, Shrubs and Flowers*

White peppered moth, *Biston betularia*

The white peppered moth is well camouflaged on pale lichen-covered tree trunks found in rural areas. It is more likely to survive to reproduce than black moths, which would be highly visible to the birds that feed on them. This leads to survival of the fittest, the light moth is camouflaged and therefore 'fittest' in country areas. The opposite applies in city areas.

Rat

2. From fossils

Camels – These appeared first in North America. Fossils show the movement and changes in camels.

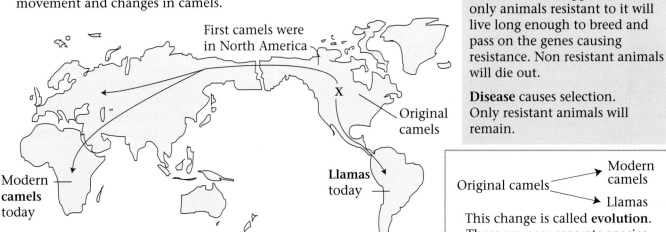

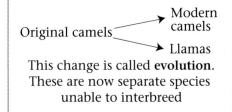

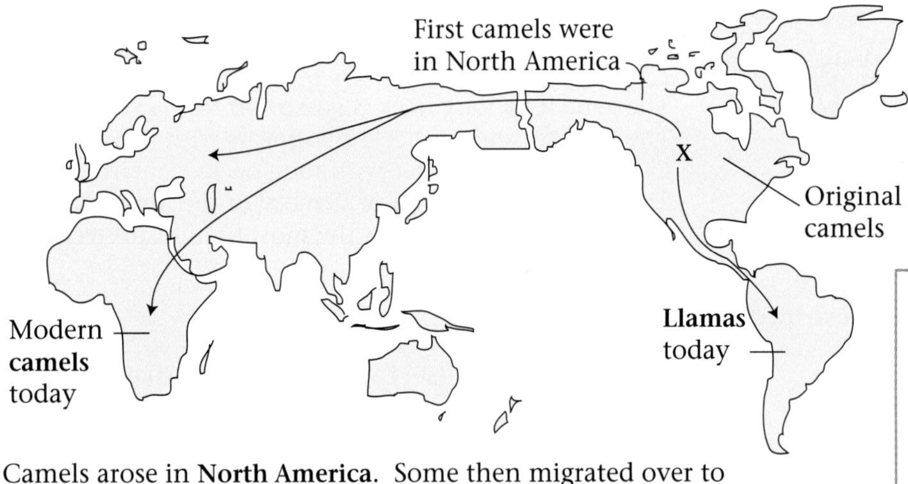

Camels arose in **North America**. Some then migrated over to **Africa** and **Asia**. Other camels moved down to **South America.**

Once the camels had been **isolated**, different features were **selected** for survival and the two camel populations became **different**. This movement is confirmed by the **fossil record** which is almost complete and can be dated.

Disease as a selection pressure
If a new disease appears, then only animals resistant to it will live long enough to breed and pass on the genes causing resistance. Non resistant animals will die out.

Disease causes selection. Only resistant animals will remain.

Original camels → Modern camels
Original camels → Llamas

This change is called **evolution**. These are now separate species unable to interbreed

Questions:

1. Why is there competition for survival?
2. Which peppered moths survive best in country areas and why?
3. What has caused the increase in the number of black peppered moths in Manchester?
4. Which two species have evolved from original camels and what is the evidence for this?
5. How can disease cause selection in animals? Which animals are more likely to survive?

SPECIES

All living organisms have two names (genus and species)

	Genus	Species	
e.g.	*Homo*	*sapiens*	(humans)
	Urtica	*dioica*	(stinging nettles)
	Lumbricus	*terrestris*	(earthworms)

Definition
Members of a species look alike and can interbreed successfully to produce fertile offspring.

Only members of the **same** species can breed and produce **fertile** offspring.

Parents (same species) *Homo sapiens*

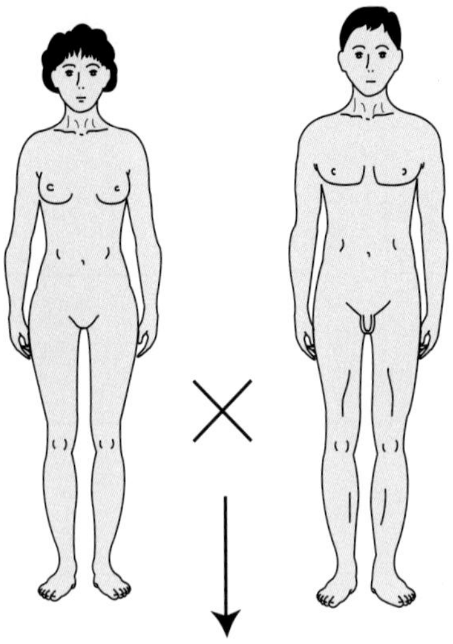

Mating possible and successful.

Children

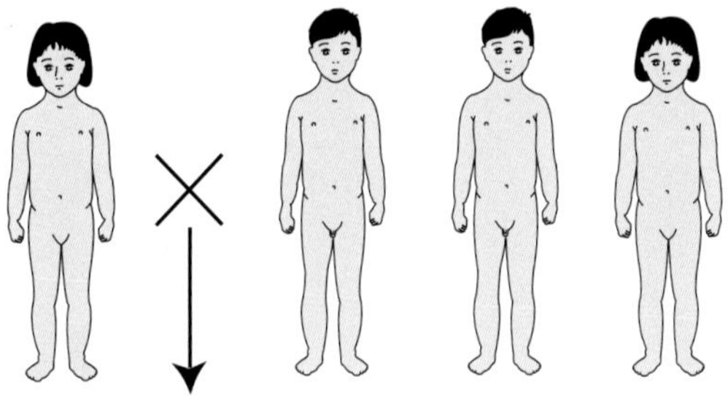

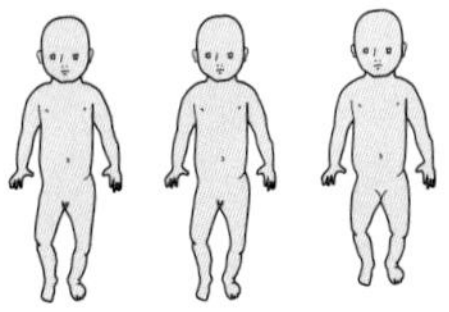

Fertile offspring

As mating is successful and produces fertile offspring they are the same species, *Homo sapiens*.

Parents (different species)

Horse   Donkey

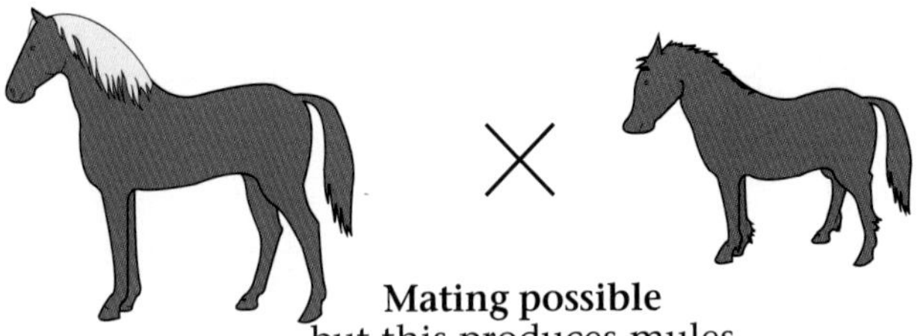

Mating possible but this produces mules.

Offspring   Mule

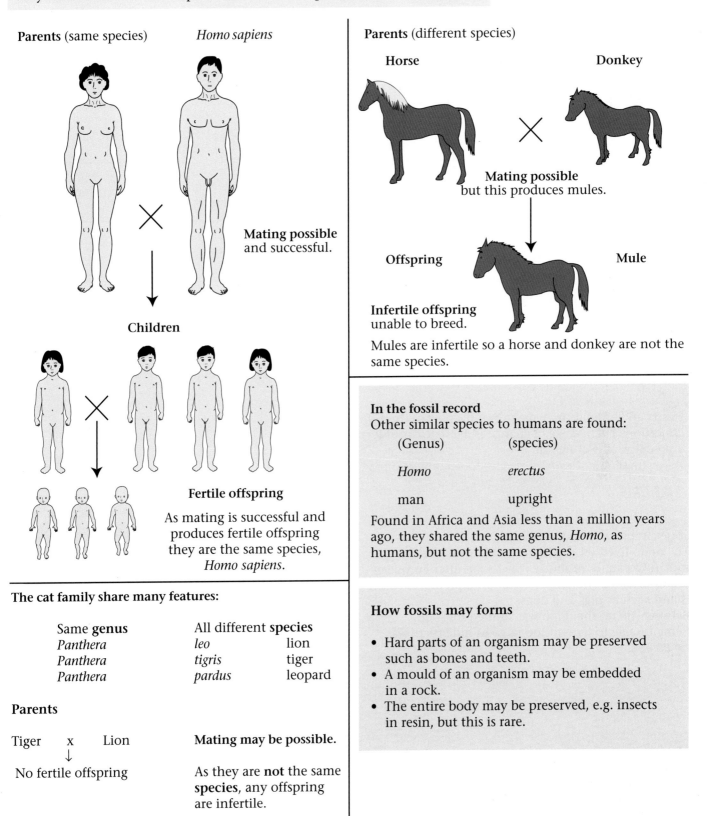

Infertile offspring unable to breed.

Mules are infertile so a horse and donkey are not the same species.

The cat family share many features:

Same **genus**	All different **species**	
Panthera	*leo*	lion
Panthera	*tigris*	tiger
Panthera	*pardus*	leopard

Parents

Tiger x Lion
↓
No fertile offspring

Mating may be possible.

As they are **not** the same **species**, any offspring are infertile.

In the fossil record
Other similar species to humans are found:

(Genus)	(species)
Homo	*erectus*
man	upright

Found in Africa and Asia less than a million years ago, they shared the same genus, *Homo*, as humans, but not the same species.

How fossils may forms

- Hard parts of an organism may be preserved such as bones and teeth.
- A mould of an organism may be embedded in a rock.
- The entire body may be preserved, e.g. insects in resin, but this is rare.

FOSSILS These are remains of plants and animals that have been preserved in rocks. A study of fossils tells us how life on earth has changed or evolved.

The fossil record is incomplete for many reasons:
- Soft tissue may not be fossilised.
- Fossils may not have been discovered.
- Fossilisation rarely occurs (animals and plants may be eaten or decomposed.

A narrowing indicates that many species died, reducing the total number. This suggests a mass **extinction**.

Here the **dinosaurs** become **extinct**. Evidence remains in the **fossil record**.

The number of mammals is increasing.

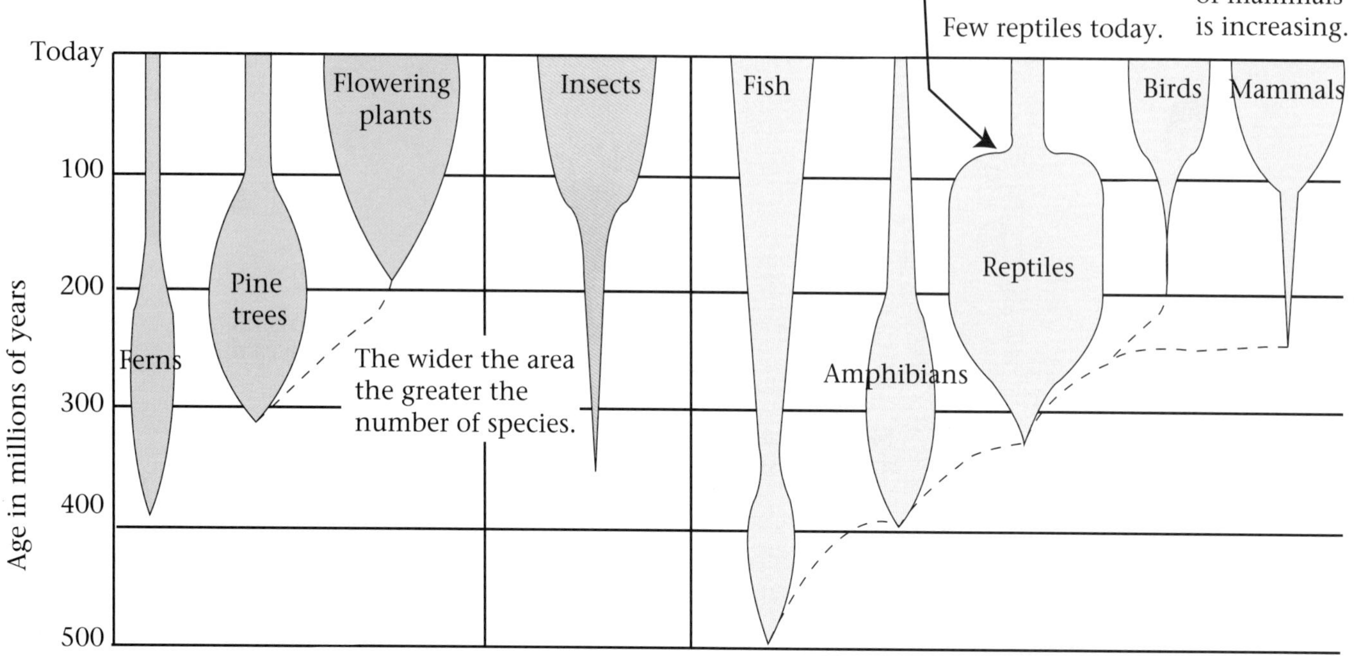

Algae and fungi present

Vertebrates evolved in the following order
Fish → Amphibians → Reptiles
↙ Mammals ↘ Birds
(evidence from the fossil record)

Plants and animals may die out or become **extinct** if conditions change, e.g. a change in climate, introduction of a predator, loss of habitat.
All of these might cause a species to become extinct.

For example, the **dodo** (extinct).
The dodo lived on the island of Mauritius in the Indian Ocean. It became extinct soon after European sailors arrived. The sailors introduced predators to the island such as pigs and dogs.
Between them, the dodo was hunted for food.
Being unable to fly and too large to hide there was no escape.

Questions:

1. What are fossils?
2. Where are fossils found?
3. Why is the fossil record incomplete?
4. Which groups were present around 400 million years ago?
5. What groups were dominant 200 million years ago, and how do we know?
6. Which two groups evolved from reptiles?
7. What does the term extinct mean?
8. Which bird became extinct in Mauritius, and why?

CLASSIFICATION OF LIVING THINGS Five kingdoms.

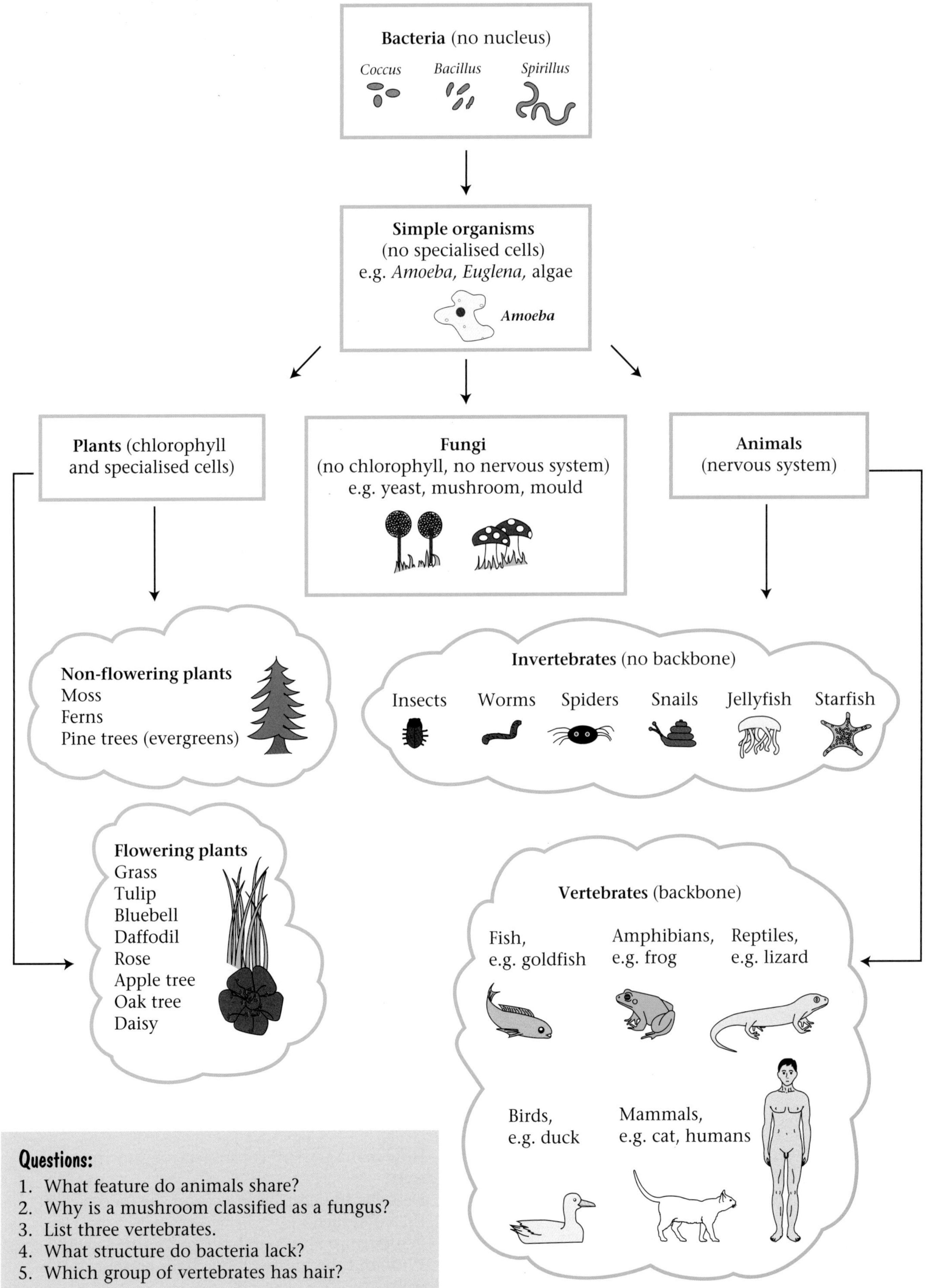

Questions:

1. What feature do animals share?
2. Why is a mushroom classified as a fungus?
3. List three vertebrates.
4. What structure do bacteria lack?
5. Which group of vertebrates has hair?

ECOLOGY

FOOD CHAINS AND FOOD WEBS

Food chain

Energy from sun	**Trophic or feeding level**	**Description**
Grass	**Producer**	They produce their own food by photosynthesis.
Caterpillars	**Primary consumer** (herbivores)	They eat or consume the producers.
Hedgehogs	**Secondary consumer** (carnivores)	These consume the primary consumers.
Foxes	**Tertiary consumer/ predator**	These eat the secondary consumers (carnivores). These are **hunters** that hunt, catch, and kill their prey for food. They are not normally killed by other animals. Top of the food chain.

A food chain is unrealistic as grass is eaten by many animals.
Foxes eat many animals not just hedgehogs.

Food web

This is more realistic than a food chain.

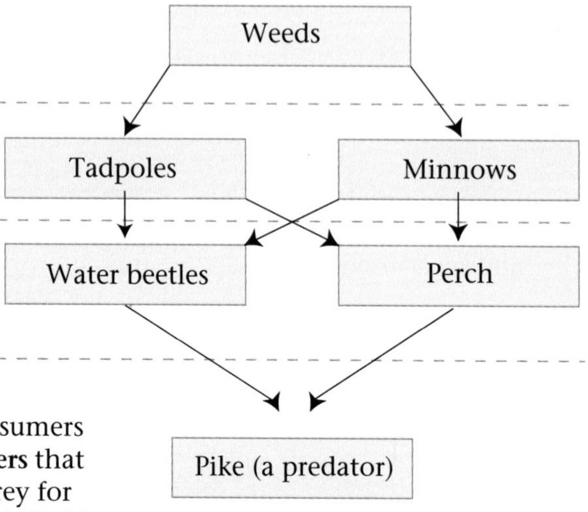

Food web in a river

Food web in a wood

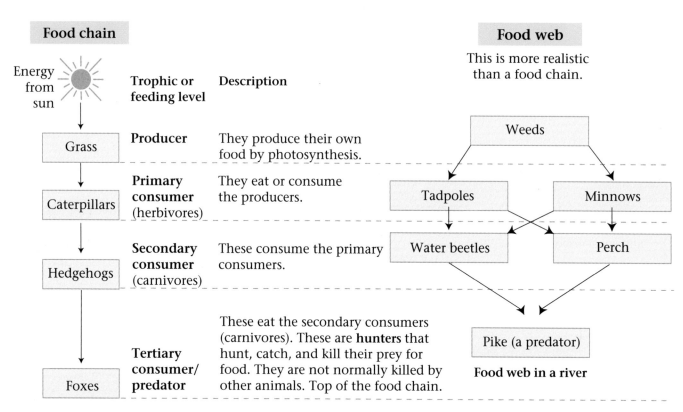

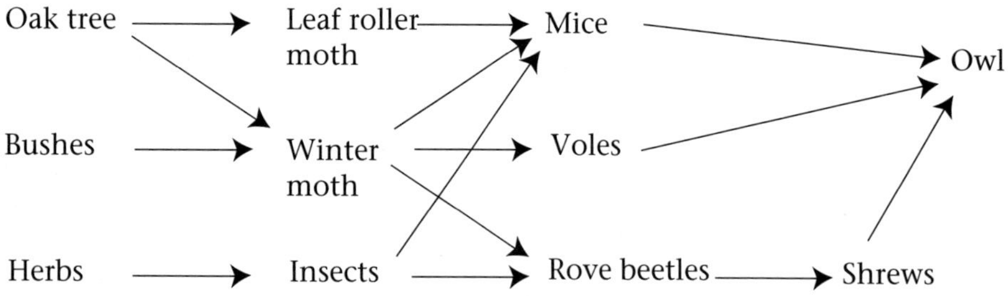

Questions:

Use the food web above to answer the following questions:

1. Name a producer. Why is it called a producer?
2. What do mice eat?
3. What eats winter moths?
4. Which is a predator? How do you know?
5. Name two primary consumers.
6. Name two secondary consumers.
7. How many tertiary consumers are shown in this web?
8. Describe the possible effects on the food web if the voles died.
9. Which animal is found at two feeding levels?
10. Write out the longest food chain present in this web.

WOODLAND HABITAT An oak wood supports a varied community.

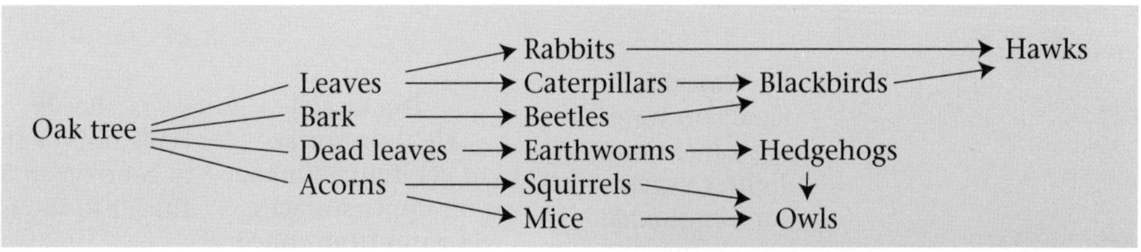

Mature trees, e.g. oak, beech. Maximum light for leaves to photosynthesise at top of canopy. **Herbivores** eat the leaves and seeds. **Carnivores** eat the herbivores.

Light

Layers in a woodland

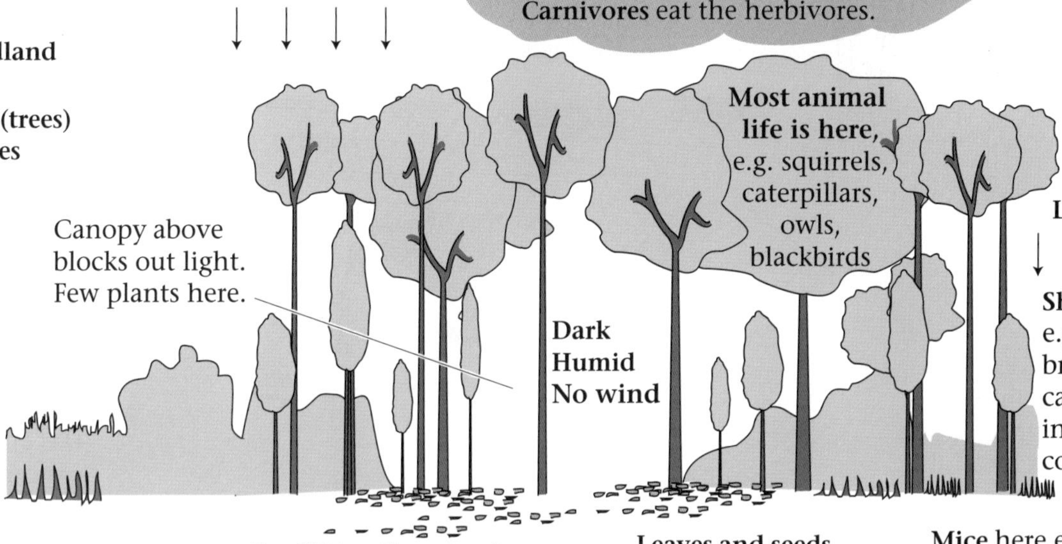

Canopy layer (trees)
Dominant species

Canopy above blocks out light. Few plants here.

Light

Shrubs, e.g. hazel, bramble, can grow in lighter conditions.

Shrub layer

Herb layer

Ground layer

Moss, fungi, earthworms, and beetles.

Leaf litter (from tree)
Earthworms and insects feed on dead leaves. Hedgehogs feed on earthworms.

Leaves and seeds fall to ground.

Fungi and bacteria feed on dead leaves here, i.e. decomposition.

Mice here eat acorns and seeds.

Death of mature trees leads to regeneration

From Bowes: *A Colour Atlas of Plant Propagation and Conservation*

Fallen trees.

Death of old tree allows light into woodland. Many plants grow.

Young tree able to grow as **light** and space are present. As it gets bigger it will block out the light below.

Light

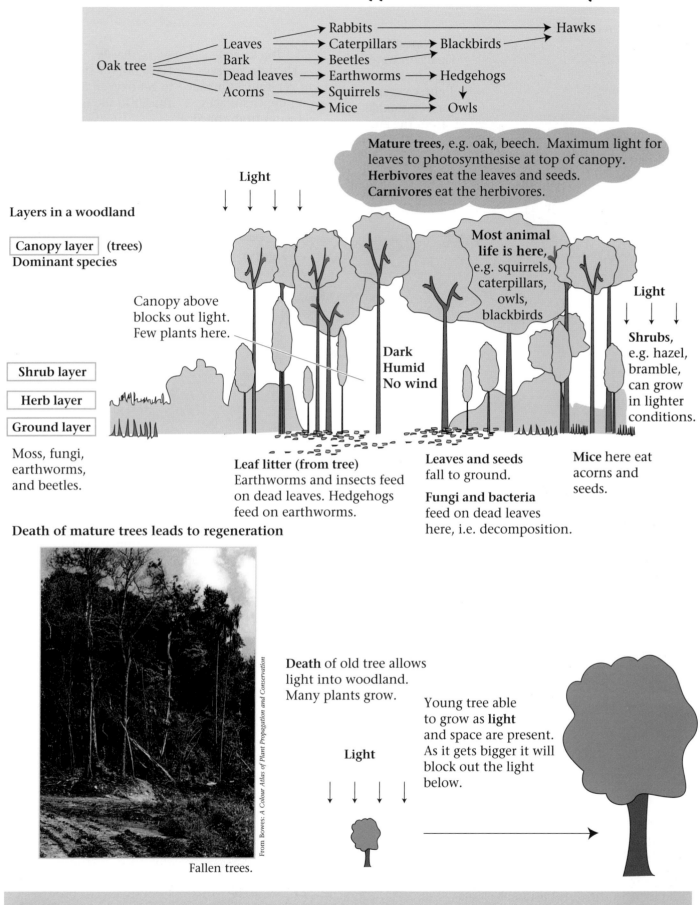

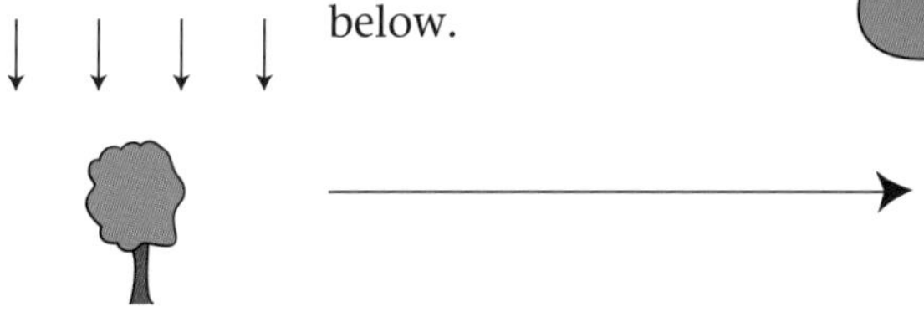

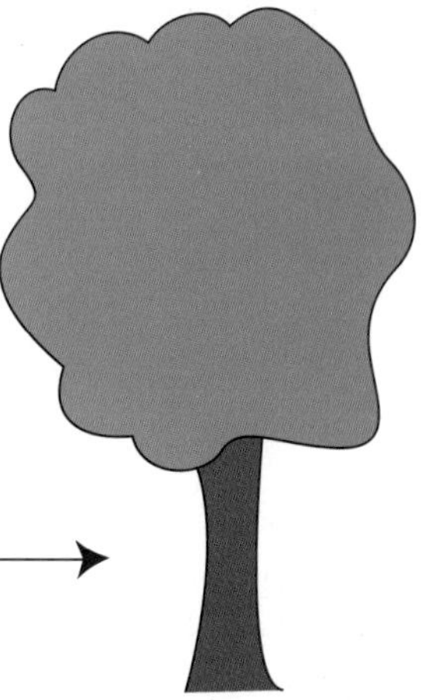

Questions:

Refer to the food web above:

1. What do squirrels eat?
2. What do blackbirds feed on?
3. Write out the longest food chain in the web.
4. If all the mice died how would it affect the food web?
5. Which organisms feed on dead leaves?

POND HABITAT How animals and plants are adapted to life in a pond.

20% oxygen in the air

Waterlily
Floats on water to get maximum light, oxygen + carbon dioxide from the air.

Duckweed
Float on surface to get light; stomata on upper surfaces get gases from the air.

Reeds
Have hollow stems to get oxygen from the air to the roots.

Pond snail
Lives in water but floats to surface to get more oxygen from the air.

Air tube

Light

No plants here
Dark

1% O_2 in water (max)

Cold

Pond weed
Grows near the top of pond, where light is present.

Mosquito larvae
Although it lives in water it gets oxygen from the air by a tube.

Trout
Has gills to obtain oxygen from the water.

Bloodworms
Feed on dead material. Have haemoglobin to collect the small amount of oxygen present.

Waterlouse
Feeds on dead material at bottom of pond.

Light Temperature O_2

Depth

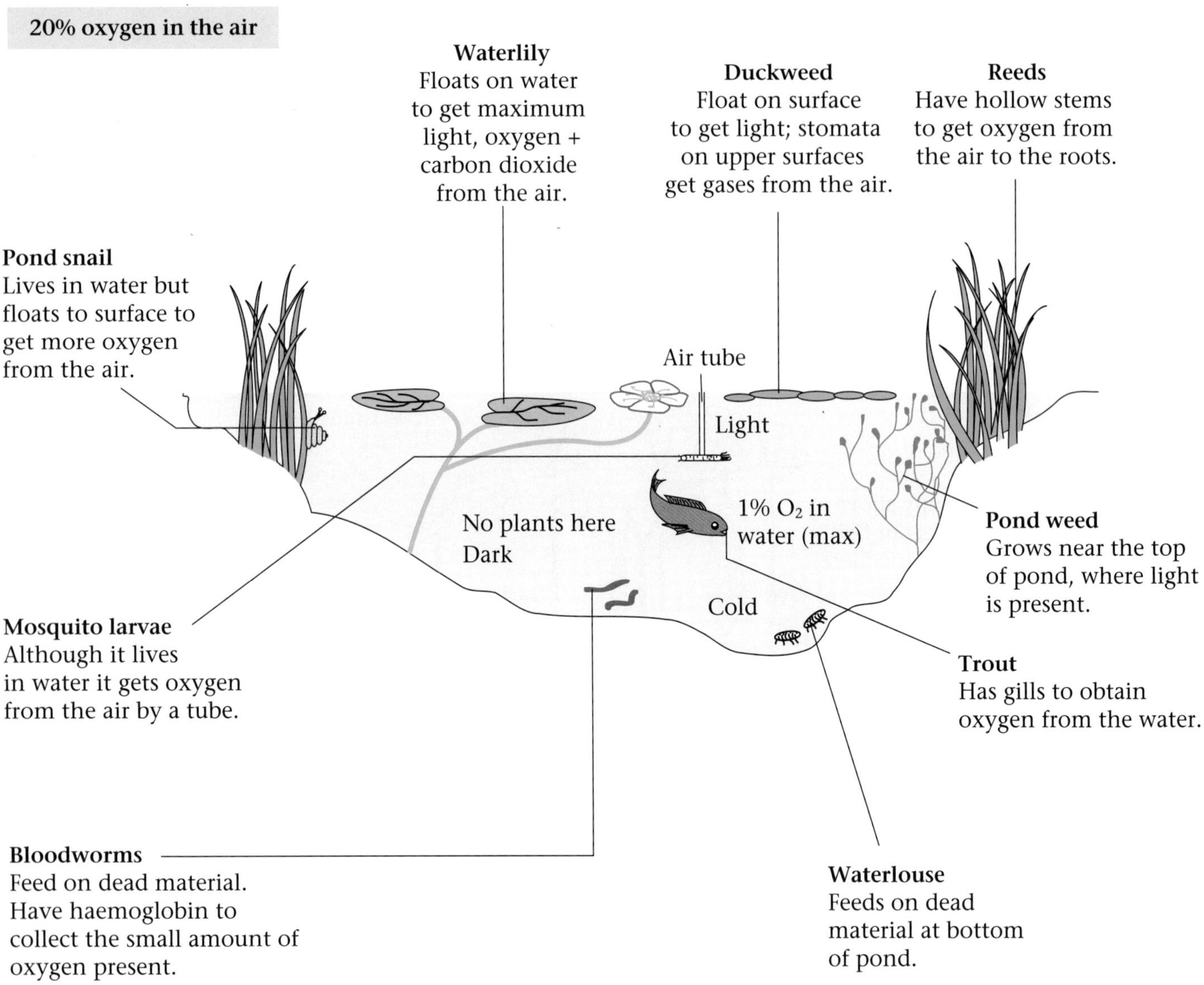

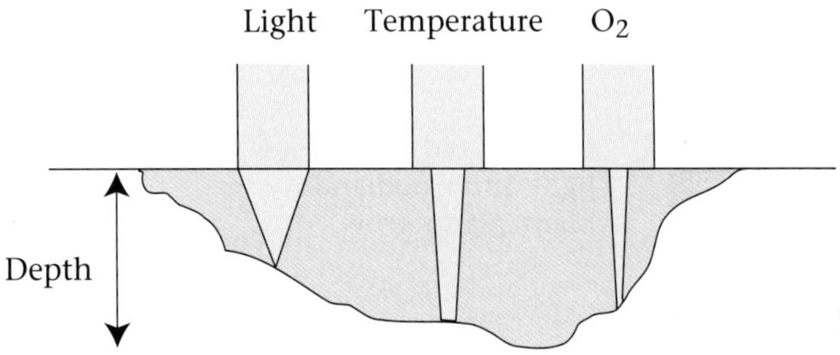

Width indicates amount present.

Questions:

1. How much oxygen is in water compared to the air?
2. Explain the distribution of plants.
3. What three factors decrease as you go deeper in the water?
4. Explain why mosquito larvae have an air tube.
5. Where are stomata usually found in aquatic plants? How does this compare with land plants?
6. How does haemoglobin help the bloodworm?
7. How do reeds get oxygen to their roots which are under water?

ANIMAL ADAPTATIONS

1. Animals in arctic conditions

e.g. **polar bear**, reindeer, wolves, Arctic fox, brown bears.

The adaptations of animals and plants determine their distribution and abundance.

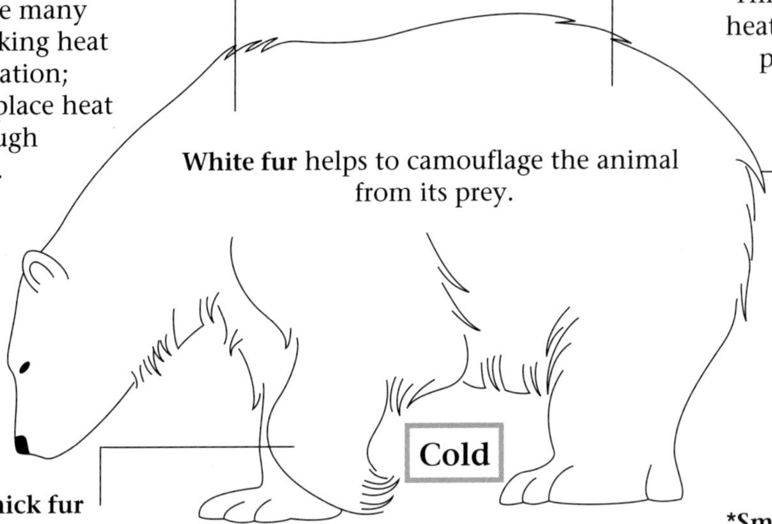

***Large size**
There are many cells making heat in respiration; these replace heat lost though the skin.

*** Thick layer of fat**
This helps to stop heat loss as fat is a poor conductor of heat.

White fur helps to camouflage the animal from its prey.

***Thick fur**
This traps a layer of warm air next to animal; insulates against cold air outside.

***Small ears and tail**
These reduce the surface area through which heat is lost. Less surface means less heat lost.

* All these features are designed to **reduce heat loss** to prevent a fall in the animal's temperature in a cold climate.

2. Animals in deserts

e.g. **Fennec fox** (smallest fox), kangaroo rat, small desert fox, desert hedgehog.

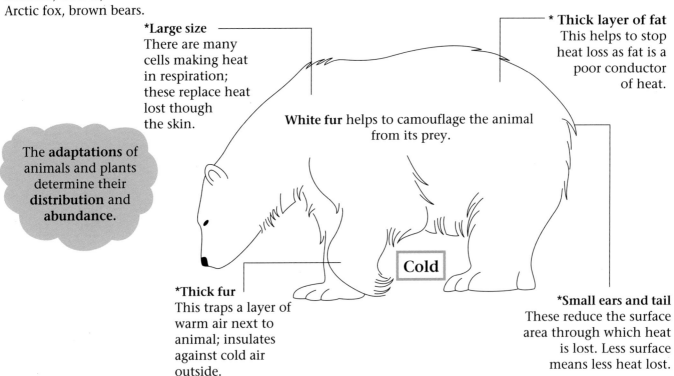

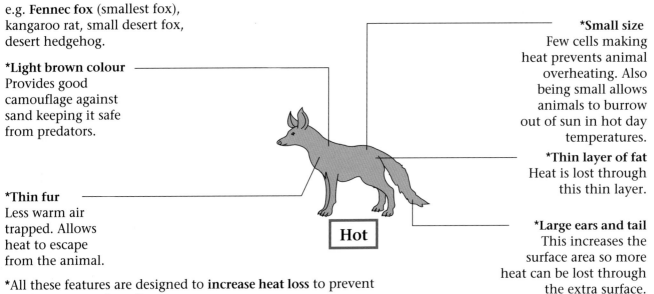

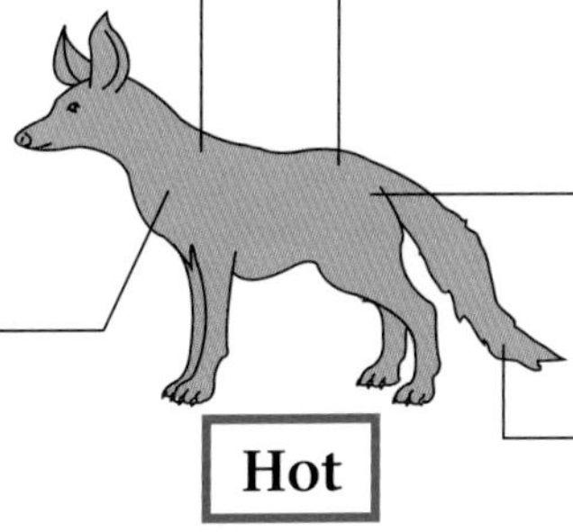

***Light brown colour**
Provides good camouflage against sand keeping it safe from predators.

***Thin fur**
Less warm air trapped. Allows heat to escape from the animal.

***Small size**
Few cells making heat prevents animal overheating. Also being small allows animals to burrow out of sun in hot day temperatures.

***Thin layer of fat**
Heat is lost through this thin layer.

***Large ears and tail**
This increases the surface area so more heat can be lost through the extra surface.

*All these features are designed to **increase heat loss** to prevent overheating in a hot climate.

Questions:

1. Animals found in arctic conditions are large. Can you suggest why?
2. How do large ears and tails help a mammal living in hot conditions?
3. Name 3 features found in mammals living in cold conditions. Explain why each is necessary.
4. How does thick fur help a mammal to keep warm in cold conditions?

ESTIMATING POPULATION SIZE

Animal populations

Most animals move making counting difficult.
The following equipment is used to collect and count small animals.

Pooters
Pooters are used to suck small insects from vegetation without harming them.

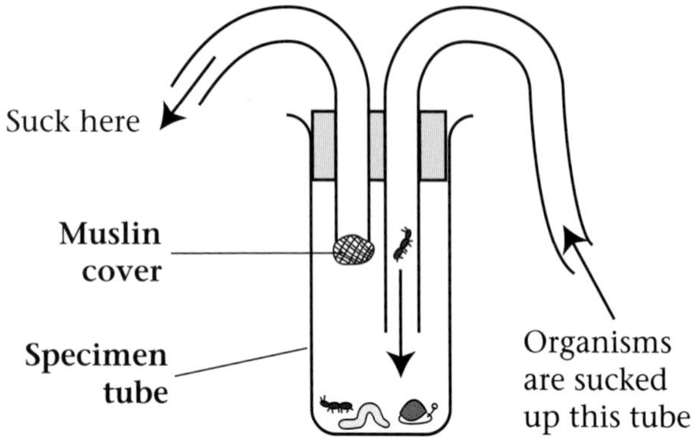

Nets e.g. sweep nets

A sweep net can be swept through grass, bushes, streams or ponds. Small insects are collected.

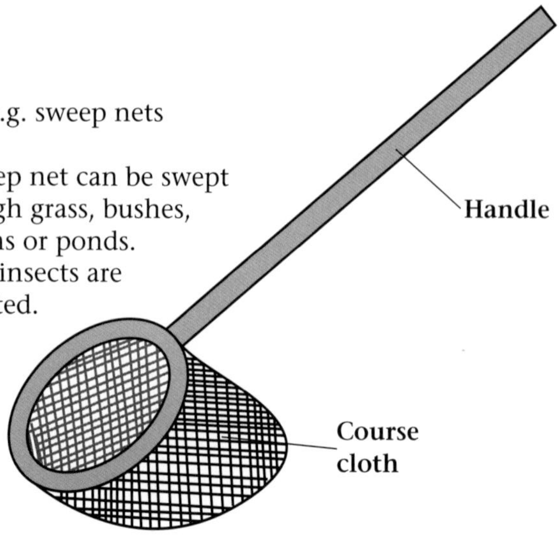

Pit-fall traps

Walking and crawling insects from the soil surface or leaf litter can be attracted to the jar with jam or meat.

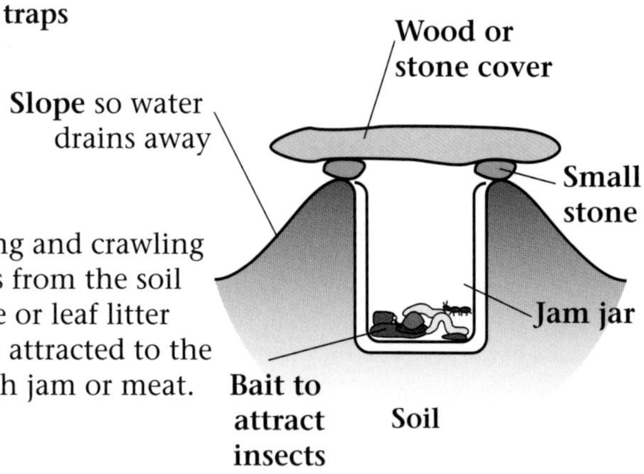

Tullgren funnel

Small insects and other arthropods move down, away from the heat towards the cooler moist conditions. They fall through holes in the metal gauze and into the alcohol below.

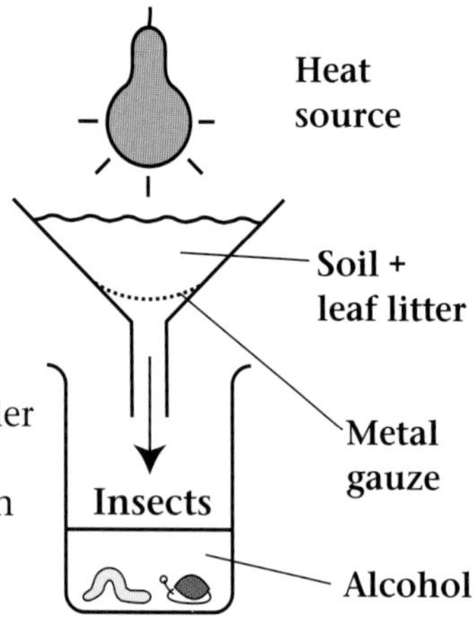

Plant populations

Counting is simpler as plants do not move. Hollow square frames called **quadrats** are used to count plants.

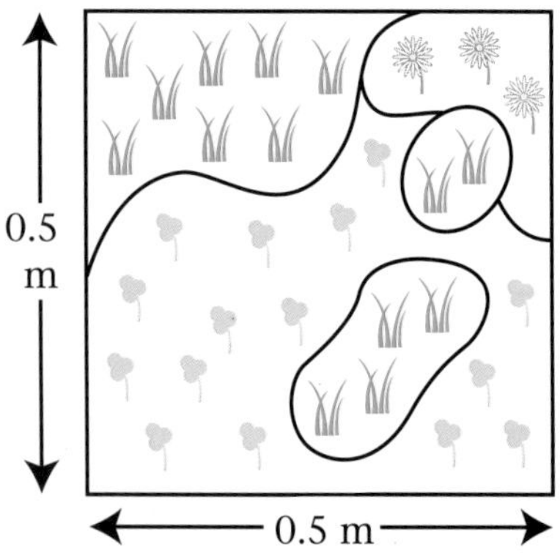

The plants can be counted or the % cover calculated

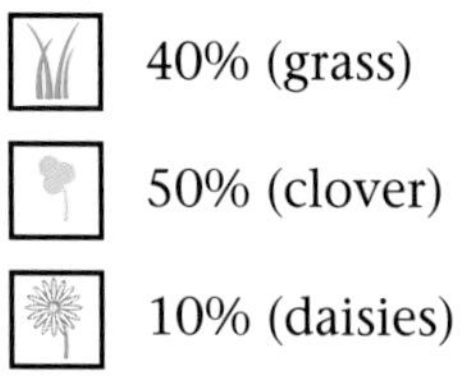

This is 0.25 m^2 quadrat frame commonly used in the field.

Gridded quadrats can also be used.

There are 100 squares.

If a plant appears in 50 squares, this can be called 50% occurrence.

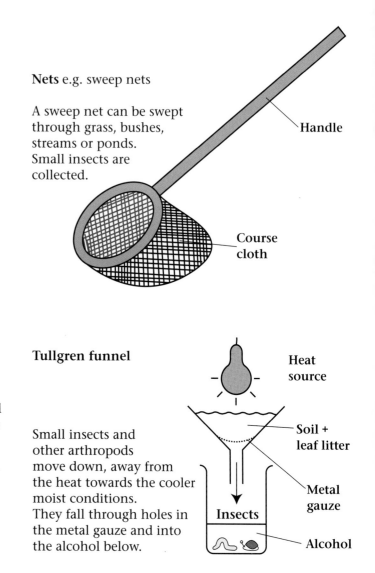

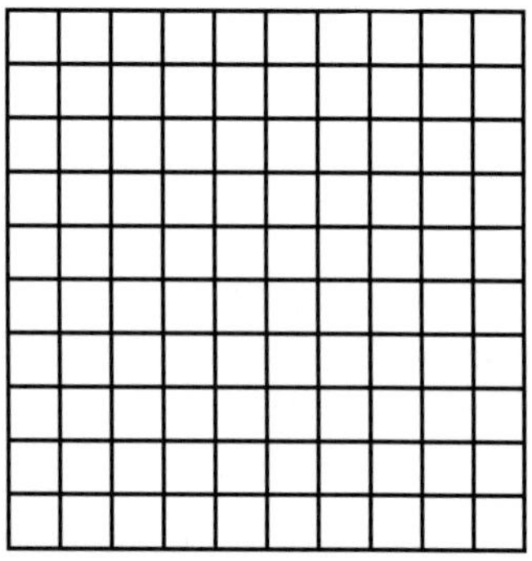

Sampling
All these methods indicate what plants and animals are present in a habitat. Normally only a few samples are taken and the total numbers can then be estimated. The more samples taken, the more reliable the estimate.

Sampling must be repeated many times at random to give a reliable indication of the plant population. The results can then be averaged. A random sampling method avoids bias.

POPULATIONS

A population is a group of organisms of the same species in one area, e.g. human population in London.

Births and **immigration** *increase* the size of the population.

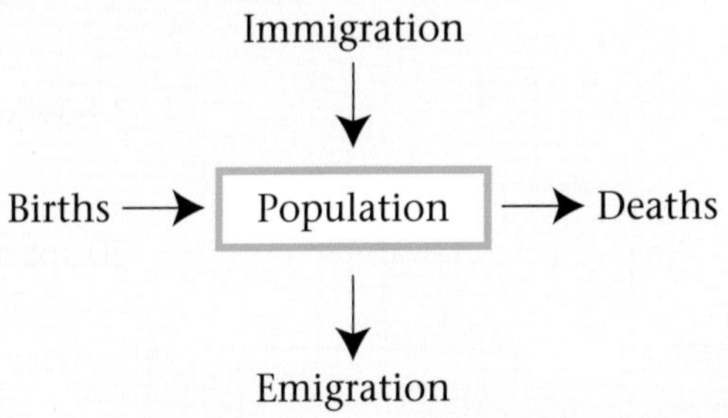

Deaths and **emigration** *reduce* the size of the population.

Paramecium and yeast in a beaker

Paramecium eats yeast.

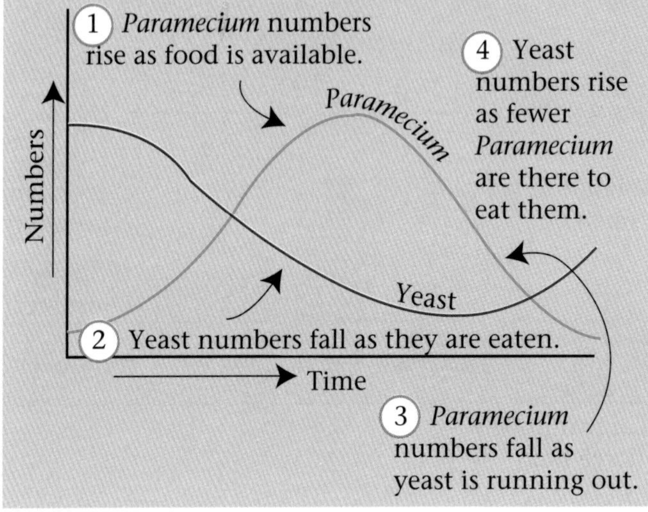

1 *Paramecium* numbers rise as food is available.
4 Yeast numbers rise as fewer *Paramecium* are there to eat them.
Paramecium
Numbers
Yeast
2 Yeast numbers fall as they are eaten.
Time
3 *Paramecium* numbers fall as yeast is running out.

Population of cane toads in Queensland, Australia

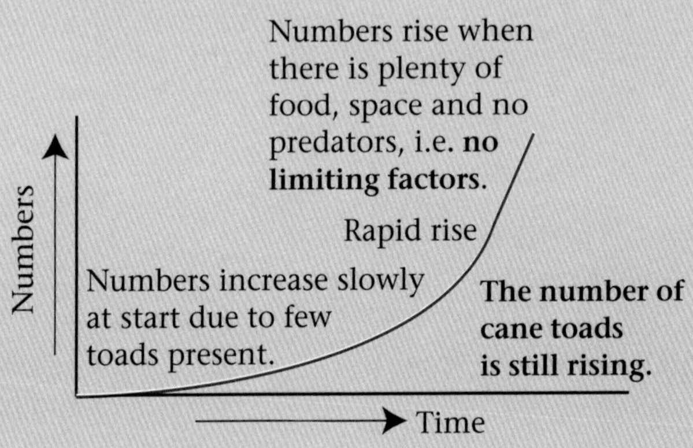

Cane toads were introduced into Australia to feed on the caterpillars spoiling the sugar cane, an example of biological control. They were alien to the country and no initial tests were carried out. Unfortunately, the cane toads did not feed on the caterpillars and instead are feeding on other, often endangered, species. The number of cane toads is now rising out of control.

Predator–prey relationship:
Rabbits (prey) ⟶ Foxes (predators)

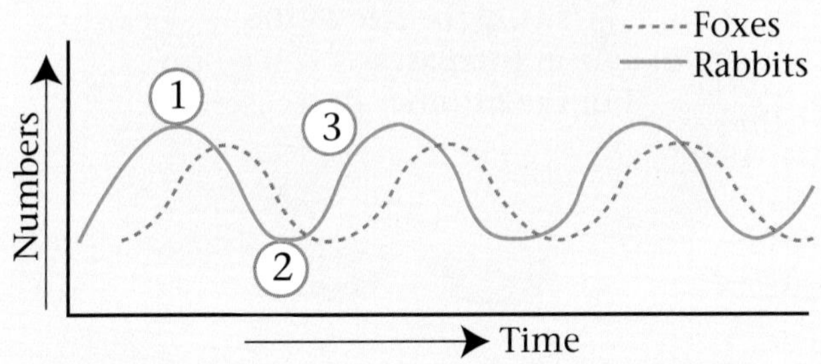

One population controls the size of the other:

1. Large numbers of rabbits provide food for foxes. So fox numbers increase, which causes rabbit numbers to decrease.
2. Few rabbits mean little food for foxes, so fox numbers decrease.
3. Few foxes allows rabbit numbers to increase again.

Yeast population in a beaker

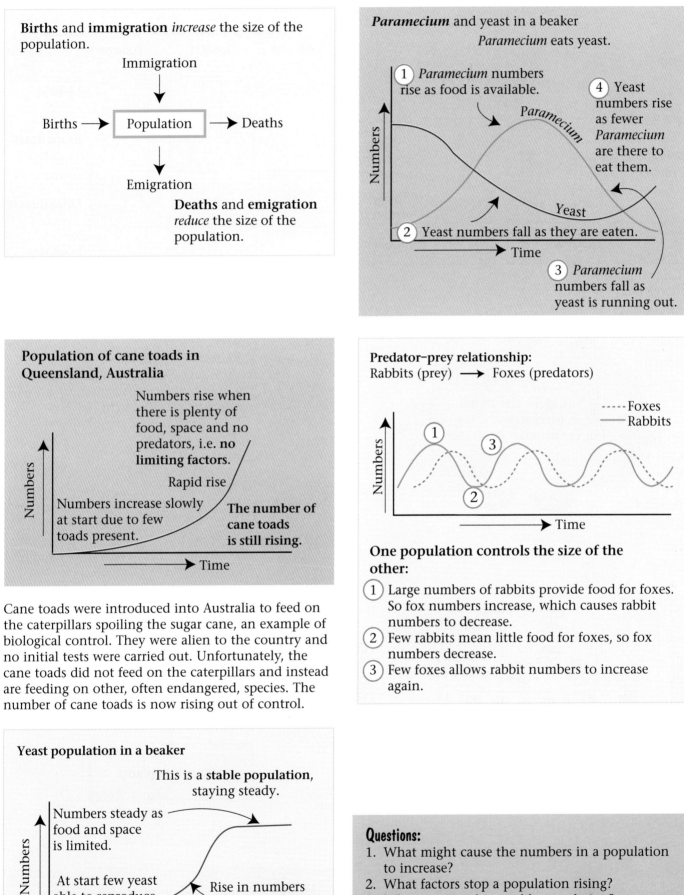

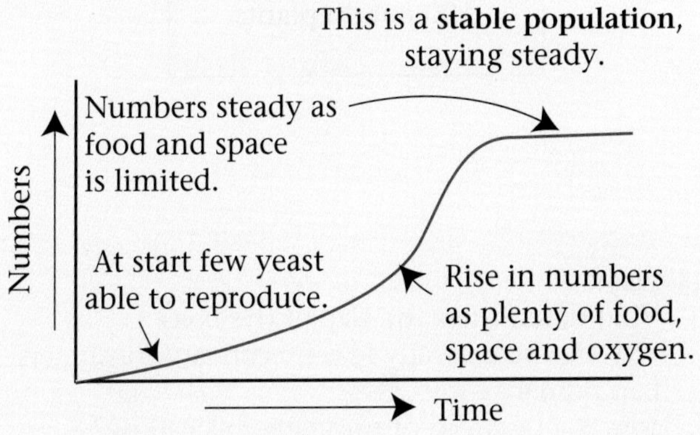

This is a **stable population,** staying steady.
Numbers steady as food and space is limited.
Numbers
At start few yeast able to reproduce.
Rise in numbers as plenty of food, space and oxygen.
Time

Questions:

1. What might cause the numbers in a population to increase?
2. What factors stop a population rising?
3. What is meant by a stable population?
4. How do rabbit numbers control fox numbers? What is this an example of?
5. What is a population?

PYRAMIDS OF NUMBERS Size of box depends on number (not mass).

The size of the box is determined by the **number** of organisms at each level.

One million e.g. grass → 10 sheep → 2 foxes

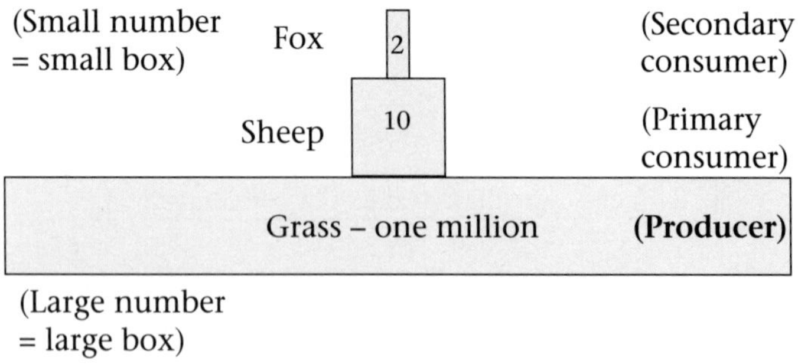

Usually the size of the organism **increases** as you go **up**, but the **number decreases**.

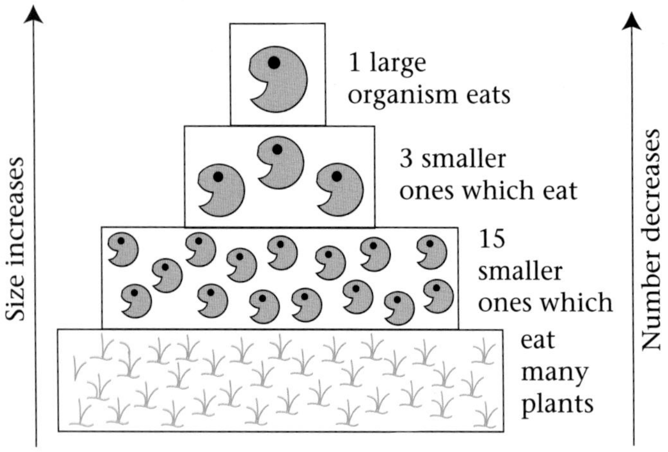

Unusual pyramids of numbers

1. 1 oak tree → 1,000 insects → 20 sparrows → 3 hawks

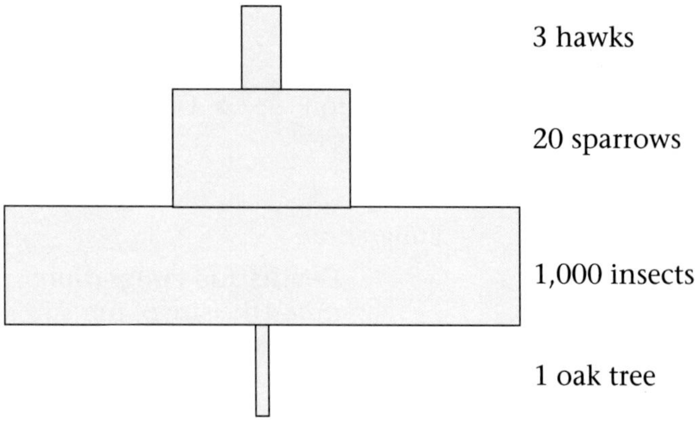

The **size of a population** depends on how successfully it **competes** for factors such as: food, water, shelter, light, minerals.

Unusual pyramid of numbers

2. 500 potato plants → 1 human → 20 fleas → 1,000 parasites

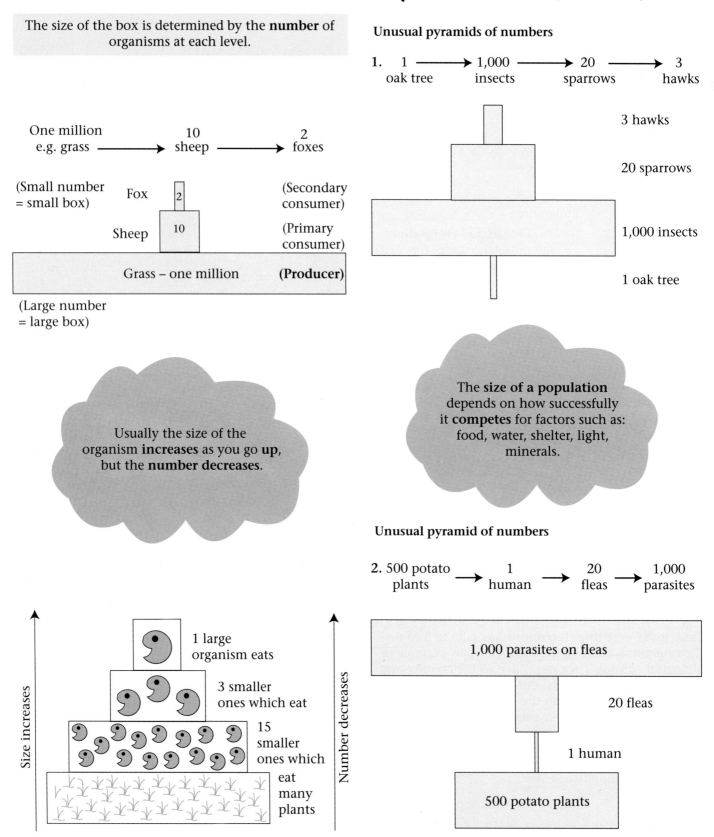

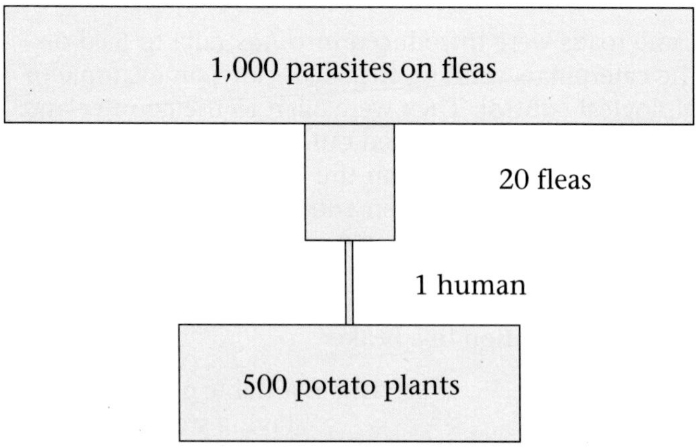

Pyramids of number give no indication of the **mass** of each organism. One grass plant is given the same area as one oak tree in the pyramid. This is misleading.

Questions:

1. What determines the size of the box?
2. Why are there usually fewer secondary consumers than primary consumers?
3. How is a pyramid of numbers misleading?

PYRAMIDS OF BIOMASS Size of box depends on mass.

The size of the box is determined by the **mass** of organisms at each level.

Oak tree ⟶ Insects ⟶ Sparrows ⟶ Hawks

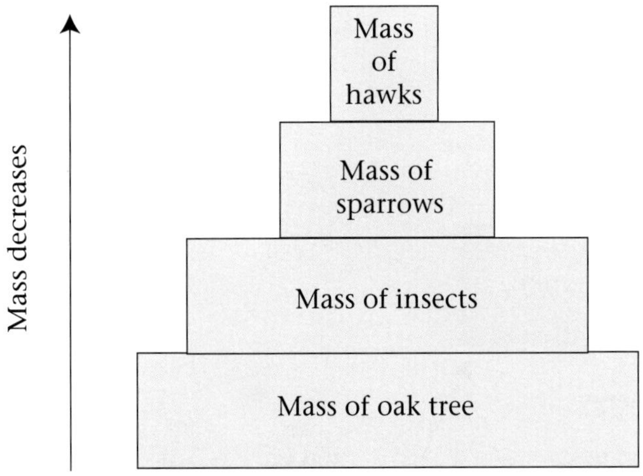

A pyramid of biomass

The total weight of an organism in a particular area is called its biomass. Normally the mass decreases as you go up.

Pyramids of biomass are usually pyramid-shaped.

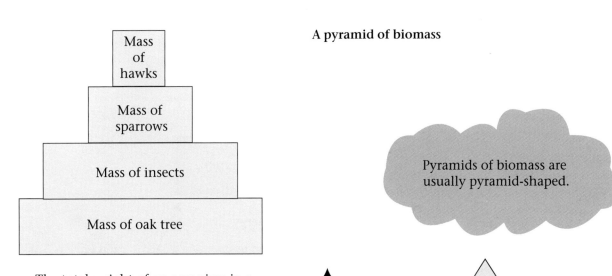

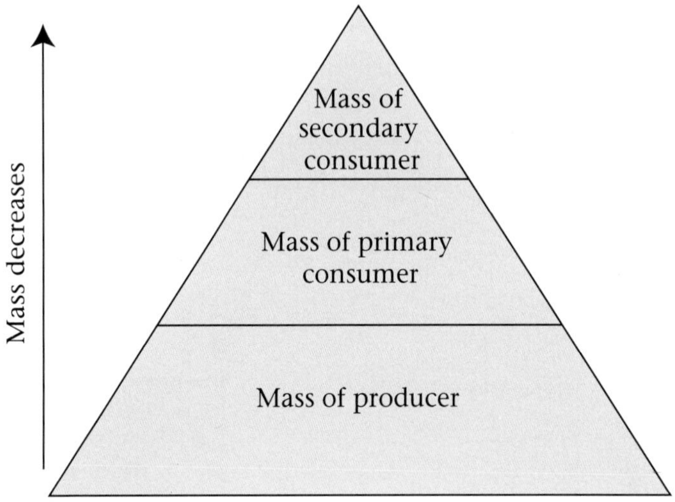

Dry mass is normally used as it is more accurate but it involves drying organisms, which kills them. Fresh mass is unreliable, especially in plants, as rain greatly increases mass for a while.

Problems

- Dry mass involves killing, which is undesirable.
- Mass varies at different times of year.
- Some masses are more productive, e.g. almost all of a grass plant produces sugar to feed herbivores; only a small percentage of an oak tree does this.

Questions:

1. What is meant by the term biomass?
2. What is dry mass and how is it measured?
3. What shape are pyramids of biomass?
4. Why does an oak tree produce less sugar per gram than grass?

PYRAMIDS OF ENERGY

The size of each box in the food chain below is determined by the **energy** flowing through each level. (Energy is measured in **kilojoules** – kJ.)

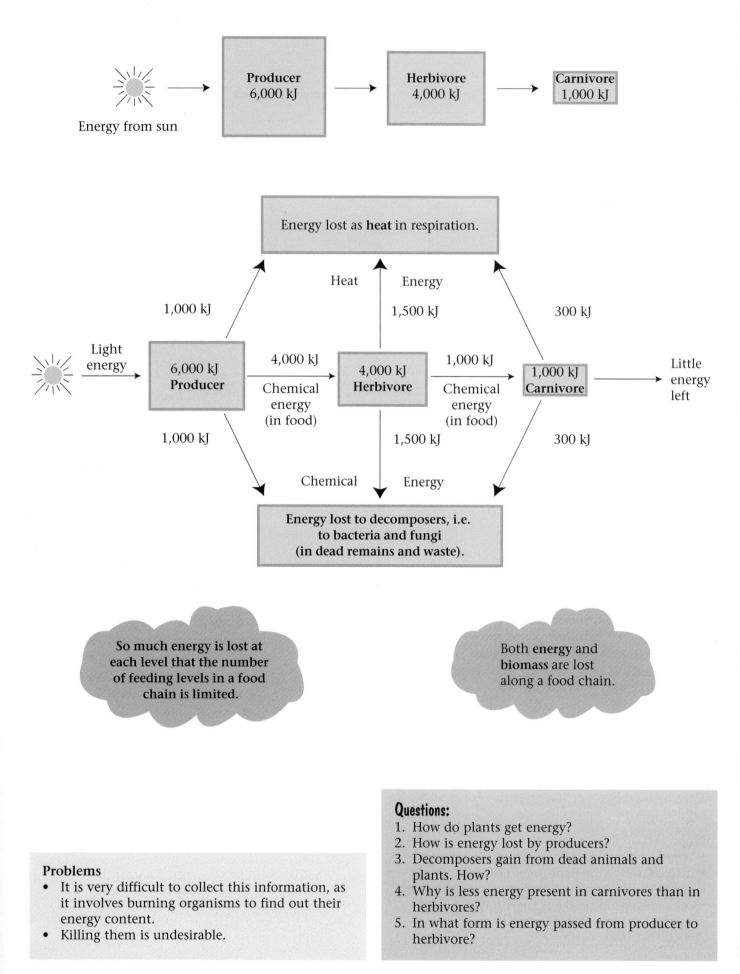

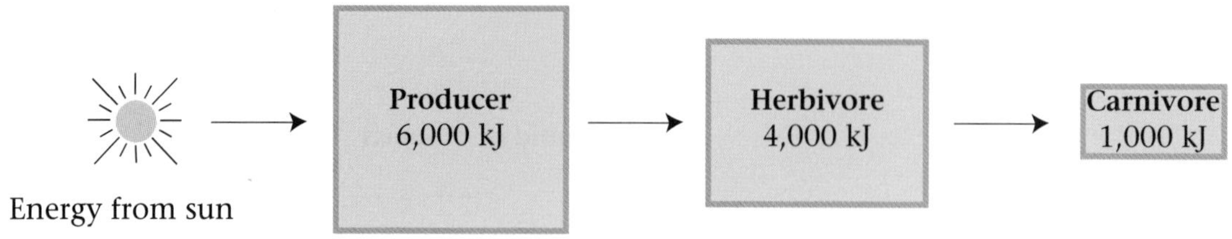

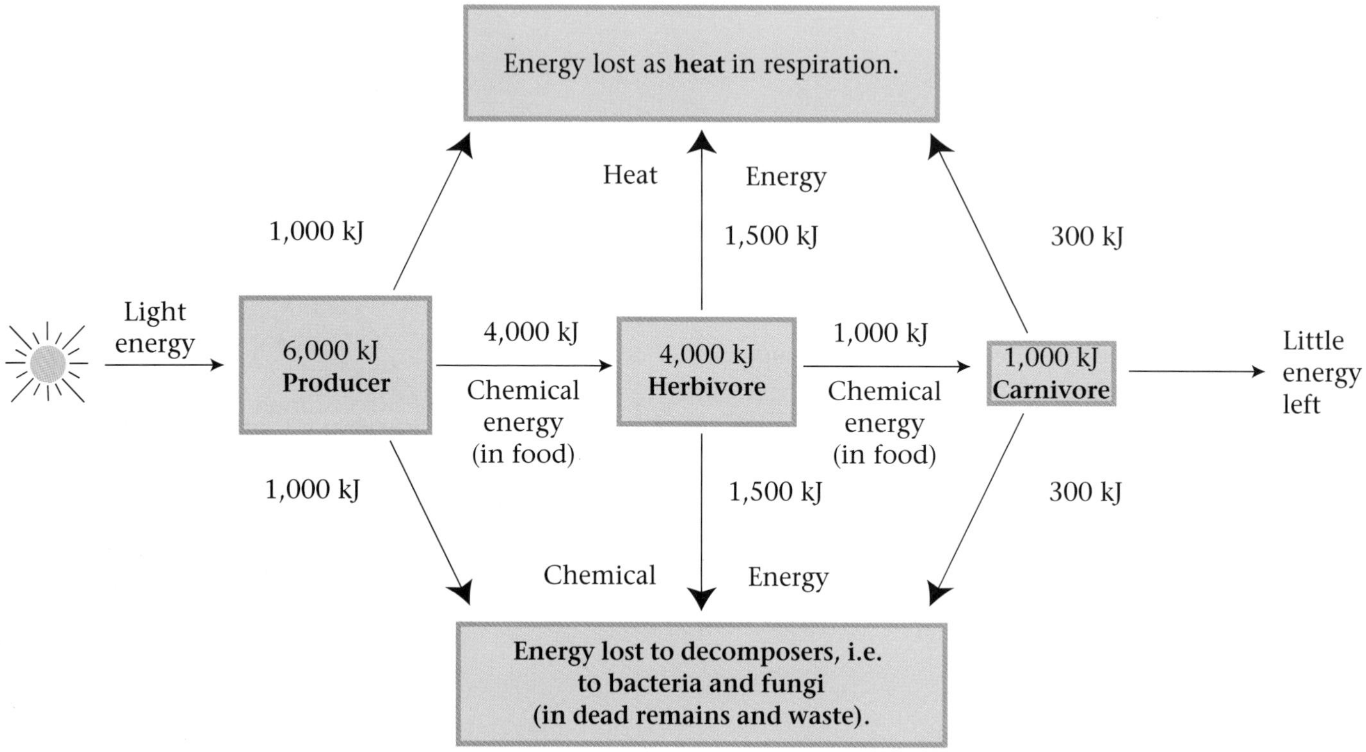

So much energy is lost at each level that the number of feeding levels in a food chain is limited.

Both **energy** and **biomass** are lost along a food chain.

Problems

- It is very difficult to collect this information, as it involves burning organisms to find out their energy content.
- Killing them is undesirable.

Questions:

1. How do plants get energy?
2. How is energy lost by producers?
3. Decomposers gain from dead animals and plants. How?
4. Why is less energy present in carnivores than in herbivores?
5. In what form is energy passed from producer to herbivore?

ENERGY LOSSES AND FOOD PRODUCTION

Energy enters the food chain from the **sun**. When a cow eats grass, the energy passes to the cow. A great deal of energy is lost at each feeding level.

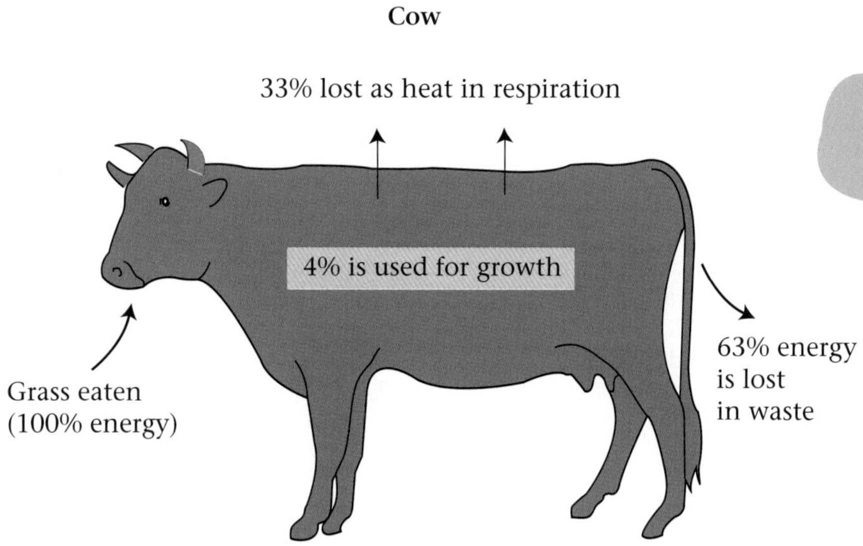

We get energy and nutrients from the food we eat.

96% of the energy in the grass taken in by a cow is **not** passed on to the next feeding level.

Energy is **lost** as it flows along a food chain. The **longer** the chain the **more** energy that is lost.

A.

Food produced
0.2 tonnes of meat

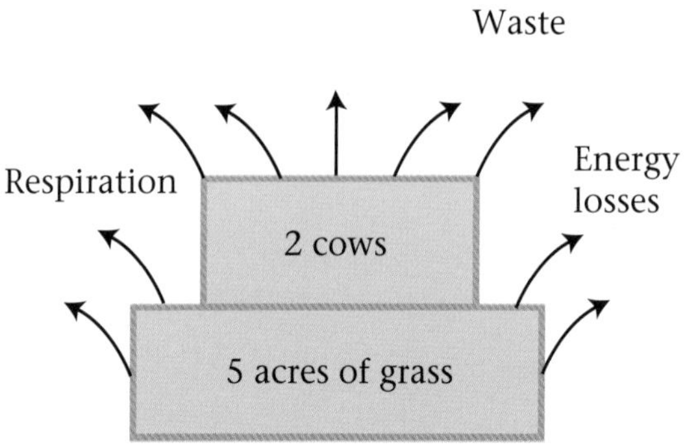

In **A** energy is lost by grass and the cows so little remains as food.

In **B** energy is lost only by potato plants, leaving more as food.

Eating **meat** is wasteful as little food remains after all the energy losses. Eating **plants** reduces the energy lost, so more remains as food. A diet with less meat and more vegetables would **increase** the food available for the world's population.

B.

Food produced
40 tonnes of potato

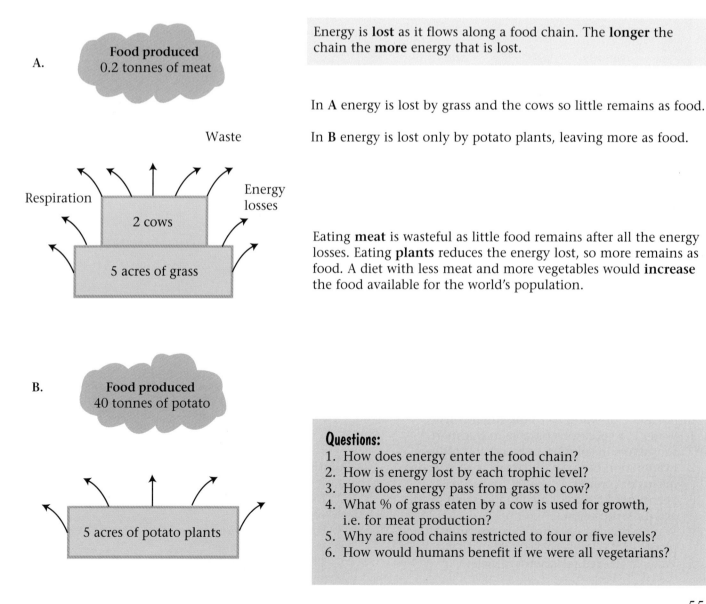

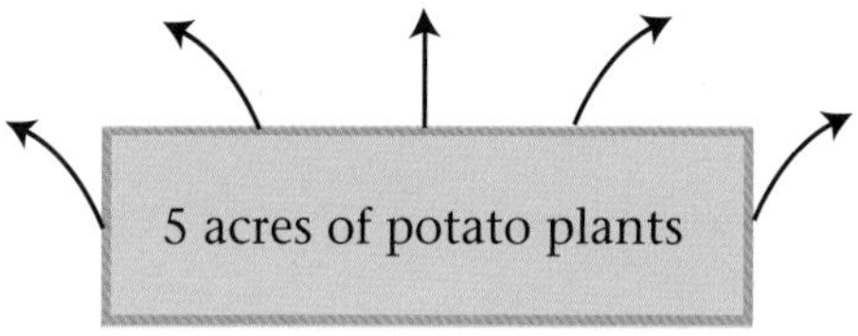

Questions:
1. How does energy enter the food chain?
2. How is energy lost by each trophic level?
3. How does energy pass from grass to cow?
4. What % of grass eaten by a cow is used for growth, i.e. for meat production?
5. Why are food chains restricted to four or five levels?
6. How would humans benefit if we were all vegetarians?

WATER CYCLE

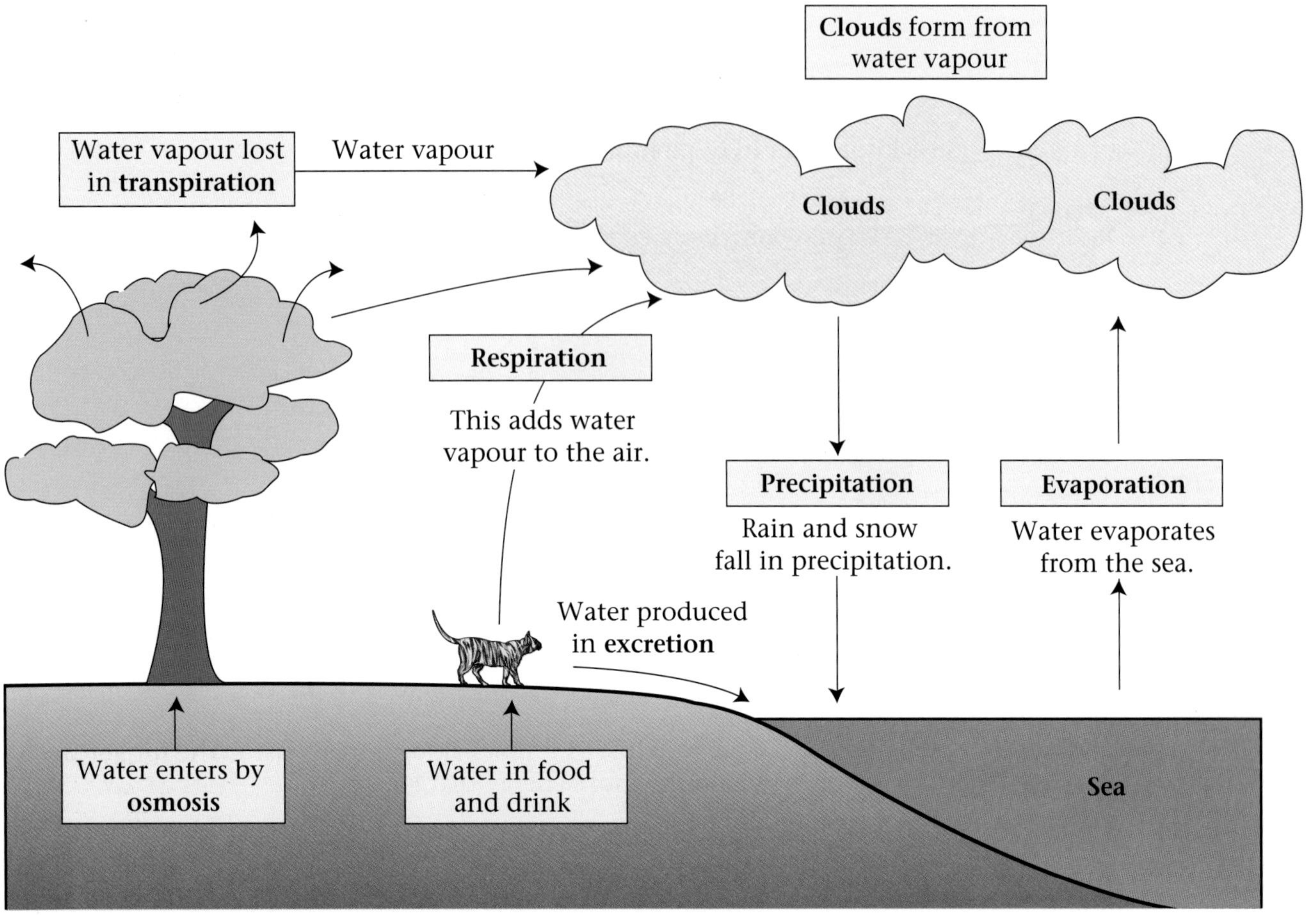

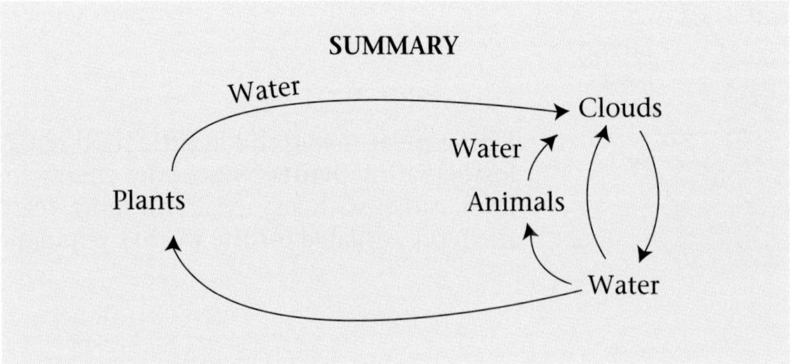

Importance of water to living organisms

1. Water is a good solvent, able to transport substances in solution in blood and phloem.
2. Evaporation of water (in sweat) is an effective means of cooling.
3. Sperm need a watery medium to swim to the egg for fertilisation.
4. Water provides a habitat for some animals and plants.
5. Water is needed for photosynthesis.

Questions:

1. By what process does water enter plant roots?
2. Name two ways in which animals produce water.
3. How do plants lose water?
4. What forms clouds?
5. What is precipitation?

CARBON CYCLE Carbon is present in protein, carbohydrate and lipid.

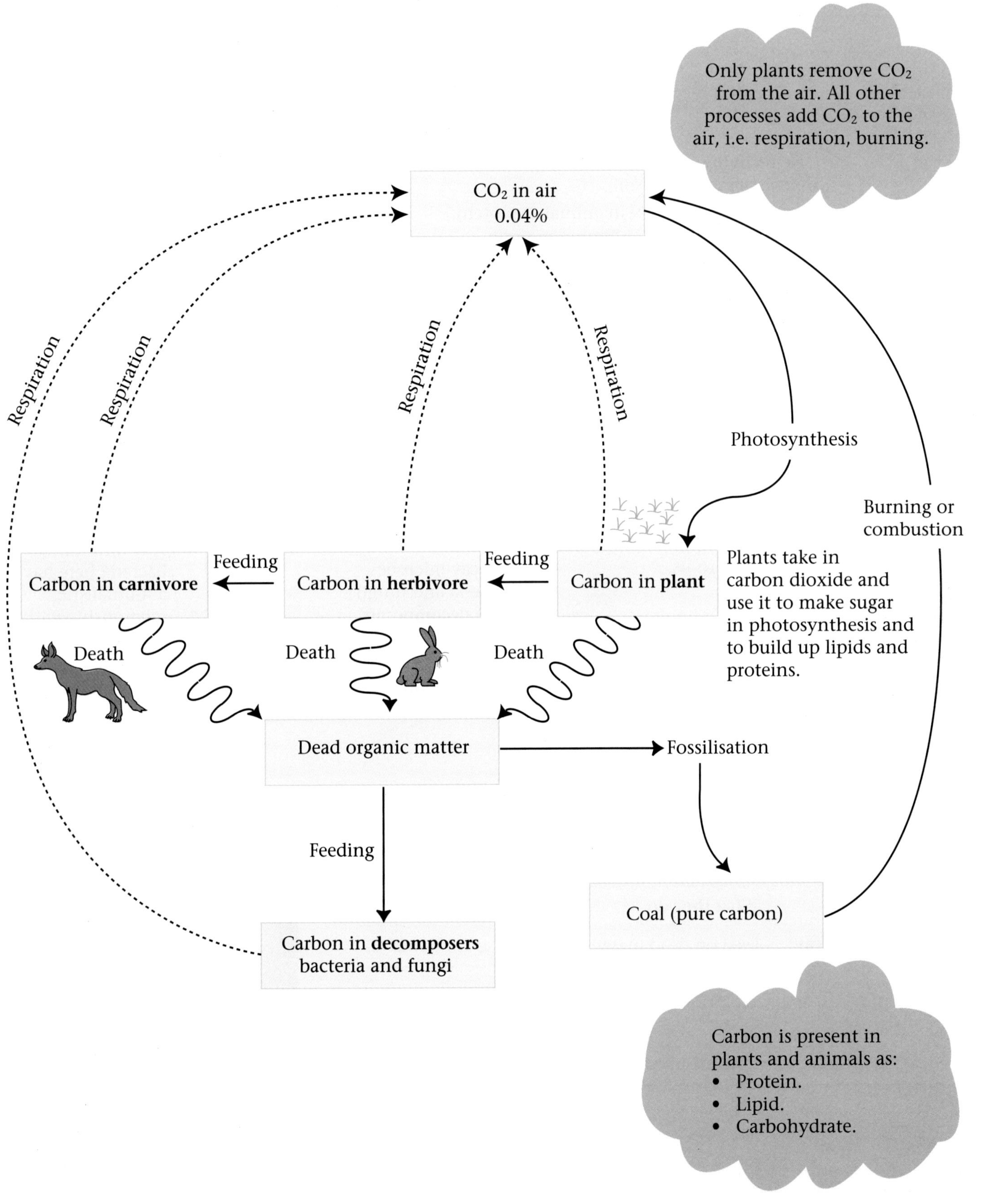

Only plants remove CO_2 from the air. All other processes add CO_2 to the air, i.e. respiration, burning.
CO_2 in air
0.04%
Respiration
Respiration
Respiration
Respiration
Photosynthesis
Burning or combustion
Carbon in carnivore
Feeding
Carbon in herbivore
Feeding
Carbon in plant
Plants take in carbon dioxide and use it to make sugar in photosynthesis and to build up lipids and proteins.
Death
Death
Death
Dead organic matter
Fossilisation
Feeding
Coal (pure carbon)
Carbon in decomposers bacteria and fungi
Carbon is present in plants and animals as:
• Protein.
• Lipid.
• Carbohydrate.

Questions:

1. For what process do plants take in CO_2?
2. What do plants use the CO_2 to make?
3. What process adds CO_2 to the air by all living organisms?
4. How does carbon pass from plants to animals?
5. What feeds on dead animals and plants?
6. How do the decomposers return carbon to the air?
7. What happens to plants when they are fossilised? How does this cause CO_2 to be added to the air?

NITROGEN CYCLE

Most plants cannot use the **nitrogen** in the air as it is **insoluble. Nitrates** are and enter roots dissolved in water.

Nitrogen ⟶ Protein ⟶ Growth

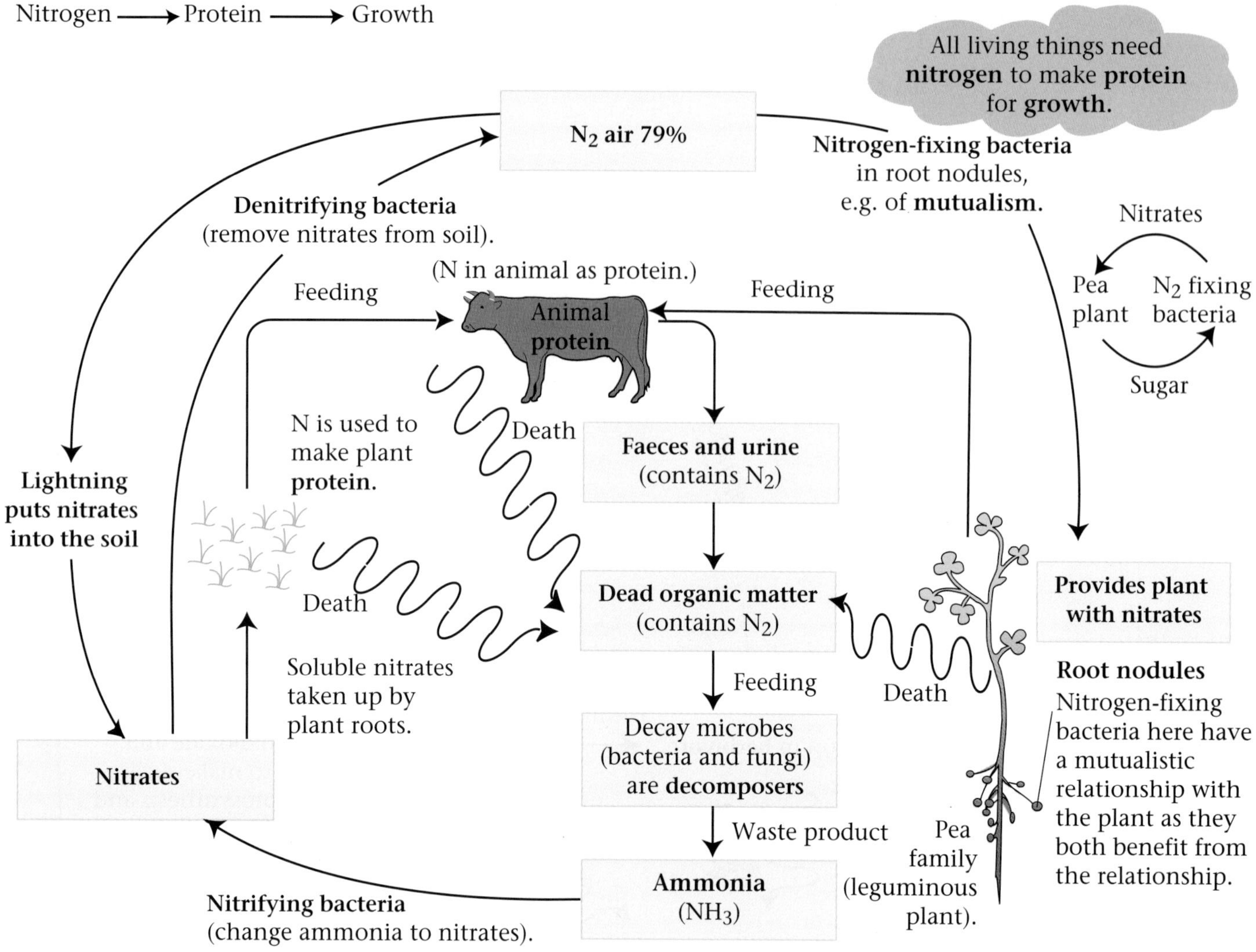

All living things need **nitrogen** to make **protein** for **growth.**
N2 air 79%
Nitrogen-fixing bacteria
in root nodules,
e.g. of mutualism.
Nitrates
Pea plant
N2 fixing bacteria
Sugar
Denitrifying bacteria
(remove nitrates from soil).
(N in animal as protein.)
Feeding
Animal protein
Feeding
N is used to make plant protein.
Death
Faeces and urine
(contains N2)
Lightning puts nitrates into the soil
Death
Dead organic matter
(contains N2)
Provides plant with nitrates
Soluble nitrates taken up by plant roots.
Feeding
Death
Root nodules
Nitrogen-fixing bacteria here have a mutualistic relationship with the plant as they both benefit from the relationship.
Decay microbes
(bacteria and fungi)
are decomposers
Nitrates
Waste product
Pea family (leguminous plant).
Ammonia
(NH3)
Nitrifying bacteria
(change ammonia to nitrates).

Bacteria in nitrogen cycle	
Type	**What they do**
Saprobiotic (decay)	Feed on dead material and release ammonia
Nitrifying	Convert ammonia to nitrates
Denitrifying	Convert nitrates to nitrogen gas
Nitrogen fixing	Convert nitrogen gas to nitrates

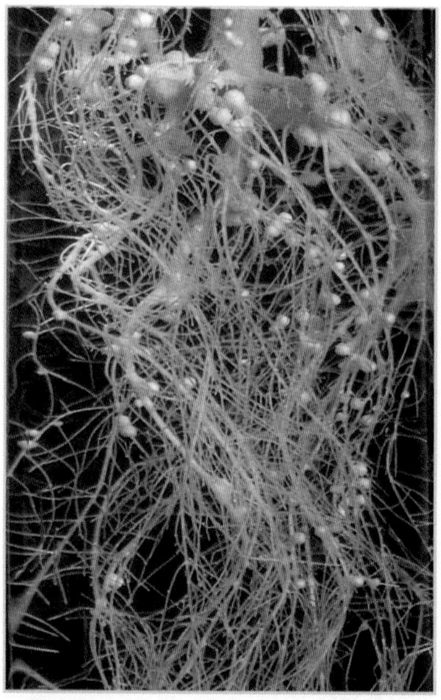

Root system of Soybean showing root nodules. The nitrogen-fixing bacteria that live here provide the plant with nitrates.

From Bowes: *A Colour Atlas of Plant Structure*

Questions:

1. Why do plants and animals need nitrogen?
2. How do plants get their nitrogen? How does this nitrogen enter plants?
3. How much nitrogen is in the air?
4. Why can't most plants make use of the nitrogen in the air?
5. What process passes nitrogen into animals from plants?
6. What are decomposers?
7. What substance is produced by the decomposers as a waste product?
8. Which bacteria produce nitrates in the soil?

HUMAN EFFECTS ON THE ENVIRONMENT

IMPORTANCE OF TROPICAL RAIN FORESTS

These are found in hot and wet areas, e.g. South America, Western Africa, Indonesia, Australia, and South-East Asia. Many trees are being cut down to provide land for **farming** and **housing**. The large scale, permanent removal of forests is called **deforestation**.

Provide a unique **habitat** for many plant and animal species. Many would face extinction without them.

Provide a **homeland** for human communities.

Canopy layer prevents leaching.

Water vapour released in transpiration causes **rain** in areas far from the sea.

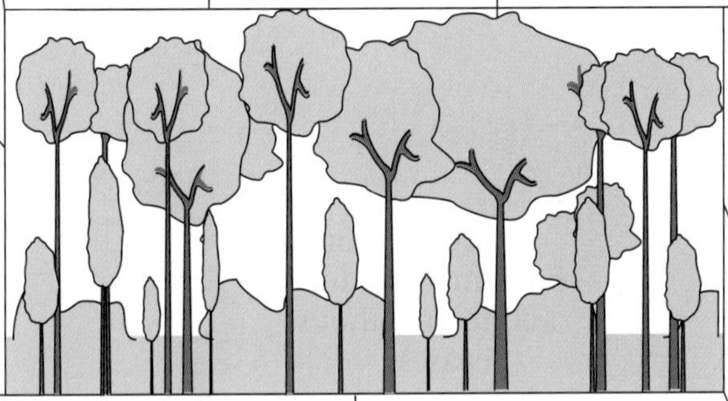

Wide variety of plants are a valuable source of **medicines.**

Carbon dioxide taken in by trees in photosynthesis reduces carbon dioxide in air, reducing the **greenhouse** effect.

Trees **anchor** the soil so fertile soil is not washed away.

Oxygen is added to the air in photosynthesis.

Trees provide **dead material** for decomposers to change into **nitrates.** This keeps the soil **fertile.**

Leaching
If heavy rain hits the soil soluble minerals such as nitrates are washed out of the top soil. They are **leached** out.

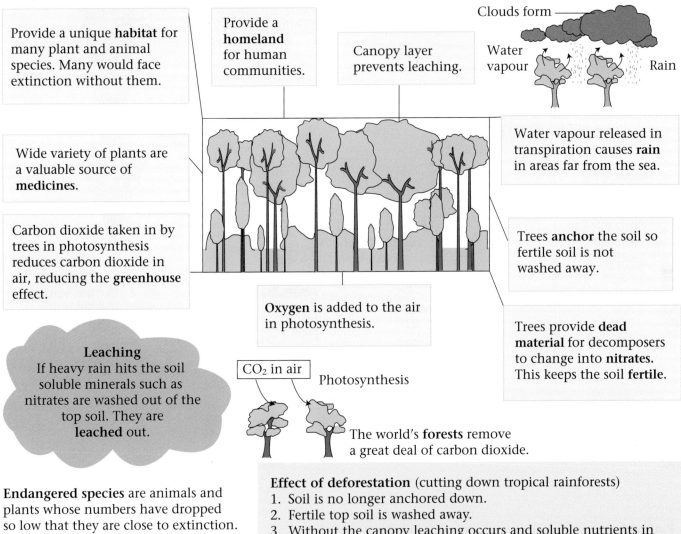

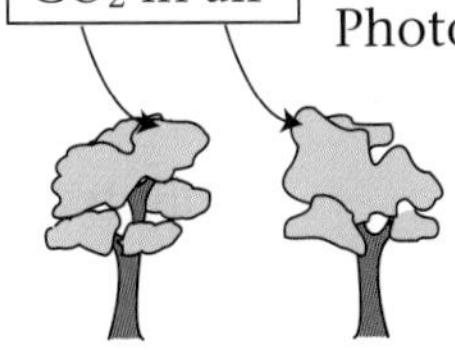

The world's **forests** remove a great deal of carbon dioxide.

Endangered species are animals and plants whose numbers have dropped so low that they are close to extinction. They can be protected in several ways:

- Protect sites in the wild – **on-site conservation** e.g. tropical forests.
- **Rare breeds parks, wild life parks and zoos.** Breeding is encouraged and predators removed.
- **Seed banks** – prevent the loss of genes in plants.
- **Legal protection** of certain species.
- **Educate** the public.

Effect of deforestation (cutting down tropical rainforests)
1. Soil is no longer anchored down.
2. Fertile top soil is washed away.
3. Without the canopy leaching occurs and soluble nutrients in the soil are washed away out of the top soil.
4. Land becomes infertile, nothing grows.
5. Burning the cut trees adds more carbon dioxide to the air – increasing the greenhouse effect.
6. Less transpiration reduces rainfall leading to drought.
7. Less trees means less photosynthesis and more carbon dioxide$_2$ in the air, adding to the greenhouse effect.
8. Loss of habitats has caused some organisms extinction, others are endangered.
9. In the short term the increase in dead material results in more decomposers respiring, so releasing more carbon dioxide into the air.

Questions:
1. Why does cutting down the trees cause the extinction of some animals?
2. How do trees help to reduce the greenhouse effect?
3. Where are tropical rain forests found in the world? Name two places.
4. How is rain produced by trees?
5. Without trees, the land may become infertile. How?

THE GREENHOUSE EFFECT This increases global warming.

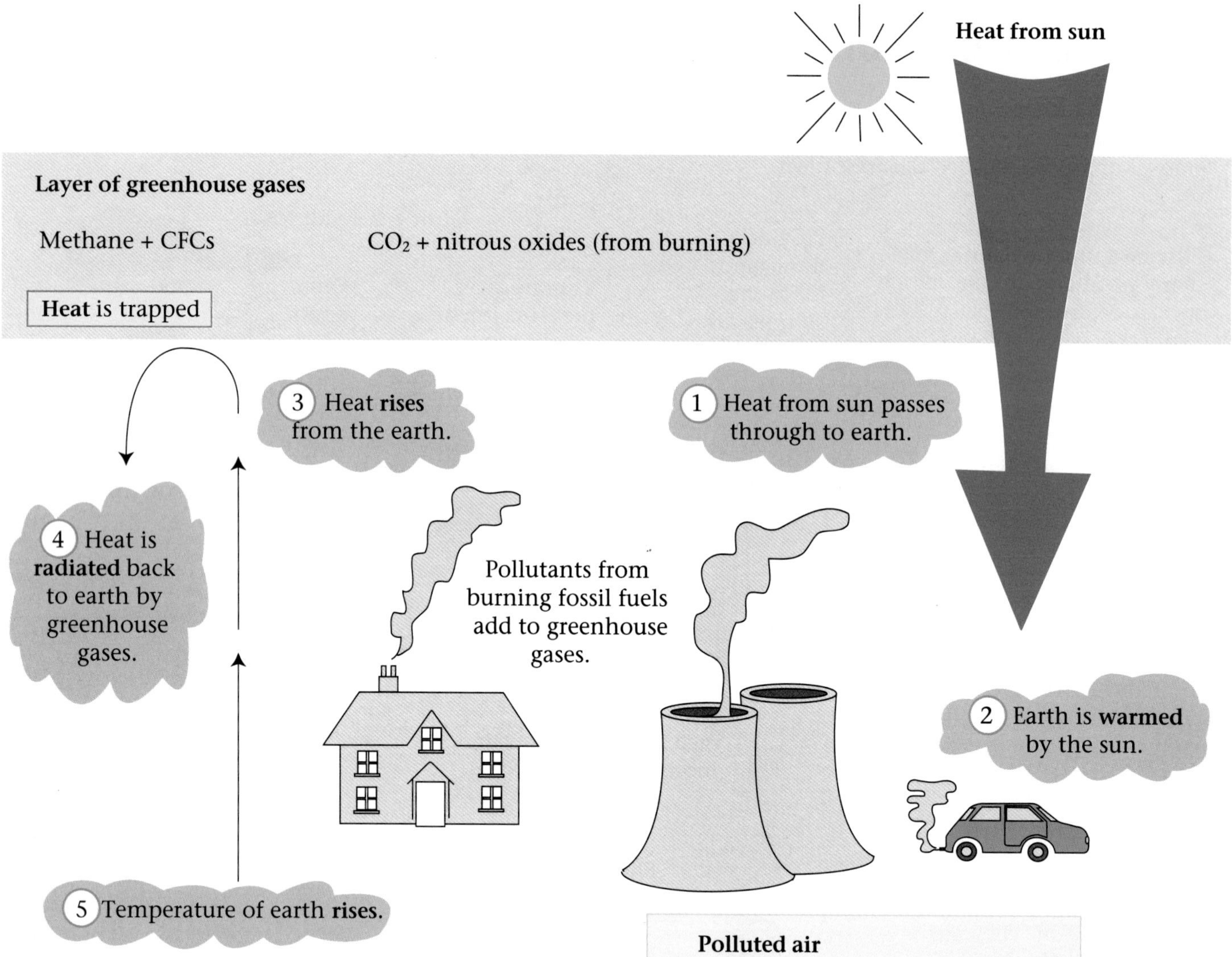

Heat from sun
Layer of greenhouse gases
Methane + CFCs
CO_2 + nitrous oxides (from burning)
Heat is trapped
3 Heat rises from the earth.
1 Heat from sun passes through to earth.
4 Heat is radiated back to earth by greenhouse gases.
Pollutants from burning fossil fuels add to greenhouse gases.
2 Earth is warmed by the sun.
5 Temperature of earth rises.

Polluted air
This contains a lot of carbon dioxide, nitrous oxides and CFCs (chlorofluorocarbons). These greenhouse gases let heat through from the sun. The warmed earth gives off heat which is radiated back to the earth by the layer of greenhouse gases.

Greenhouse effects
1. Earth temperature is rising.
2. Ice-caps are melting.
3. Sea level is rising.
4. Lowland areas will be flooded.

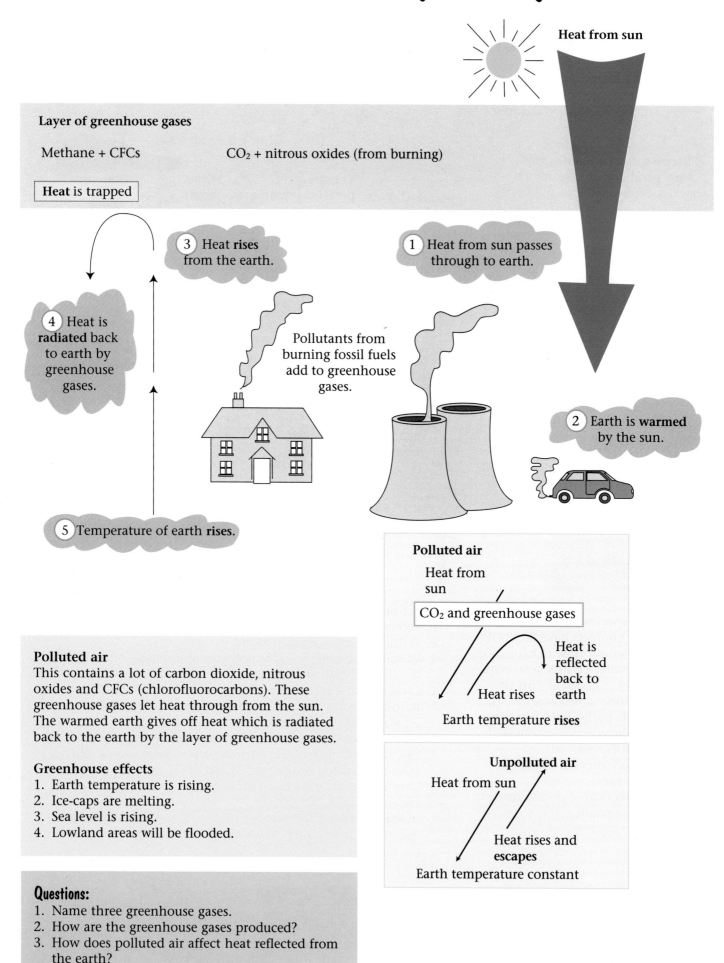

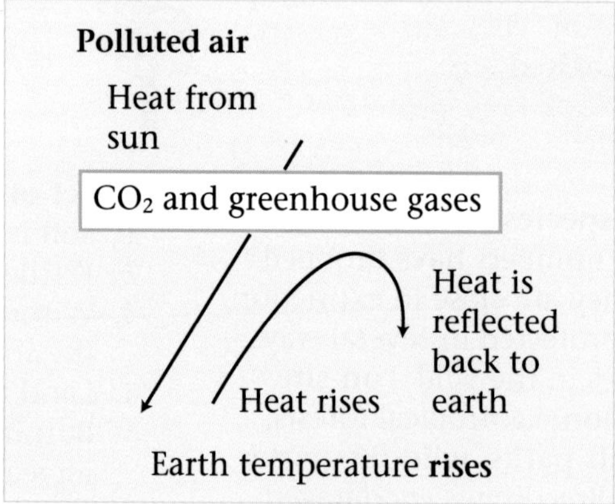

Polluted air
Heat from sun
CO_2 and greenhouse gases
Heat is reflected back to earth
Heat rises
Earth temperature rises

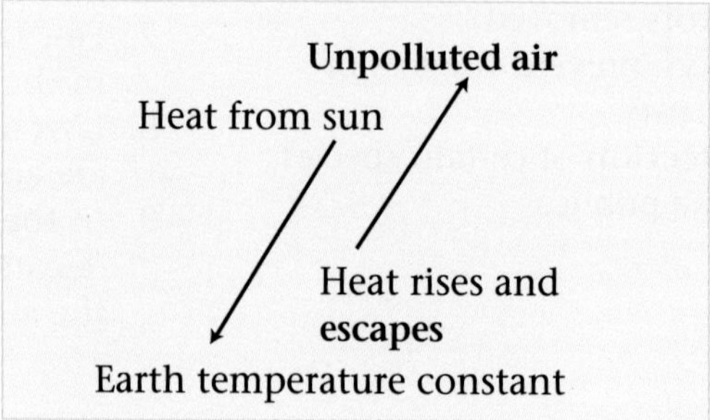

Unpolluted air
Heat from sun
Heat rises and escapes
Earth temperature constant

Questions:
1. Name three greenhouse gases.
2. How are the greenhouse gases produced?
3. How does polluted air affect heat reflected from the earth?
4. Name two effects of global warming.

AIR POLLUTION

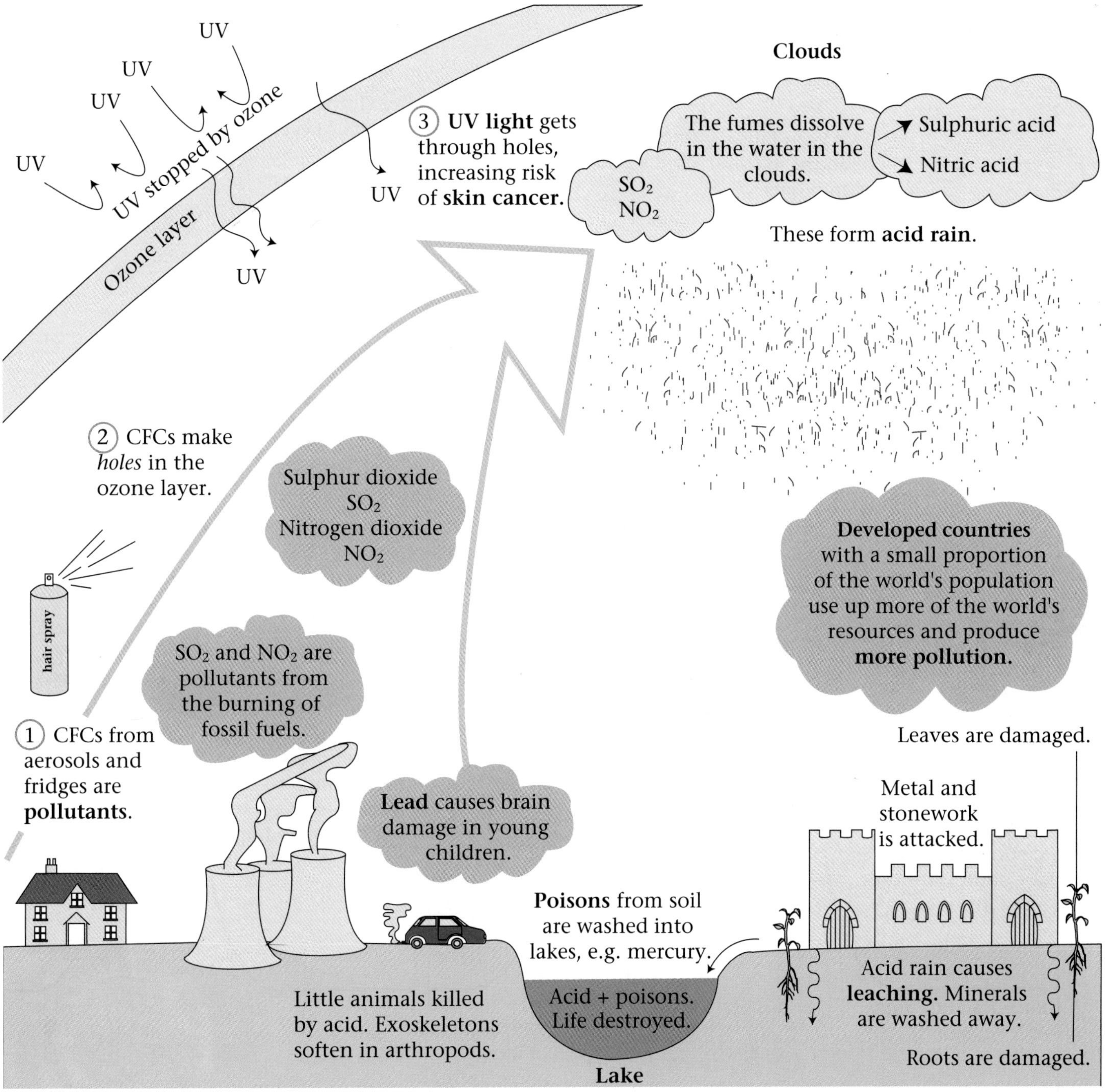

Prevention of acid rain

The amount of sulphur dioxide and nitrogen oxides produced in burning can be reduced if we:

1. Use a catalytic converter in cars to reduce the pollutants released.
2. Design furnaces so that some pollutants are trapped, e.g. a wet scrubber removes sulphur dioxide.
3. Clean coal and oil before burning so that less sulphur is present.

CFC = chlorofluorocarbons

Lichens as indicator species

Lichens are unusual organisms, as they are a mixture of two different species, algae and fungi. They form the crusty pale surface on trees in rural areas. Also the yellow powder and black 'paint' looking crusts on rocky shores. Lichens are indicators of air quality. If SO_2 levels are low, lichens are abundant. With rising SO_2 levels, lichens are scarce.

Questions:

1. List three effects of acid rain.
2. How does the ozone layer protect humans?
3. How do chlorofluorocarbons affect the ozone layer?
4. Why is soil often infertile in acid rain regions?
5. How does lead from burning fuels affect human health?

POLLUTION IN A RIVER

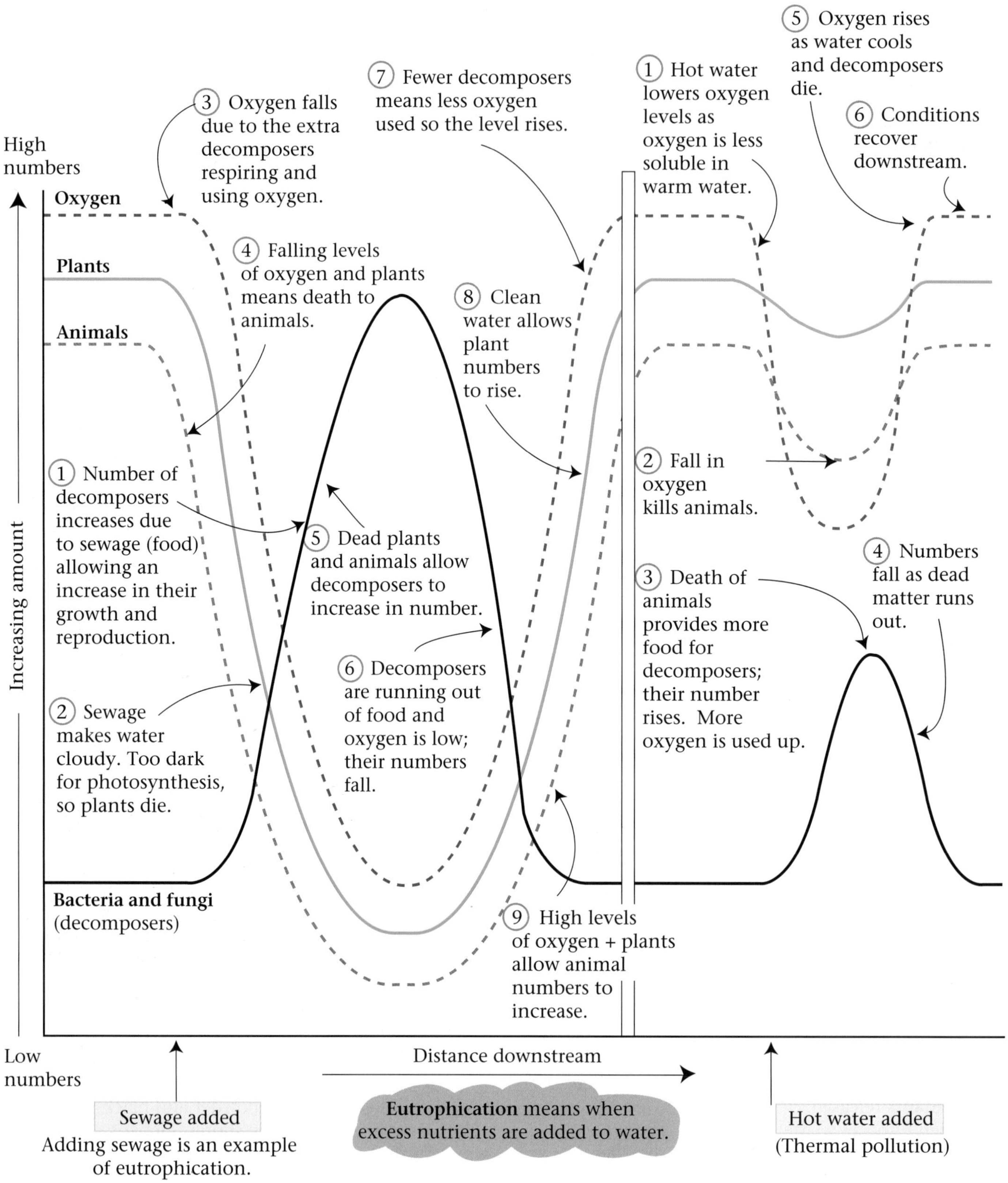
High numbers
Low numbers
Increasing amount
Oxygen
Plants
Animals
Bacteria and fungi (decomposers)
Distance downstream
1 Number of decomposers increases due to sewage (food) allowing an increase in their growth and reproduction.
2 Sewage makes water cloudy. Too dark for photosynthesis, so plants die.
3 Oxygen falls due to the extra decomposers respiring and using oxygen.
4 Falling levels of oxygen and plants means death to animals.
5 Dead plants and animals allow decomposers to increase in number.
6 Decomposers are running out of food and oxygen is low; their numbers fall.
7 Fewer decomposers means less oxygen used so the level rises.
8 Clean water allows plant numbers to rise.
9 High levels of oxygen + plants allow animal numbers to increase.
1 Hot water lowers oxygen levels as oxygen is less soluble in warm water.
2 Fall in oxygen kills animals.
3 Death of animals provides more food for decomposers; their number rises. More oxygen is used up.
4 Numbers fall as dead matter runs out.
5 Oxygen rises as water cools and decomposers die.
6 Conditions recover downstream.
Sewage added
Adding sewage is an example of eutrophication.
Eutrophication means when excess nutrients are added to water.
Hot water added
(Thermal pollution)

Questions:

1. Why do the numbers of bacteria and fungi rise when sewage is added?
2. What gas are these decomposers taking in for respiration?
6. What is eutrophication?
7. Explain why oxygen levels fall when hot water is added to a river.

Indicator species are organisms whose presence indicates particular levels of oxygen or pH in the water. For example, the presence of trout indicates high levels of oxygen.

EUTROPHICATION The sudden increase in the nutrient content of a lake or river

e.g. when excess nitrates are washed out of the soil and into a lake (leaching).

Farmer adds nitrate fertiliser to farmland. Nitrates increase plant growth and profit.

→ Rain pours on the soil and soluble nitrates are washed into a lake.

→ Nitrates in the lake cause the rapid growth of algae on the lake surface.

→ A thick 'blanket' of algae covers the lake surface and blocks light to the plants below. This is an **algal bloom.**

→ The blanket of algae die due to their short life cycle.

→ There is an increase in the dead material in the lake due to the dead surface algae and the plants below.

→ The **decomposers**, bacteria and fungi, feed on the dead material and their number increases.

→ Decomposers are **aerobic**, using up oxygen in respiration. Large numbers of decomposers respiring, reduce the oxygen level of the lake.

→ Oxygen level in the lake falls.

→ Aerobic organisms (those requiring oxygen) die.

→ **Dead lake**

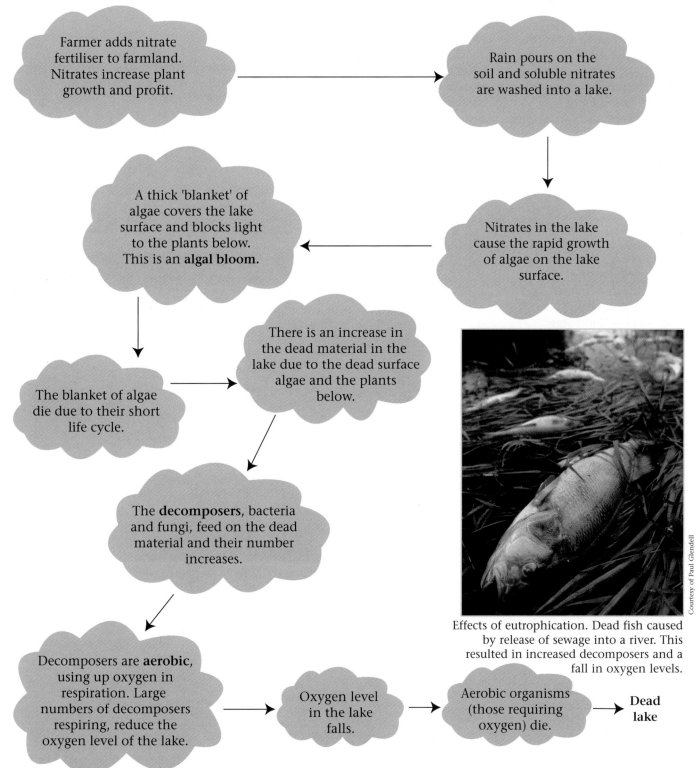

Courtesy of Paul Glendell

Effects of eutrophication. Dead fish caused by release of sewage into a river. This resulted in increased decomposers and a fall in oxygen levels.

Questions:

1. Why do farmers add nitrates to the soil?
2. What causes the nitrate to pass into rivers and lakes?
3. How does nitrate cause an algal bloom and what is it?
4. Algae only live for a short time. What feeds on the dead algae?
5. What happens to the oxygen level in the lake and why?
6. Which organisms are affected by the change in oxygen level.
7. What does the word eutrophication mean?
8. Can eutrophication be prevented?

FISH FARMING A managed ecosystem (aquaculture).

The North Sea was a rich fishing area with plenty of cod, haddock, herring, plaice, and mackerel. Overfishing has now reduced the supplies and fishing restrictions have been brought in to protect fish numbers in two main ways:

- Restricting the numbers of fish caught; a quota system.
- Only catching larger fish. Increasing the mesh size allows smaller fish to escape and live long enough to breed.

Fishermen furious at planned restrictions
A 40,000 square mile area of the North Sea, almost $^1/_5$ of its entire area may be off limits to cod, haddock and whiting fishermen as part of a desperate attempt to ensure survival of the cod stock.

SALMON fish farming developed during the 1970s to provide large quantities of salmon, without depleting the wild salmon numbers. By 1980, 800 tonnes of salmon was produced on the west coast of Scotland, providing a rich source of protein. The salmon are farmed in cages in sea lochs and in sheltered inlets of the sea, where conditions can be controlled. Although huge numbers are farmed, pollution problems are causing serious concerns.

A wild salmon leaping up a river to reach the pool where it hatched out. Here it will mate and produce more salmon.

Courtesy of Scotland in Focus/L. Campbell.

EUTROPHICATION
Faeces from the fish and uneaten food, fall to the bottom of the loch, causing an increase in the number of decomposers that feed on it. The decomposers use up oxygen so oxygen levels fall on the sea bottom, causing death to organisms here.

CHEMICAL POLLUTION
A large number of salmon in a confined space are more likely to have and to spread disease. The fish louse, which feeds on the fish, is widespread and is treated with the chemical **dichlorvos**. This chemical affects other organisms in the food chain killing lobsters, prawns, and other fish.

The pollution and parasites from the salmon farms have resulted in a dramatic fall in the number of wild salmon. Wild salmon have been found with up to 500 lice feeding on them.

Questions:
1. Why does fishing need to be regulated?
2. In what two ways are fish numbers being protected?
3. When and where did salmon fish farming develop?
4. What problems are caused by the vast numbers of salmon kept in cages?
5. What effect have farmed salmon had on wild salmon?

FARMED SALMON — Stages in production.

Newly hatched salmon in fresh water **hatchery**

↓ Controlled food provided

Controlled light, with many longer periods of light mimicking summers, speed up growth of fish.

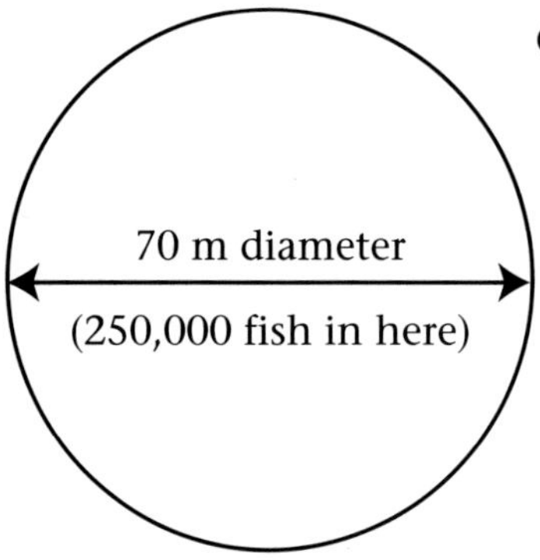

Cage for salmon

Fish moved to **cages** in sea loch

↓

Fish packed tight, up to 250,000 in each cage. Disease and parasites a problem, e.g. sea lice.

Pesticides used to kill lice → Chemical seeps out to rest of sea

People expect salmon to be pink. In the wild this is due to salmon eating pink crustaceans, such as prawn and shrimp. Farmed salmon is given pink pigment in the diet, adding about 15% to the cost of food.

Salmon **grow** quickly

↓

Salmon fed pellets of concentrated small fish, vaccines, growth hormones, and pink colouring.

Salmon starved for 10 days

This empties the gut so increasing the storage life of the fish.

By 2000 the West Coast of Scotland produced 127,000 tonnes of salmon.

↓

Salmon reach the shops

Pollution from the fish farms here is comparable to the sewage from 9.4 million people.

If salmon farms were situated in areas with good tidal flow, most of these problems would have been prevented.

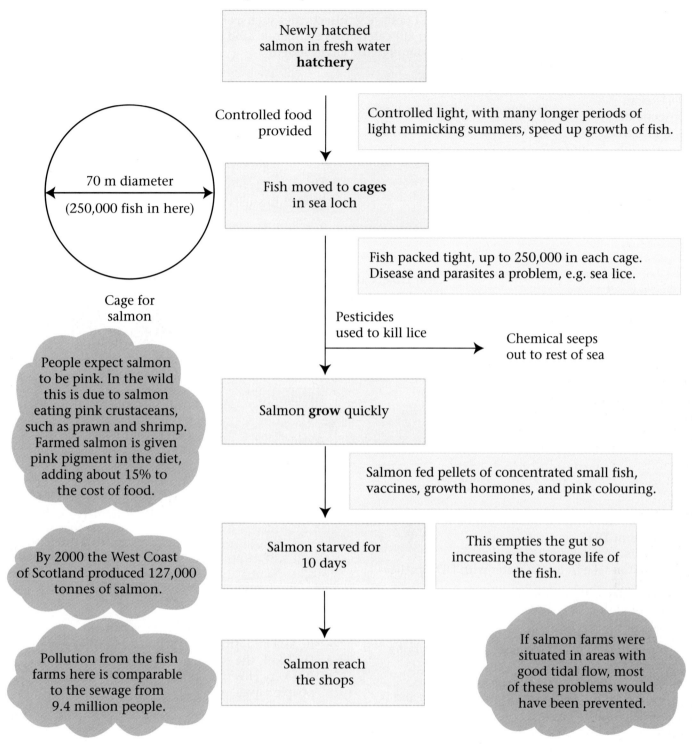

From Summerhayes and Thorpe: *Oceanography, An Illustrated Guide*, courtesy of Dr Jim Buchanan.

Salmon farming.

PESTICIDES Poisonous chemicals which kill pests.

Pesticides are poisonous chemicals which kill pests, e.g. insecticides such as DDT kill insects. Fungicides kill fungi. Pesticides are used to kill weeds and to kill caterpillars eating our food crops. These can **improve** our crop production but unfortunately they can also **damage** the environment.

DDT is an insecticide – it kills insects

DDT in ppm

Plankton (0.04) → Fish (2) → Grebes (25)

ppm = parts per million

Pesticides like DDT are harmful to wildlife as they do not break down. Unchanged they pass up the food chain causing death.

High DDT levels cause bird eggs to have thin shells and break easily.

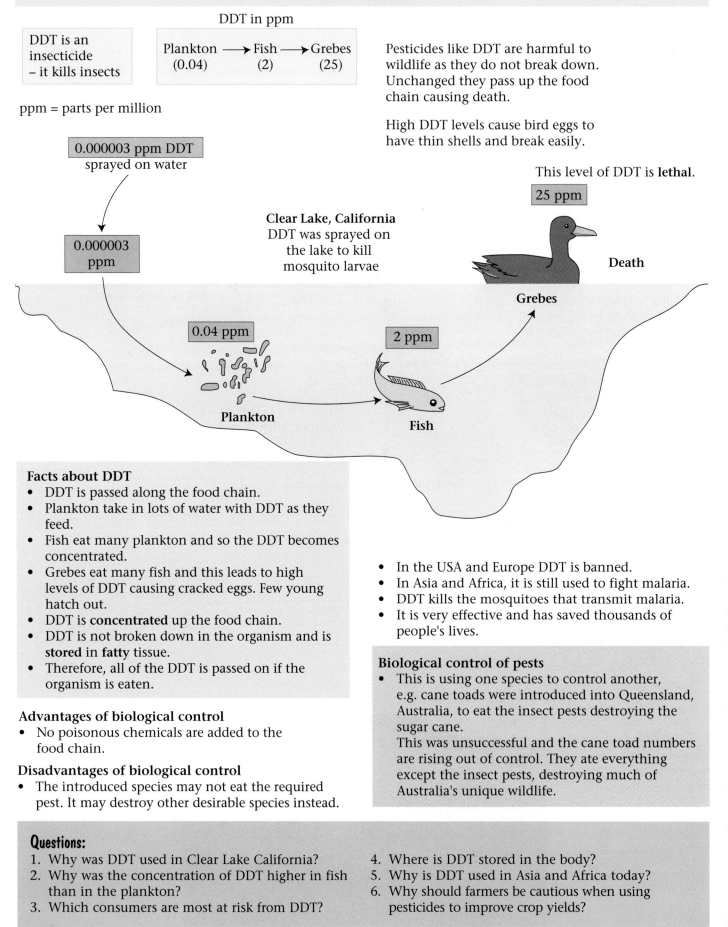

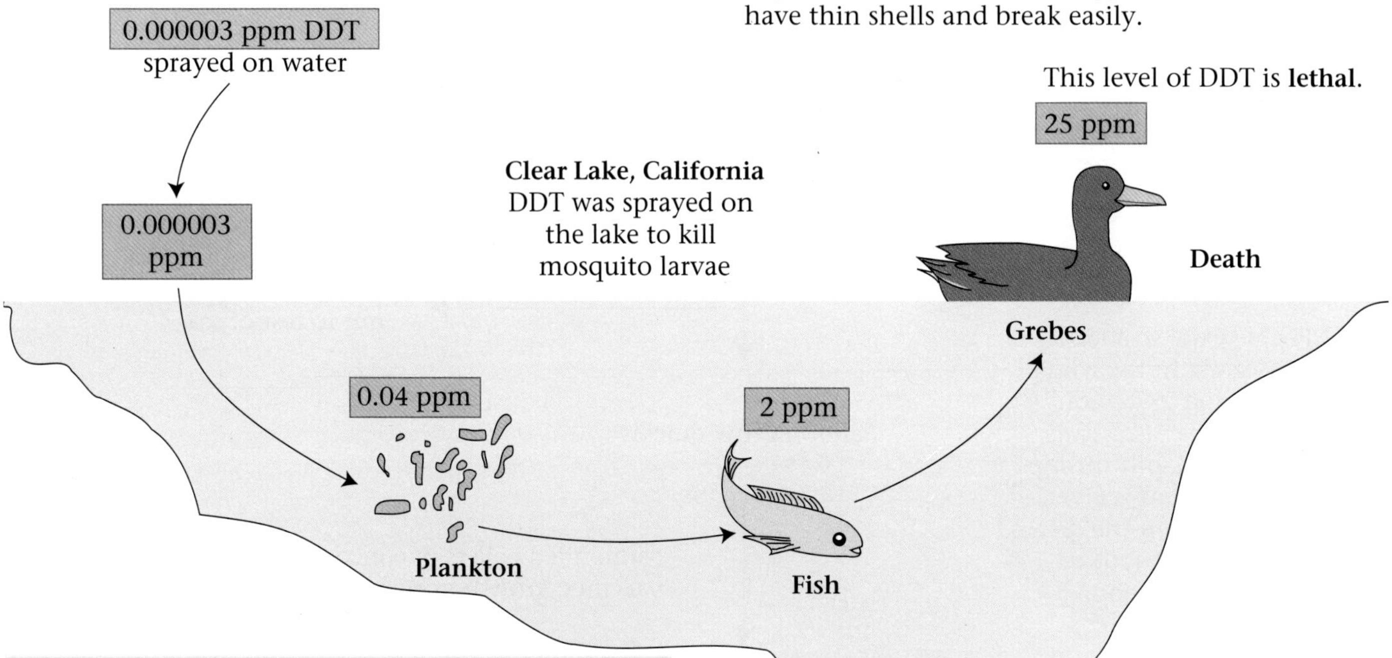

Facts about DDT

- DDT is passed along the food chain.
- Plankton take in lots of water with DDT as they feed.
- Fish eat many plankton and so the DDT becomes concentrated.
- Grebes eat many fish and this leads to high levels of DDT causing cracked eggs. Few young hatch out.
- DDT is **concentrated** up the food chain.
- DDT is not broken down in the organism and is **stored** in **fatty** tissue.
- Therefore, all of the DDT is passed on if the organism is eaten.

- In the USA and Europe DDT is banned.
- In Asia and Africa, it is still used to fight malaria.
- DDT kills the mosquitoes that transmit malaria.
- It is very effective and has saved thousands of people's lives.

Biological control of pests

- This is using one species to control another, e.g. cane toads were introduced into Queensland, Australia, to eat the insect pests destroying the sugar cane.
  This was unsuccessful and the cane toad numbers are rising out of control. They ate everything except the insect pests, destroying much of Australia's unique wildlife.

Advantages of biological control

- No poisonous chemicals are added to the food chain.

Disadvantages of biological control

- The introduced species may not eat the required pest. It may destroy other desirable species instead.

Questions:

1. Why was DDT used in Clear Lake California?
2. Why was the concentration of DDT higher in fish than in the plankton?
3. Which consumers are most at risk from DDT?
4. Where is DDT stored in the body?
5. Why is DDT used in Asia and Africa today?
6. Why should farmers be cautious when using pesticides to improve crop yields?

MICROBES

USEFUL AND HARMFUL MICROBES

	Bacteria	Fungi	Viruses
Structure			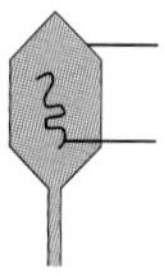
Useful	• Yoghurt making • Breaking down dead material to release useful substances, e.g. nitrates • Making insulin in genetic engineering • Breaking down sewage • Making vinegar	• Producing penicillin (an antibiotic) • Cheese making • Yeast is used in baking and wine making • Mushrooms provide food • Fungi are used to make mycoprotein – a food source	• Used in genetic engineering to make useful products
Harmful	**Cause diseases** • Tuberculosis • Food poisoning *(Salmonella)* • Tonsillitis • Whooping cough • Tetanus Bacteria feed on our food and cause **decay**	**Cause diseases** • Athlete's foot • Ringworm Fungi feed on our food and cause **decay**	**Cause diseases** • Influenza • Chicken pox • Small pox • Mumps • Rabies • Polio • AIDS
Examples	• *Bacillus tuberculosi* • *Salmonella*	• Penicillin • Bread mould • Yeast • Mushroom	• HIV(human immunodeficiency virus) The AIDS virus attacks the white blood cells which help our bodies fight disease.

Antibiotics

Antibiotics are chemicals used to kill bacteria. Penicillin was the first antibiotic discovered, in London, by Sir Alexander Fleming in 1928. It was a huge breakthrough in the fight against bacterial diseases.

Find out how Fleming's research led to this important discovery.

Over use of antibiotics can lead to the evolution of resistant bacteria, such as MRSA, which has caused the death of many hospital patients.

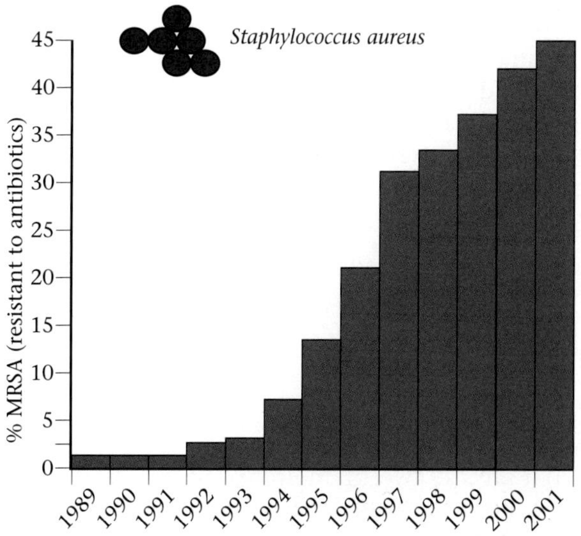

The percentage of *Staphylococcus aureus* resistant to methicillin (MRSA), in England and Wales 1989–2001.

Questions:

1. How are fungi useful to us? Name two ways.
2. Name three diseases caused by viruses.
3. Give three examples of fungi.
4. Which microbe is made of hyphae threads?

DECOMPOSERS The decay organisms.

There are two main groups, **bacteria** and **fungi**. They feed on dead material and return useful nutrients back into the soil to be used again, e.g. nitrates.

Bacteria and fungi are called **Saprobiotic** as they feed on **dead**, organic matter, such as dead animals and plants. They also may cause decay in our food.

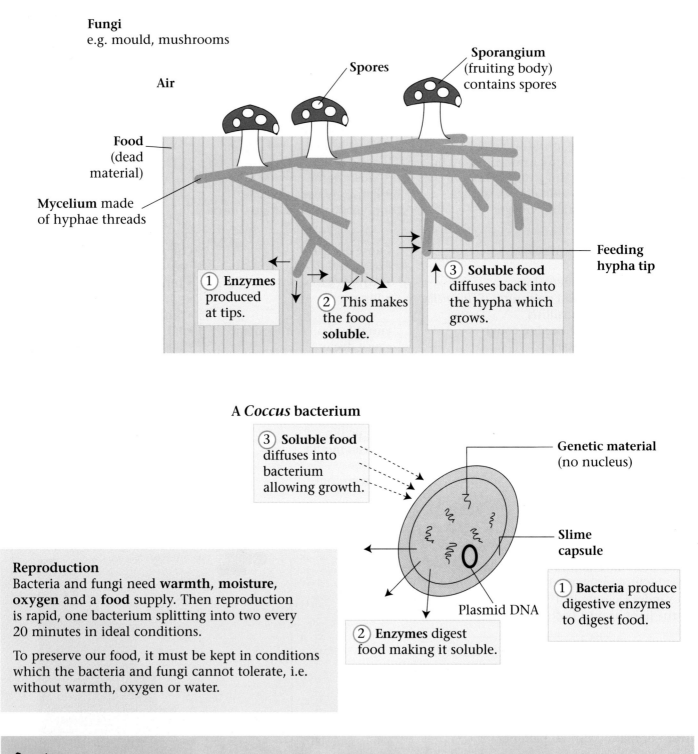

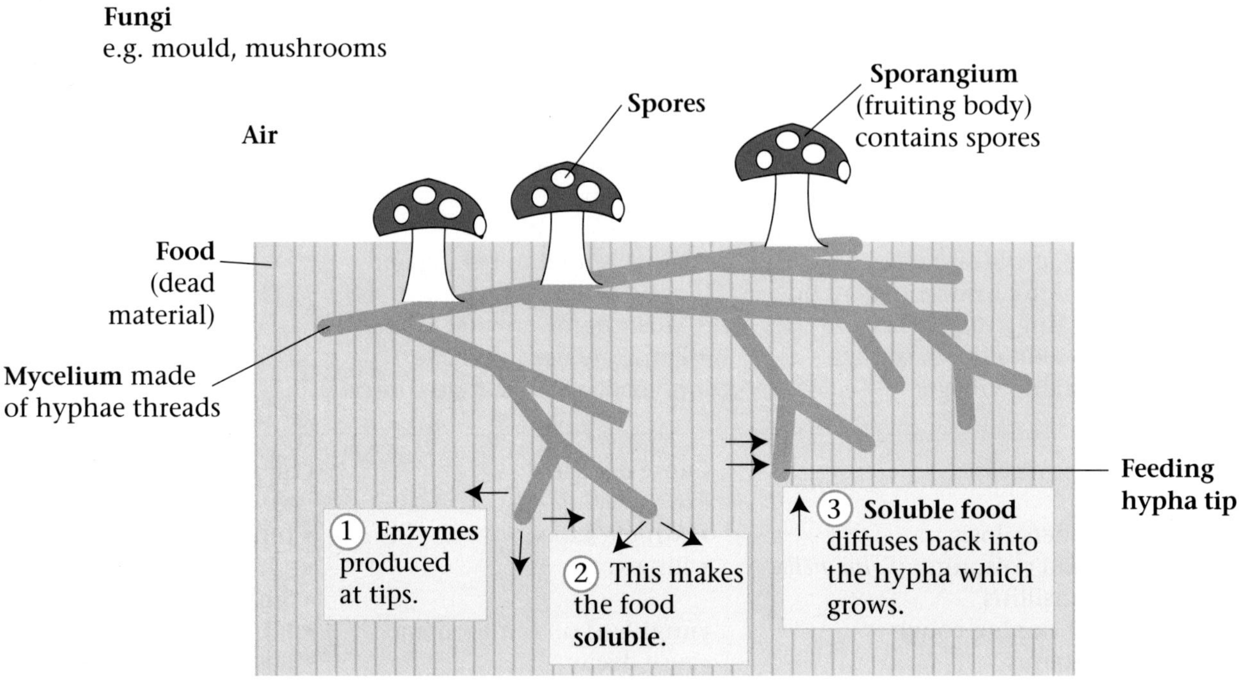

Fungi
e.g. mould, mushrooms
Air
Spores
Sporangium
(fruiting body)
contains spores
Food
(dead
material)
Mycelium made
of hyphae threads
Feeding
hypha tip
1 **Enzymes** produced at tips.
2 This makes the food **soluble.**
3 **Soluble food** diffuses back into the hypha which grows.

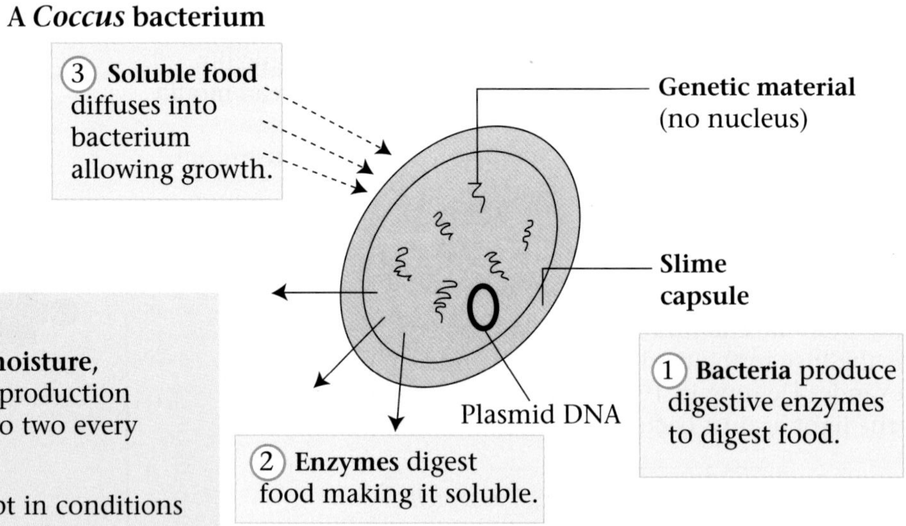

A *Coccus* bacterium
3 **Soluble food** diffuses into bacterium allowing growth.
Genetic material
(no nucleus)
Slime
capsule
Plasmid DNA
1 **Bacteria** produce digestive enzymes to digest food.
2 **Enzymes** digest food making it soluble.

Reproduction
Bacteria and fungi need **warmth, moisture, oxygen** and a **food** supply. Then reproduction is rapid, one bacterium splitting into two every 20 minutes in ideal conditions.

To preserve our food, it must be kept in conditions which the bacteria and fungi cannot tolerate, i.e. without warmth, oxygen or water.

Questions:
1. What are the two main groups of decomposers?
2. How often can bacteria reproduce in ideal conditions?
3. What causes our food to decay?
4. What are the three types of bacteria?
5. What would happen if there were no decomposers?

TREATMENT OF SEWAGE

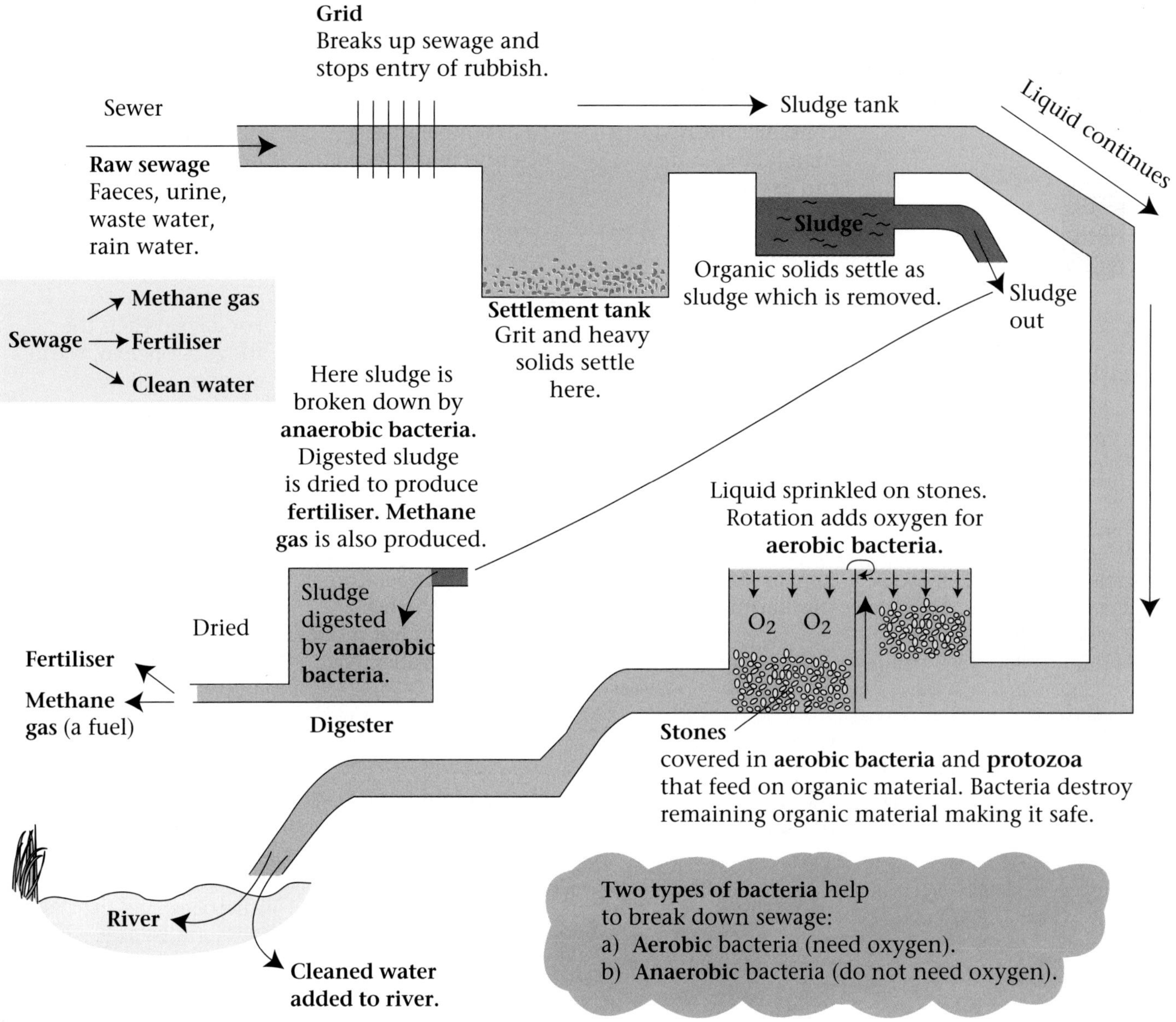
Grid
Breaks up sewage and stops entry of rubbish.
Sewer
Sludge tank
Liquid continues
Raw sewage
Faeces, urine, waste water, rain water.
Sludge
Organic solids settle as sludge which is removed.
Sludge out
Settlement tank
Grit and heavy solids settle here.
Sewage → Methane gas
Sewage → Fertiliser
Sewage → Clean water
Here sludge is broken down by anaerobic bacteria. Digested sludge is dried to produce fertiliser. Methane gas is also produced.
Liquid sprinkled on stones. Rotation adds oxygen for aerobic bacteria.
O2 O2
Dried
Sludge digested by anaerobic bacteria.
Fertiliser
Methane gas (a fuel)
Digester
Stones
covered in aerobic bacteria and protozoa that feed on organic material. Bacteria destroy remaining organic material making it safe.
River
Cleaned water added to river.

Two types of bacteria help to break down sewage:
a) **Aerobic** bacteria (need oxygen).
b) **Anaerobic** bacteria (do not need oxygen).

Bacteria are used to break down sewage to clean water, methane gas and fertiliser. The **aerobic** bacteria feed on the organic material in the liquid sewage, resulting in clean water. The solid organic material, sludge, is digested by the **anaerobic** bacteria, producing **fertiliser** and **methane gas**, both useful products.

Problems caused by an increasing human population

- increased consumption of limited resources, e.g. coal, oil, wood
- build-up of greenhouse gases
- sewage disposal
- air and water pollution
- loss of habitats leading to loss of species, i.e. reduced biodiversity
- waste disposal

Recycling of glass, tins and paper is now routine in most areas of the UK. The use of biodegradable materials is also on the rise.

Questions:

1. Which two types of bacteria are needed to break down sewage?
2. Why is sewage treatment necessary?
3. What useful products are gained from sewage treatment?

FOOD PRESERVATION

Food goes **bad** because **bacteria** and **fungi** feed on it causing **decay**. Bacteria and fungi need **warmth**, **oxygen**, and **water**. If any of these are removed the food will stay fresh as microbes cannot survive without these conditions.

Example	Type	What is done	How it works	How long it stays fresh
Baked beans 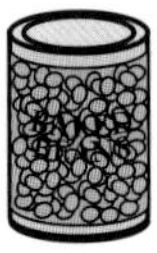	Tinned foods	Sealed in airtight container. Heated to a high temperature (no oxygen).	No oxygen present. High temperature destroys microbes.	For years, unless the tins are damaged or punctured.
Milk Margarine 	Fridge	Kept at 2°C (low temperature).	Cool temperature slows down growth and reproduction of microbes. Slows rate of decay.	For days. When the fridge is opened, warm air enters, allowing reproduction of microbes.
Ice-cream Peas 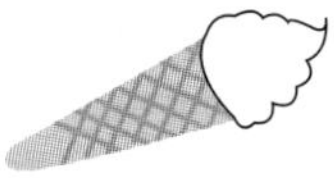	Frozen foods	Kept at -18°C (no warmth).	Very low temperature stops growth and reproduction but it does not kill microbes. No decay.	For years. Food will come out with the same quantity of microbes as went into the freezer.
Dried potato Dried milk	Dried foods	All the water is removed (no water).	No water means no decay.	For years, unless packet is opened. Damp air will then allow entry and growth of microbes.
Jam Pickled onions Bacon  Pickled onions	Chemicals	Placed in chemicals. **Salt:** bacon, ham (no water).	Salt **dries** the food, so no water, no decay.	Weeks.
		Sugar: marmalade, jam (removes water).	Sugar **dries** microbes.	Months, if unopened.
		Vinegar: pickles, chutney (acid conditions).	Acid **destroys** bacteria and fungi.	Months.

Food labelling

Processed food often contains a high proportion of fat and salt.

Too much salt in the diet can lead to increased blood pressure for about 30% of the population.

Choose 3 highly processed foods, e.g. baked beans, crisps and pot noodles. Write down the energy content of each and list all the additives present.

Use the internet to find out why these additives are added.

Questions:

1. What conditions do bacteria and fungi need?
2. How does canning foods stop decay?
3. What effect does freezing have on the microbes in food?
4. Why does bacon last for several weeks without going bad?
5. Pickles last for months. How does their treatment preserve them?

HOW DISEASES SPREAD

Direct contact
- Touching infected person spreads microbes causing disease, e.g.
  - athlete's foot
  - ringworm
  - boils.
- Body fluid contact through sexual intercourse, e.g.
  - AIDS
  - gonorrhoea
  - syphilis.
- Blood contact from transfusions or infected needles, e.g.
  - AIDS
  - hepatitis.

In droplets of water
- Coughs and sneezes spread diseases.
- Bacteria spread in water droplets from one person to another.
- Most bacteria spread in crowded damp rooms.
- Viruses spread both in damp and dry air.
- Diseases spread in this way:
  - colds
  - influenza
  - tuberculosis
  - pneumonia
  - whooping cough.

Mosquitoes may bite a person who suffers from malaria. Human blood, with the malarial parasite therefore passes to the mosquito.When this mosquito bites its next victim, the parasite causing malaria is passed on.

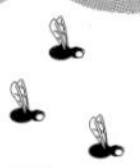
Mosquitoes

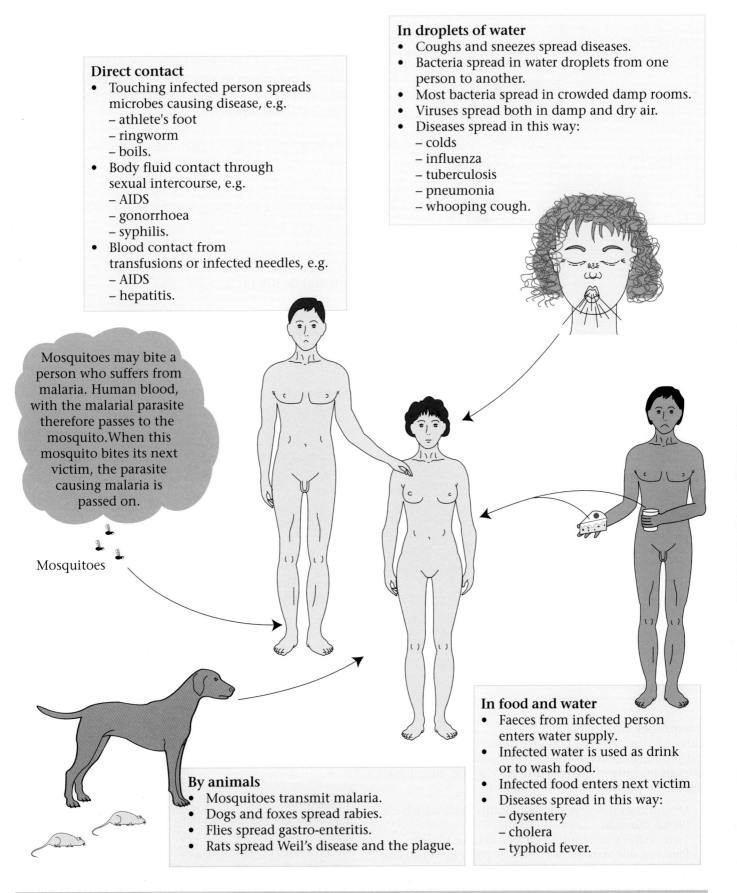

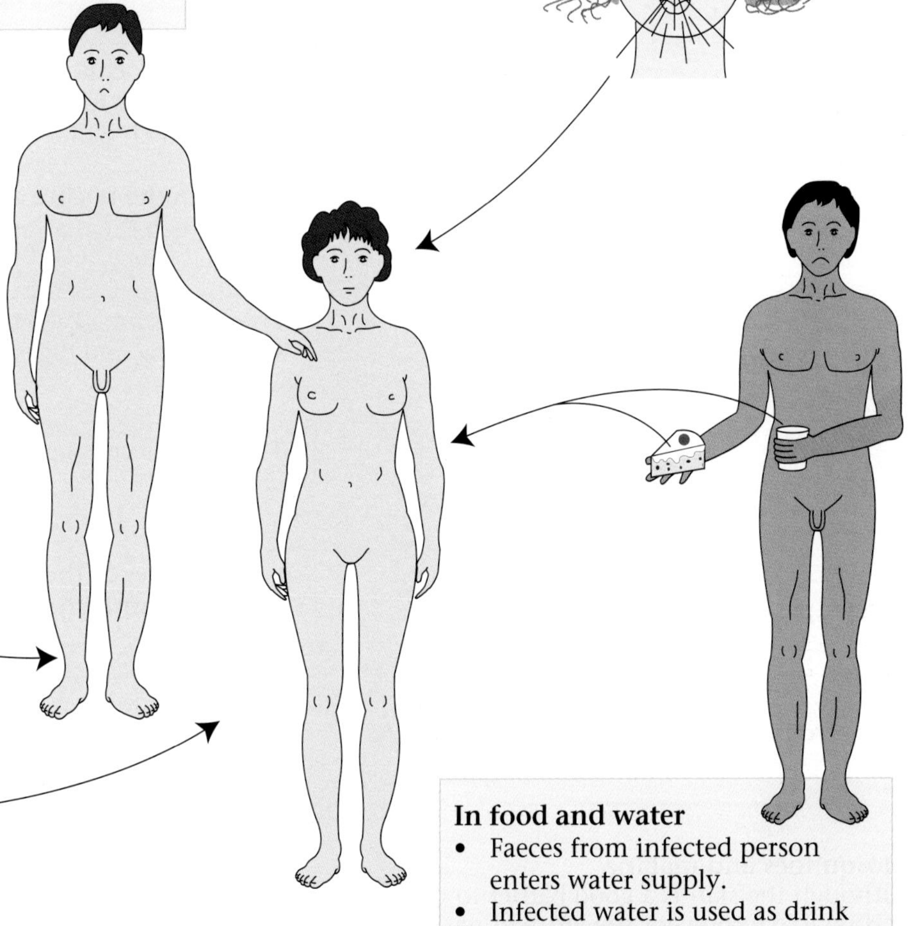

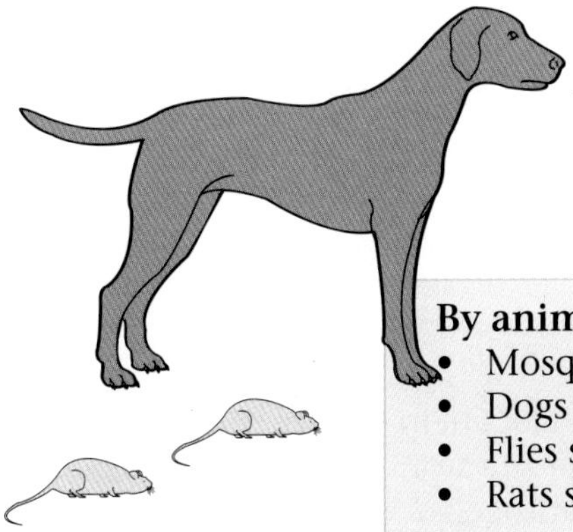

By animals
- Mosquitoes transmit malaria.
- Dogs and foxes spread rabies.
- Flies spread gastro-enteritis.
- Rats spread Weil's disease and the plague.

In food and water
- Faeces from infected person enters water supply.
- Infected water is used as drink or to wash food.
- Infected food enters next victim
- Diseases spread in this way:
  - dysentery
  - cholera
  - typhoid fever.

Questions:
1. How is malaria spread?
2. What diseases are spread if sewage is not properly treated?
3. Why should someone with whooping cough keep away from healthy people?
4. In what conditions do bacterial infections spread most rapidly? How can this be prevented?

DEFENCES OF THE BODY TO PATHOGENS

A **pathogen** is an organism that causes disease.

Part of body		How it works
Skin	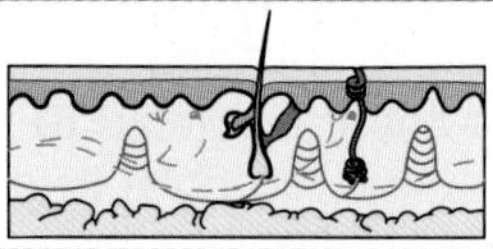	Skin is a barrier to pathogens except where it is damaged. Sweat contains the enzyme lysozyme which kills bacteria by breaking their cell wall open.
Blood clot		Blood clots stop entry of germs (pathogens) at cuts.
Eyes (tears)	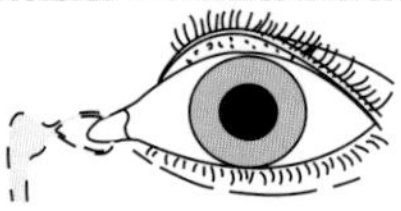	Tears also contain the enzyme lysozyme to kill bacteria.
Senses	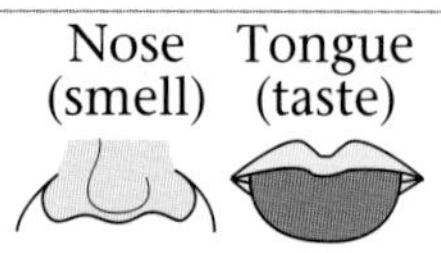	Food that is bad may smell or taste 'off'. We find it unpleasant and do not eat it.
Stomach	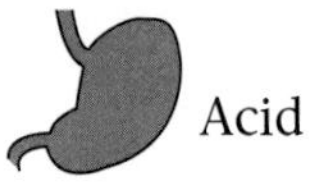	Strong hydrochloric acid (pH2) kills most bacteria that are present in our food. Contaminated food may make us vomit, which will remove the harmful bacteria, and food.
Respiratory passages	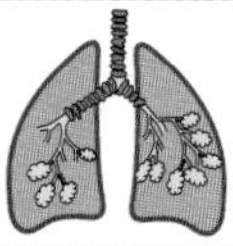	Mucus in the trachea and bronchi traps bacteria, and cilia hairs sweep them away from the lungs.
Lungs (alveoli)	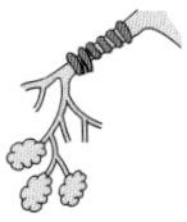	Phagocytes here keep the surface of the alveoli clean by surrounding and digesting any bacteria present.
Phagocytes		These white cells in the blood surround and digest invading bacteria.
Lymphocytes	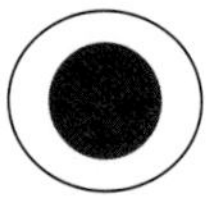	Lymphocytes in the blood produce antibodies to kill the germs and antitoxins to destroy their poisons.
Reproductive system	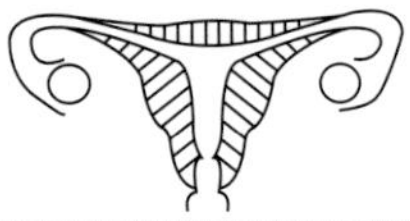	Little defence here. Vagina and urethra (in penis) have acid conditions which destroys some bacteria.

Mosquitoes and malaria
Although the skin is a good barrier to most pathogens, it is unable to prevent the entry of **pathogens** from animals and insects which **bite** us, e.g. the **mosquito** carries the **pathogen** causing **malaria**.

- Mosquitoes have specialised mouth parts which can pierce our skin.
- This enables mosquitoes to feed on our blood.
- Piercing our skin allows entry of pathogens causing malaria.
- Mosquitoes can pass on pathogens from one person to another when they bite.
- Malaria can be spread quickly and easily by mosquitoes.

If travelling to malarial regions of the world, it is essential to take anti-malaria tablets and use mosquito nets and creams to prevent their bites.

Questions:
1. How do blood clots help to stop disease?
2. What feature of the stomach helps to prevent the entry of pathogens?
3. How do our senses of smell and taste protect us from pathogens?
4. What does the enzyme lysozyme do and where is it produced?
5. Why do mosquitoes bite us?
6. What disease can mosquitoes spread?
7. How does the mucus in the breathing tubes help to protect us from pathogens?

ANTIBODIES AND IMMUNITY

Disease is caused by invading bacteria and viruses. These are **pathogens** (disease-causing organisms). White blood cells called **lymphocytes** produce antibodies which are specific to the pathogen.

Antibodies attack both the organism and the poisons (toxins) they produce. Antibodies that attack toxins are called antitoxins

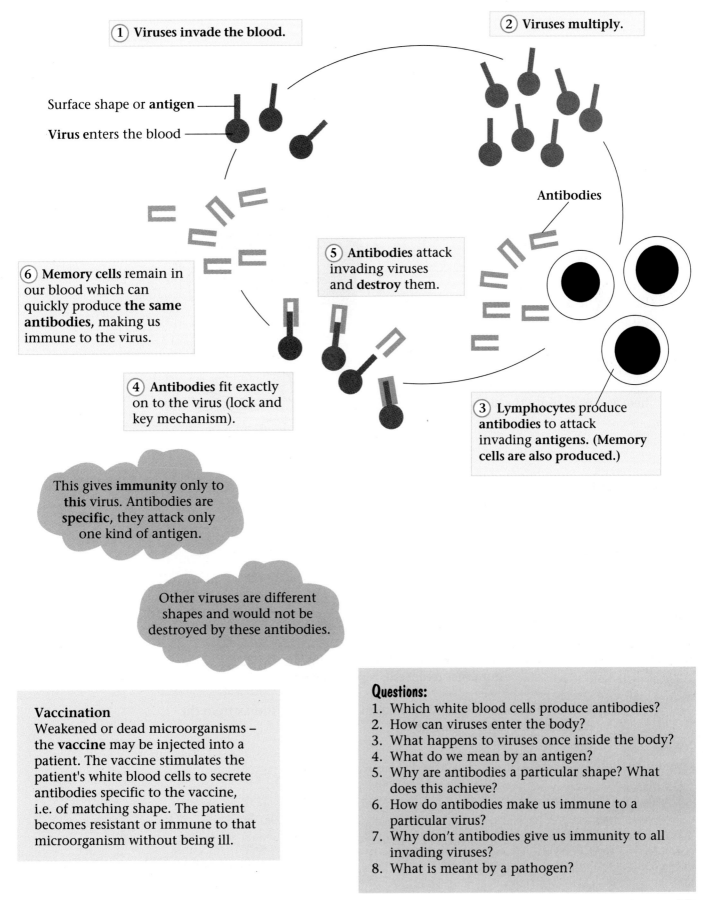

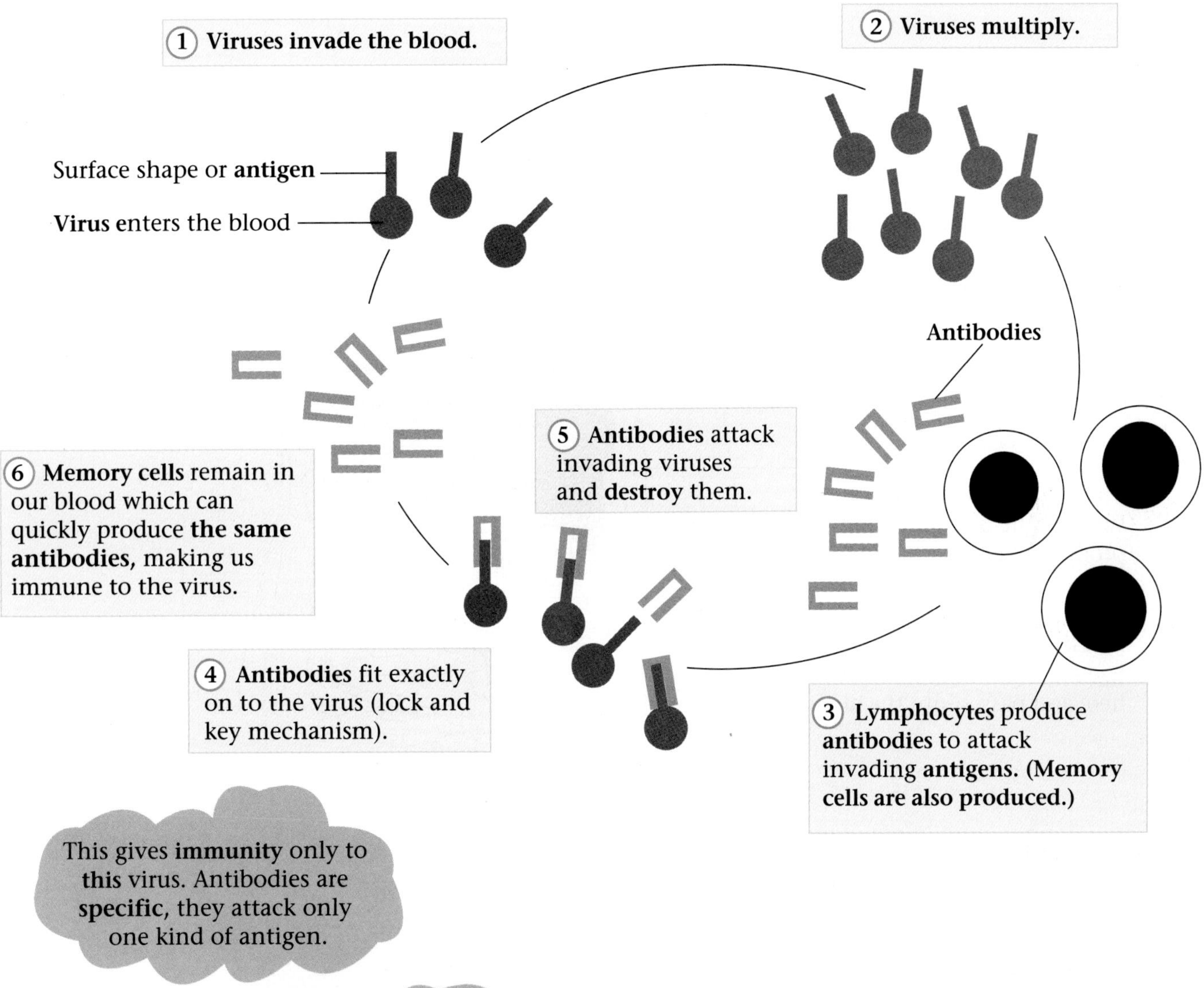

1 Viruses invade the blood.
2 Viruses multiply.
Surface shape or antigen
Virus enters the blood
Antibodies
3 Lymphocytes produce antibodies to attack invading antigens. (Memory cells are also produced.)
4 Antibodies fit exactly on to the virus (lock and key mechanism).
5 Antibodies attack invading viruses and destroy them.
6 Memory cells remain in our blood which can quickly produce the same antibodies, making us immune to the virus.

This gives **immunity** only to **this** virus. Antibodies are **specific**, they attack only one kind of antigen.

Other viruses are different shapes and would not be destroyed by these antibodies.

Vaccination
Weakened or dead microorganisms – the **vaccine** may be injected into a patient. The vaccine stimulates the patient's white blood cells to secrete antibodies specific to the vaccine, i.e. of matching shape. The patient becomes resistant or immune to that microorganism without being ill.

Questions:

1. Which white blood cells produce antibodies?
2. How can viruses enter the body?
3. What happens to viruses once inside the body?
4. What do we mean by an antigen?
5. Why are antibodies a particular shape? What does this achieve?
6. How do antibodies make us immune to a particular virus?
7. Why don't antibodies give us immunity to all invading viruses?
8. What is meant by a pathogen?

THE HUMAN BODY

Brain
This co-ordinates body activities.

Lungs
Here oxygen passes into the blood.

Heart
This pumps blood around the body.

Stomach
Food is digested here.

Liver
This regulates the amount of sugar in the blood. Urea is produced here.

Intestines
Food is digested and absorbed here and passes into the blood.

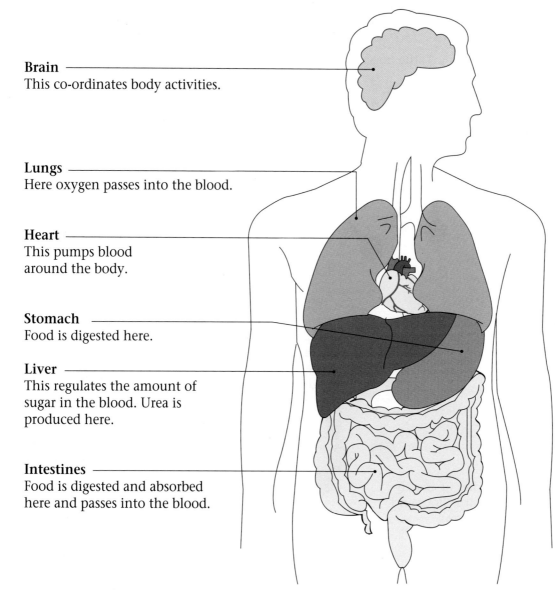

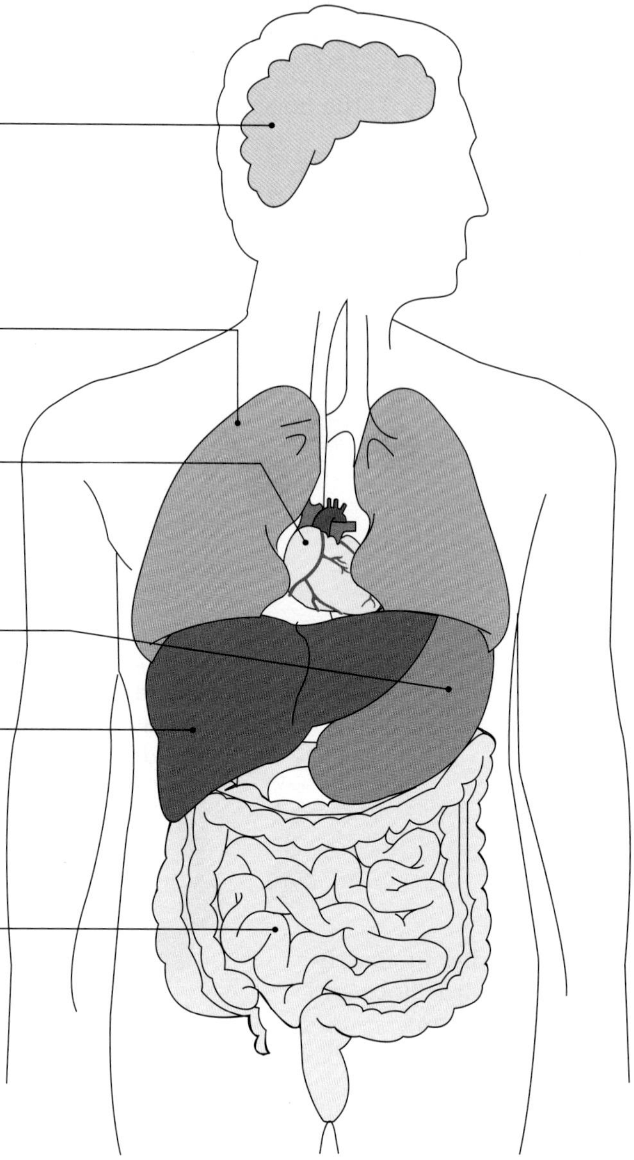

The human body is divided into 3 parts:

Part	Organs present
Head	Brain, eyes, ears
Thorax	Lungs, heart
Abdomen	Liver, stomach, intestines, kidneys (only visible when intestines removed), reproductive organs (not shown)

A sheet of muscle called the diaphragm separates the thorax from the abdomen (see pages 77 and 88).

Questions:
1. In which organs is food digested?
2. Where does oxygen pass into the blood?
3. Which two major organs are found in the thorax?
4. Moving from the head to the abdomen, list the organs that are visible in the diagram.

NUTRITION AND CIRCULATION

HUMAN TEETH

Humans have two sets of teeth. The first set are called **milk teeth** and these are replaced by **permanent teeth** from the age of five.

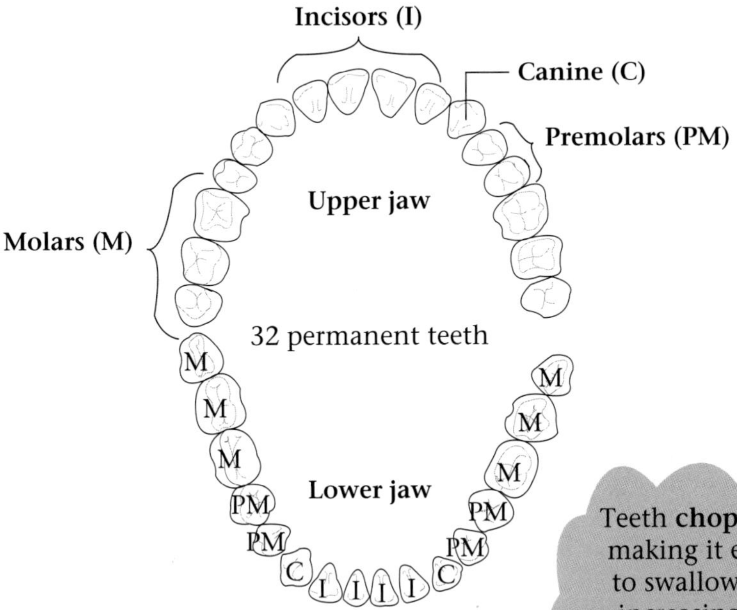

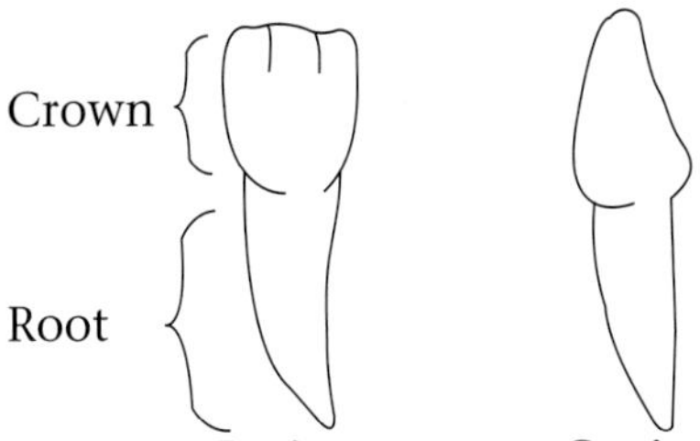

Incisors
Sharp edges to cut off food.

Canines
Large and pointed for tearing food.

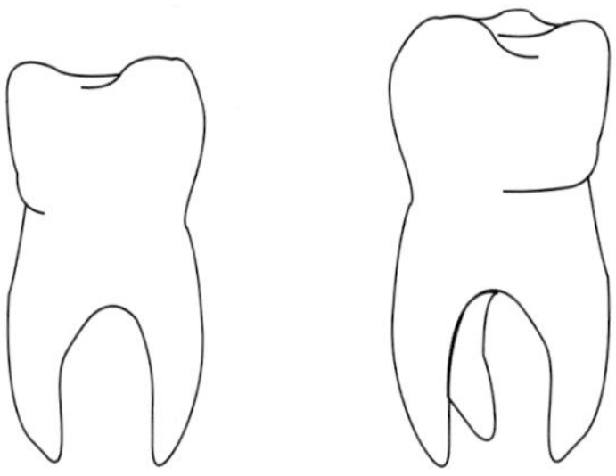

Premolars **Molars**

These crush and grind food so that it is easier to swallow.

Teeth **chop food** making it easier to swallow and increasing the surface area for enzyme action.

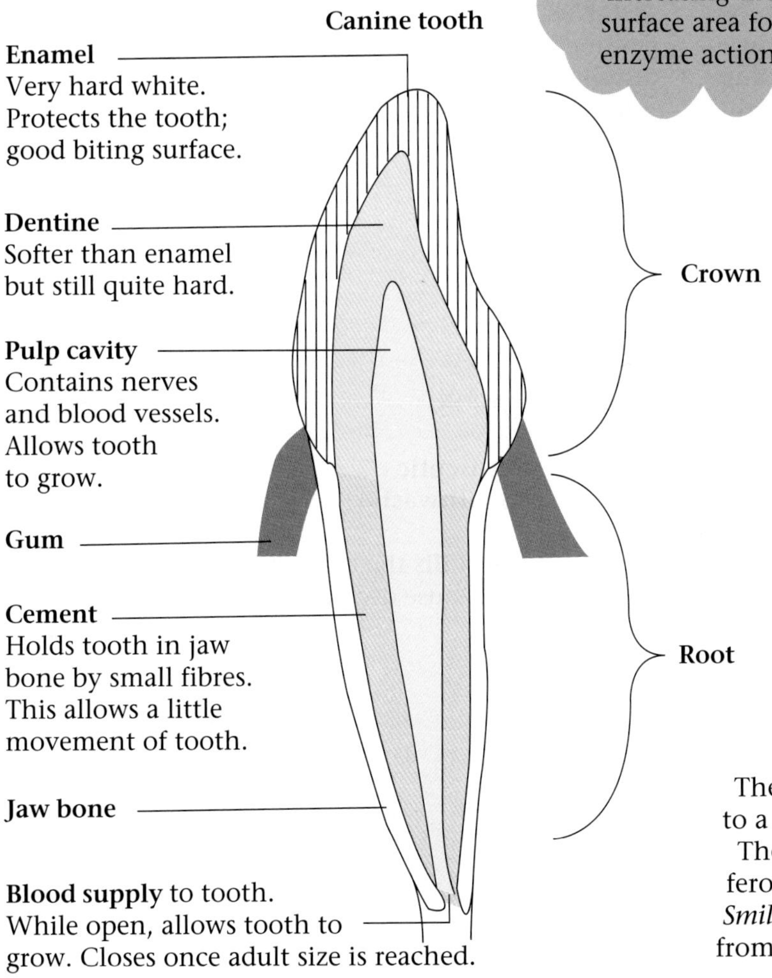

Animals that eat meat, **carnivores**, have large canine teeth to stab and kill their prey.

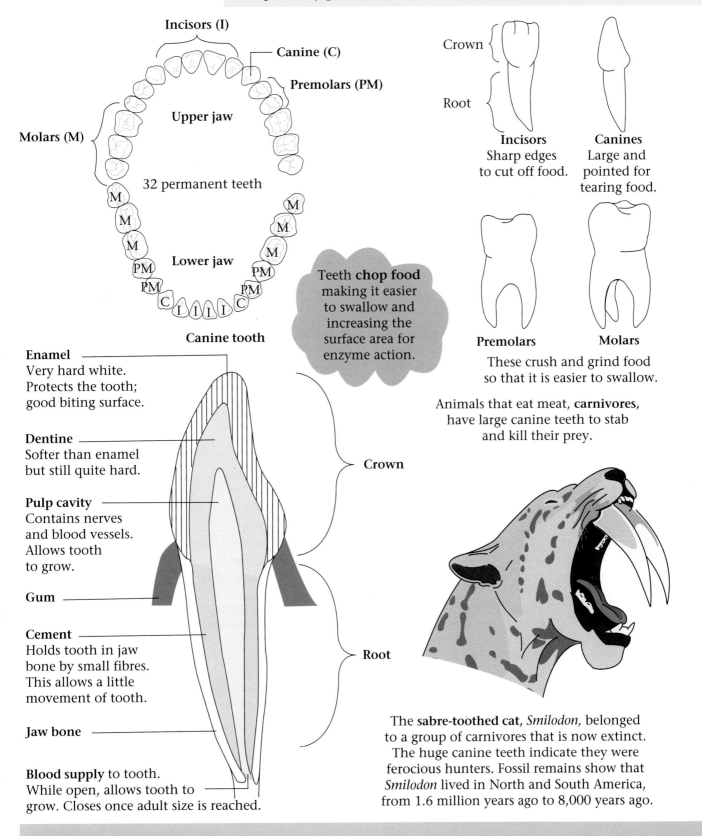

The **sabre-toothed cat**, *Smilodon,* belonged to a group of carnivores that is now extinct. The huge canine teeth indicate they were ferocious hunters. Fossil remains show that *Smilodon* lived in North and South America, from 1.6 million years ago to 8,000 years ago.

Questions:

1. What are the four types of teeth?
2. What protects our teeth from damage?
3. Why do molars and premolars have rough edges?
4. Which teeth are pointed?
5. Why do sheep not have canine teeth?
6. Herbivores like sheep spend a lot of time chewing. Why do their teeth continue to grow all their life?
7. Which part of the tooth contains nerves and blood?
8. What do incisors do?

TOOTH DECAY

After eating, sugar is left in the mouth. Bacteria feed on the sugar and produce an **acid** that causes decay. The white, sticky mixture of food and bacteria is called **plaque**. This should be removed regularly to avoid tooth decay and damage to the gums.

Stages in tooth decay

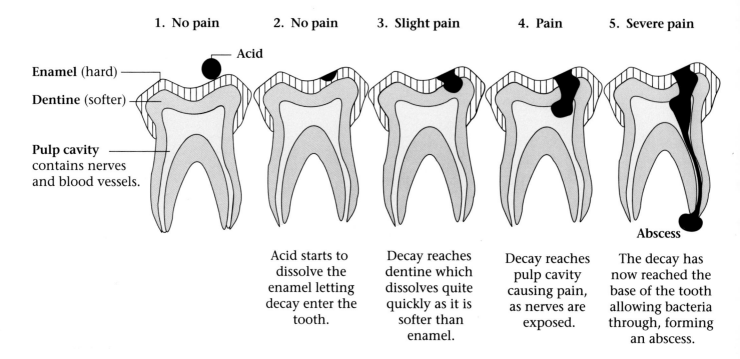

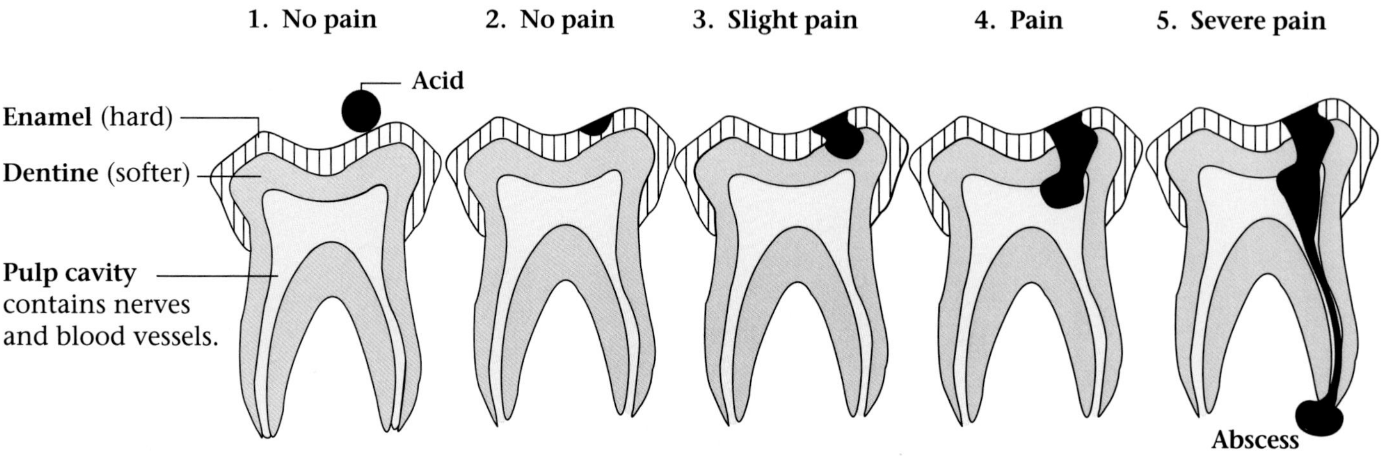

2. Acid starts to dissolve the enamel letting decay enter the tooth.

3. Decay reaches dentine which dissolves quite quickly as it is softer than enamel.

4. Decay reaches pulp cavity causing pain, as nerves are exposed.

5. The decay has now reached the base of the tooth allowing bacteria through, forming an abscess.

Prevention of tooth decay – five ways				
Diet	**Fluoride**	**Brushing teeth**	**Antiseptic mouthwash**	**Dentist**
Calcium and **Vitamin D** harden teeth, giving greater resistance to decay.	1. **Fluoride** hardens enamel, reducing effect of acid. 2. **Fluoride** is an **alkali** and neutralises the acid causing decay.	This **removes** trapped food which is a source of sugar for bacteria.	This **kills** the bacteria that cause decay.	**Dentists** can identify holes caused by decay and fill them, thus preventing further entry of bacteria.

Questions:

1. What causes tooth decay?
2. Which part of the tooth is affected by acid first?
3. When will pain first appear and why?
4. Why should the dentist be visited before pain starts?
5. How can tooth decay be prevented?

THE HUMAN DIGESTIVE SYSTEM (I)

This is where the digestion and absorption of food takes place.

Human digestive system

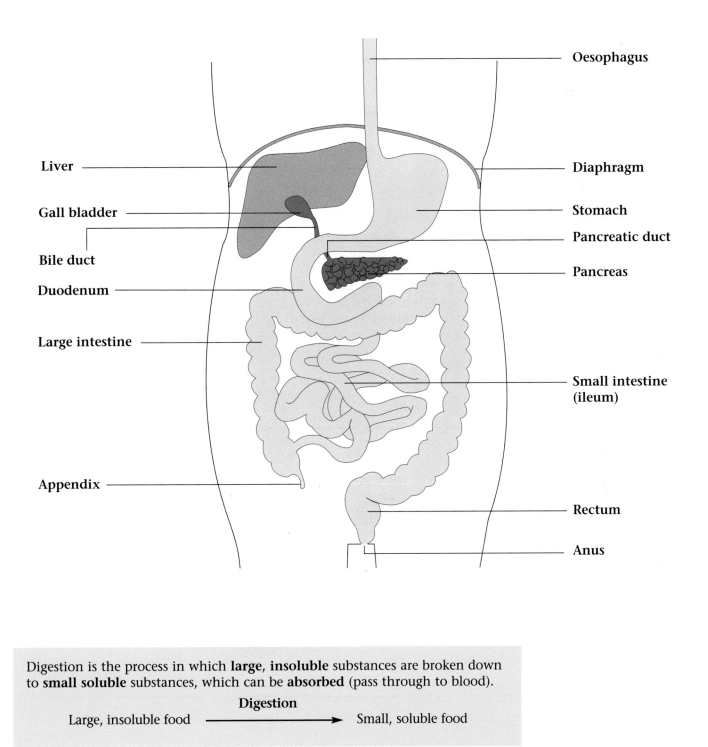

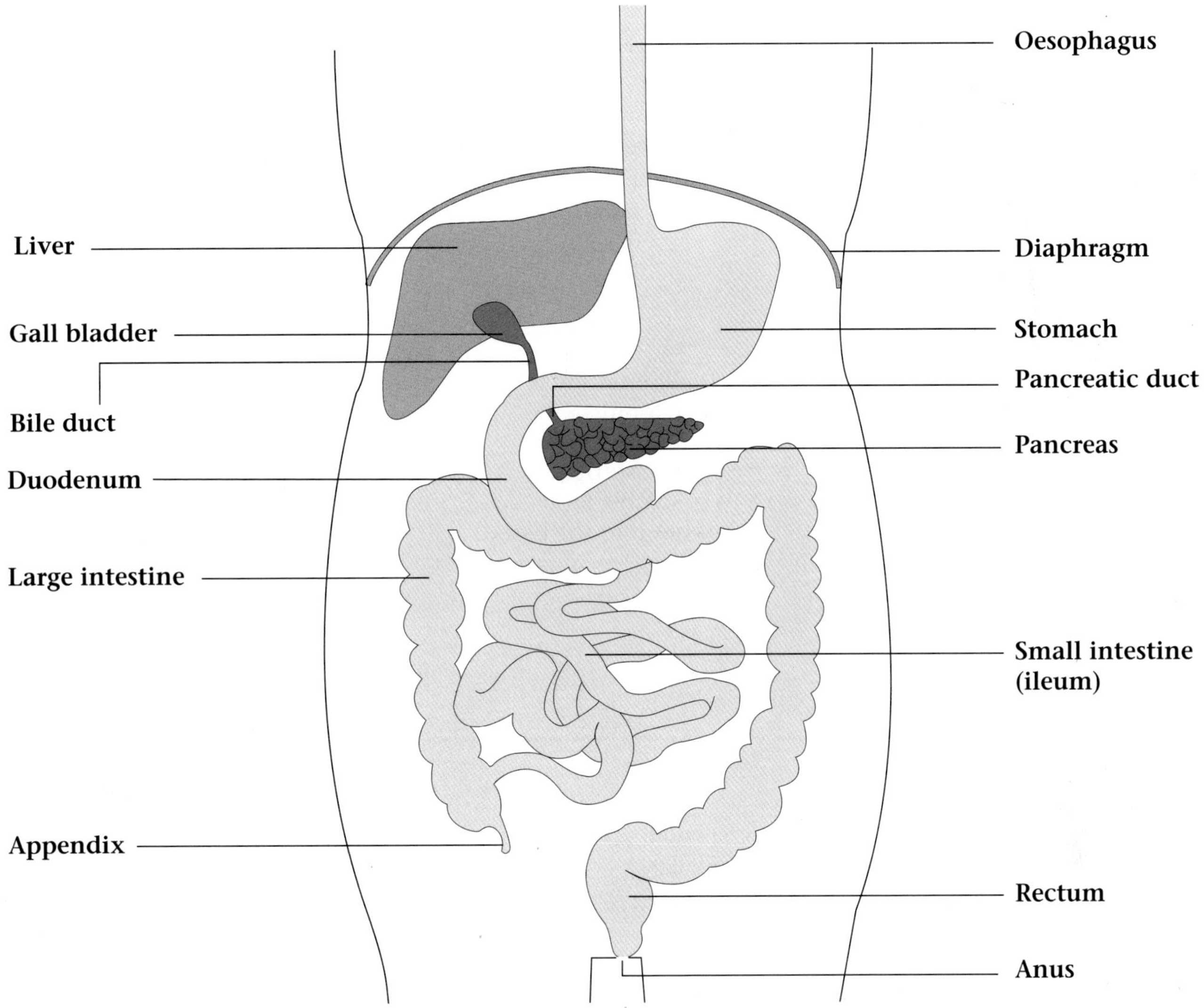

Digestion is the process in which **large, insoluble** substances are broken down to **small soluble** substances, which can be **absorbed** (pass through to blood).

Large, insoluble food —**Digestion**→ Small, soluble food

Questions:

1. What are the tubes that food passes through from the mouth to anus? List them in the correct order.
2. Where is the gall bladder found?
3. What separates the thorax from the abdomen?
4. Why does food need to be digested?

THE HUMAN DIGESTIVE SYSTEM (II)

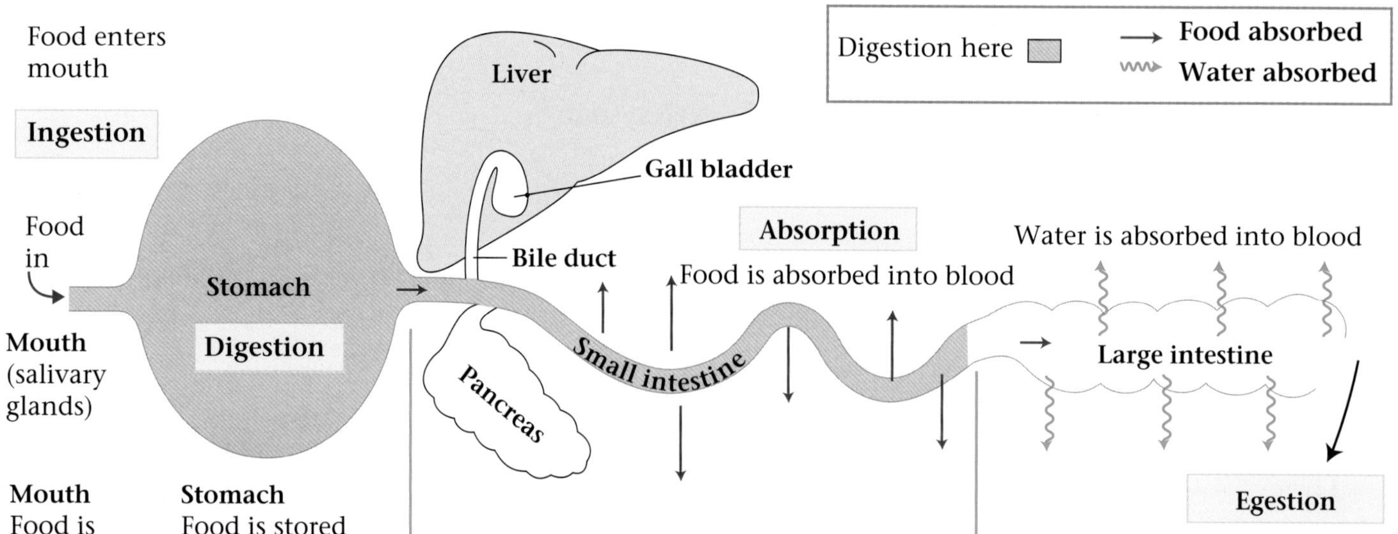

Undigested food comes out of anus

Mouth
Food is chopped into small pieces by teeth. The salivary glands add saliva to the food. This digests starch and moistens the food. Moist, chopped food is easier to swallow.

Stomach
Food is stored here for several hours while it is digested (broken down).

Small intestine
Here food is digested and then absorbed, i.e. soluble food passes into the blood.

Large intestine
Here water is absorbed so only a little water is lost in the faeces.

Food is pushed along by **peristalsis** (muscle contraction)

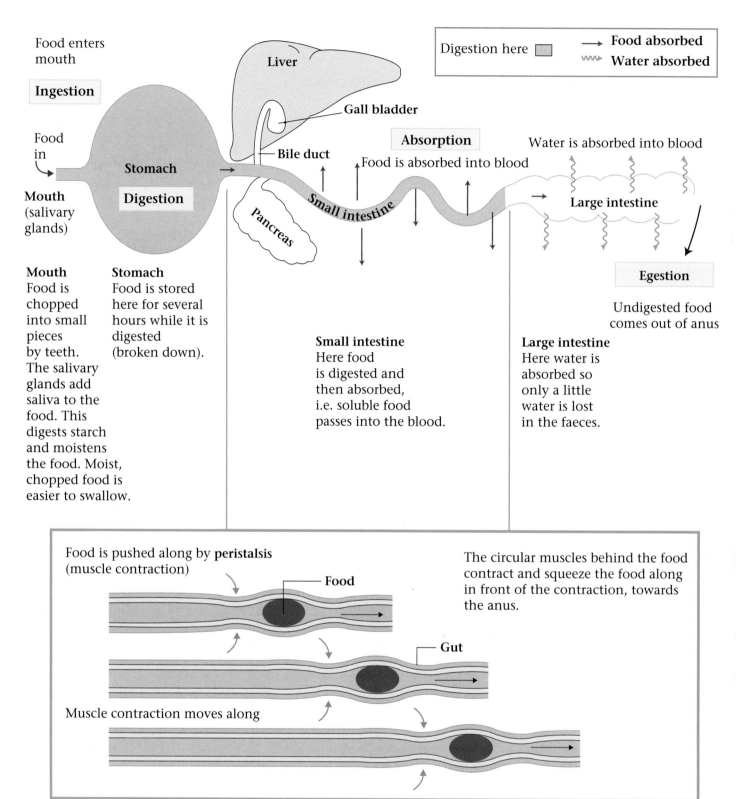

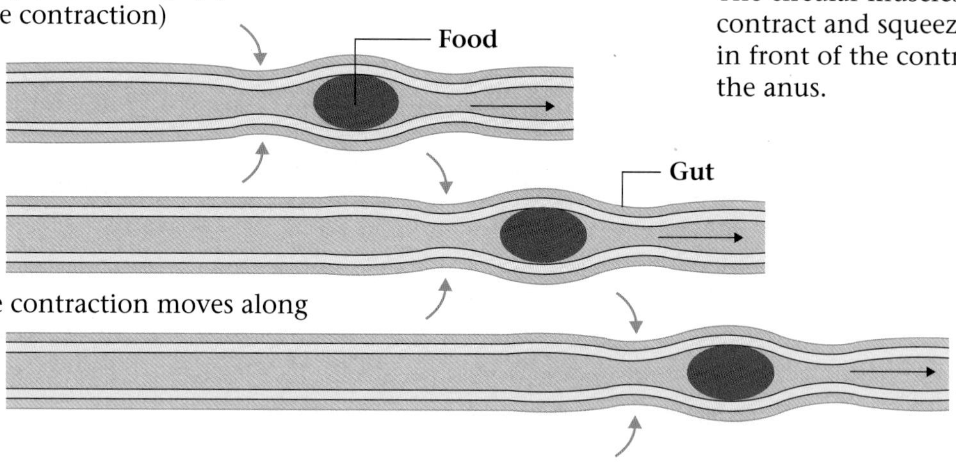

Muscle contraction moves along

The circular muscles behind the food contract and squeeze the food along in front of the contraction, towards the anus.

DUODENUM

Bile • Emulsifies fats.
• Neutralises acid.

Pancreas - secretes enzymes:
- protease
- amylase
- lipase.

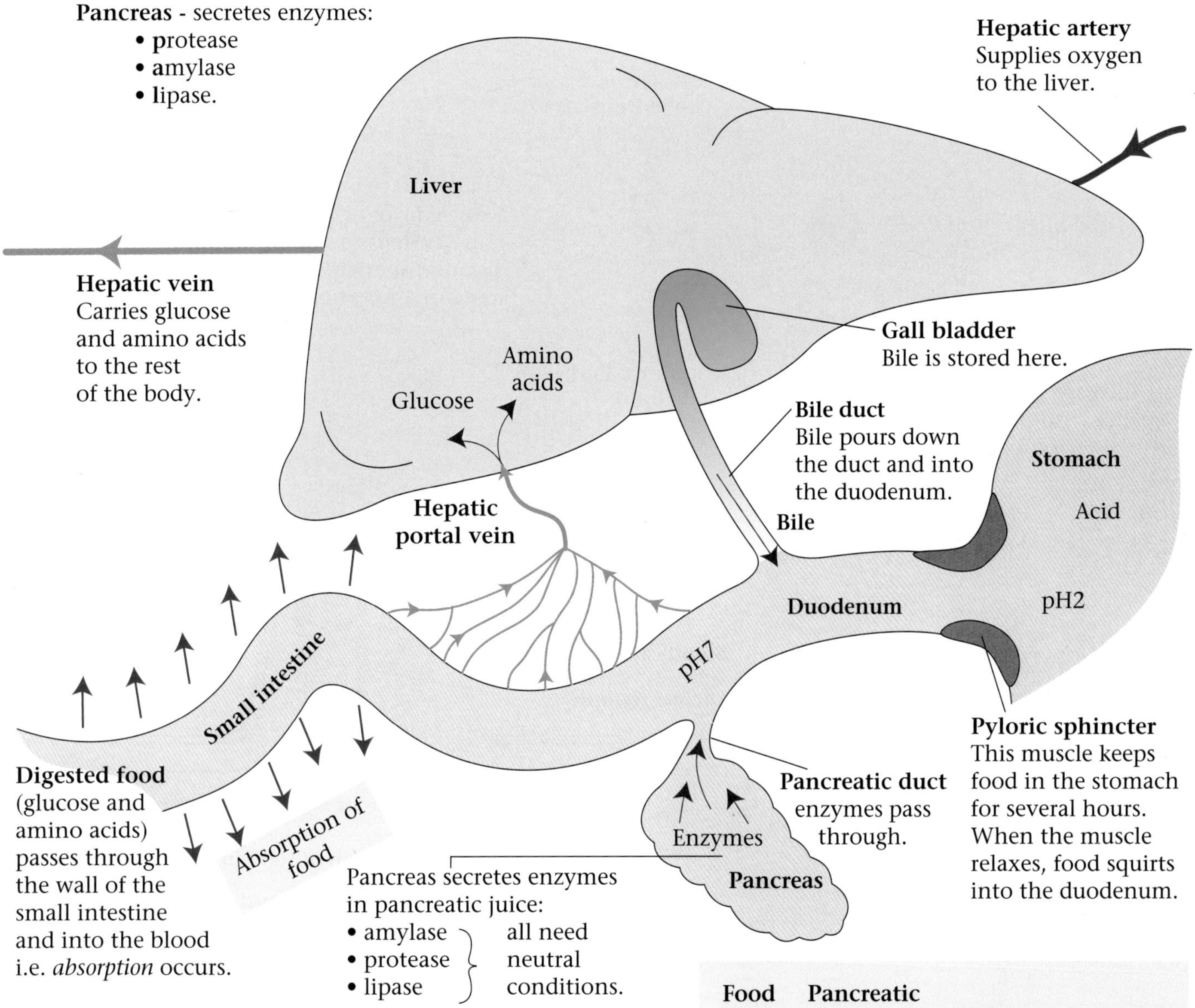

Hepatic artery
Supplies oxygen to the liver.
Liver
Hepatic vein
Carries glucose and amino acids to the rest of the body.
Gall bladder
Bile is stored here.
Amino acids
Glucose
Bile duct
Bile pours down the duct and into the duodenum.
Stomach
Acid
Hepatic portal vein
Bile
Duodenum
pH2
pH7
Small intestine
Pyloric sphincter
This muscle keeps food in the stomach for several hours. When the muscle relaxes, food squirts into the duodenum.
Digested food
(glucose and amino acids) passes through the wall of the small intestine and into the blood i.e. *absorption* occurs.
Absorption of food
Pancreatic duct
enzymes pass through.
Enzymes
Pancreas
Pancreas secretes enzymes in pancreatic juice:
• amylase
• protease
• lipase
all need neutral conditions.

The pancreas produces pancreatic juice. This contains **three enzymes** to digest the main food groups.

Food group	Pancreatic enzyme	Product
Protein	protease →	Amino acids
Starch	amylase →	Maltose
Lipids	lipase →	Fatty acids and glycerol

Bile
Bile emulsifies fats into small droplets.

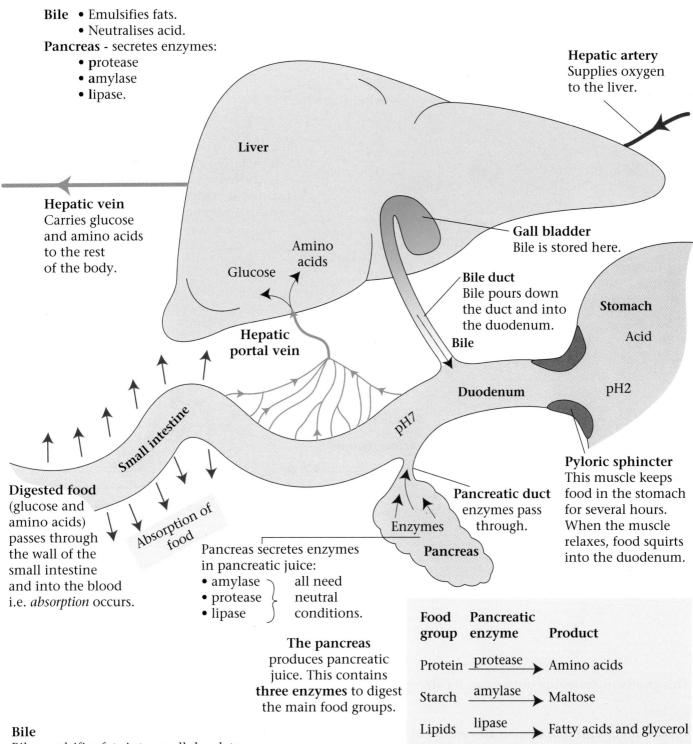

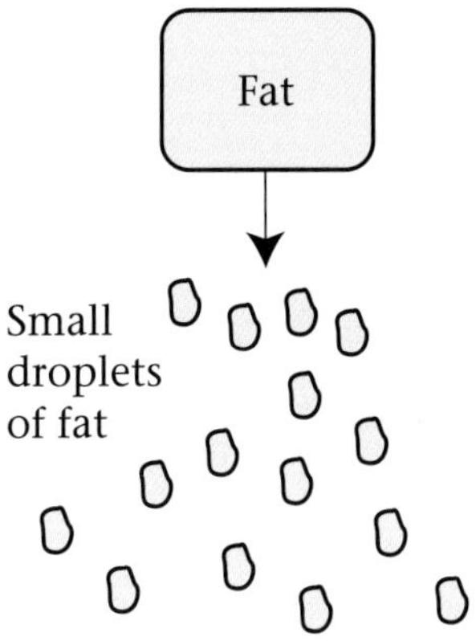

Fat
Small droplets of fat

This increases the surface area for enzymes to digest and speeds up the rate of digestion.

Bile contains alkali to **neutralise** acid from the stomach

Questions:
1. What are the three main food groups?
2. What enzymes are released by the pancreas?
3. What pH do the pancreatic enzymes need?
4. What are the two main functions of bile?
5. How does bile help the pancreatic enzymes to work?
6. Where is food absorbed?
7. Which organ is the digested food taken to and in what blood vessel?
8. What happens to food that has not been digested?

ABSORPTION

This takes place through the **villi** of the small intestine.

Transverse section (TS) of the small intestine

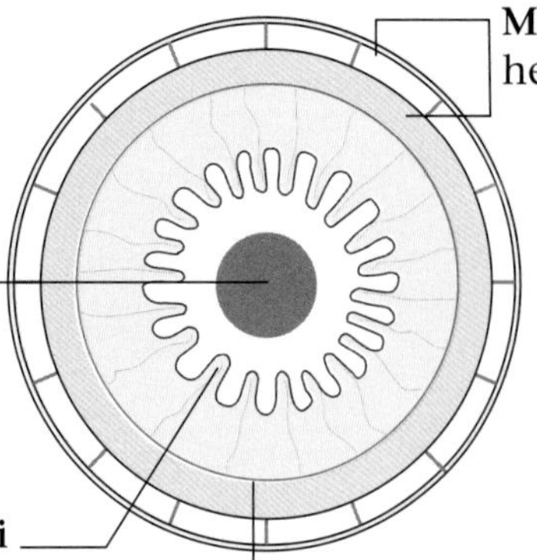
Muscle here
helps peristalsis.
Food in gut.
Villi
increase surface
area for absorption
of food.
Blood
takes absorbed
food away.

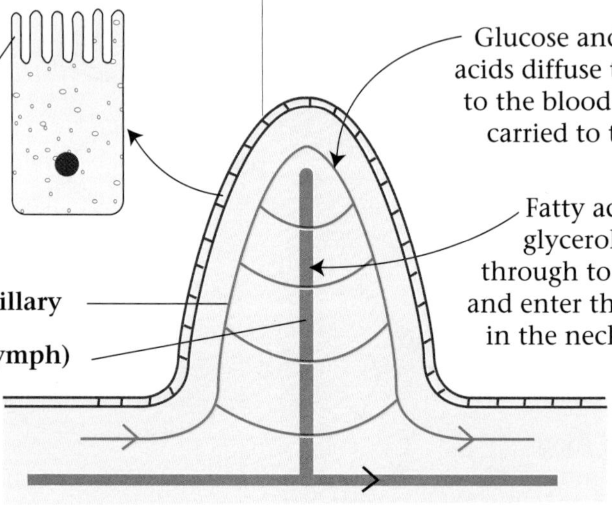
Villus surface is made up of
a single layer of cells.
This shortens the diffusion
distance, so increasing the
speed of absorption of food.
One villus
One cell
Microvilli
Glucose and amino
acids diffuse through
to the blood and are
carried to the liver.
Fatty acids and
glycerol diffuse
through to lymph,
and enter the blood
in the neck region.
Blood capillary
Lacteal (lymph)

Low concentration of food in blood

Blood carries **glucose** and **amino** *acids* away.

The wall of the small intestine is folded into villi and microvilli.

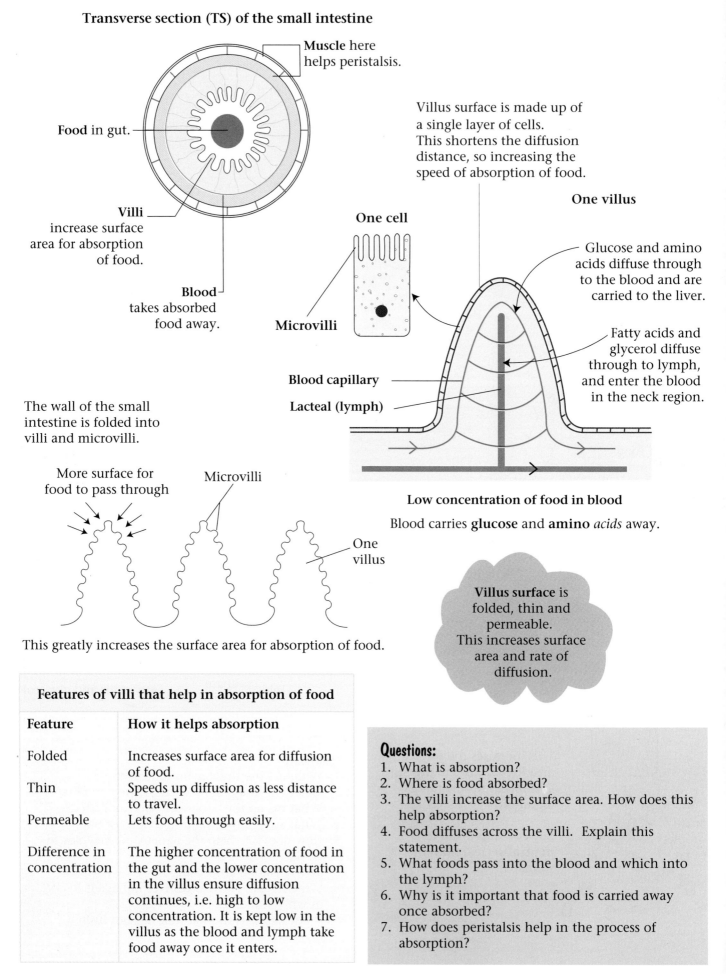

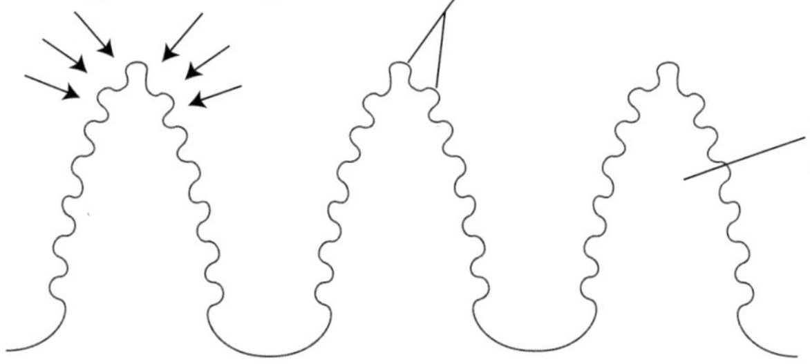
More surface for
food to pass through
Microvilli
One
villus

This greatly increases the surface area for absorption of food.

Villus surface is folded, thin and permeable.
This increases surface area and rate of diffusion.

Features of villi that help in absorption of food	
Feature	**How it helps absorption**
Folded	Increases surface area for diffusion of food.
Thin	Speeds up diffusion as less distance to travel.
Permeable	Lets food through easily.
Difference in concentration	The higher concentration of food in the gut and the lower concentration in the villus ensure diffusion continues, i.e. high to low concentration. It is kept low in the villus as the blood and lymph take food away once it enters.

Questions:

1. What is absorption?
2. Where is food absorbed?
3. The villi increase the surface area. How does this help absorption?
4. Food diffuses across the villi. Explain this statement.
5. What foods pass into the blood and which into the lymph?
6. Why is it important that food is carried away once absorbed?
7. How does peristalsis help in the process of absorption?

STRUCTURE OF BLOOD

Blood

(Liquid part)

Plasma
A yellow liquid, 90% water, 10% dissolved substances (solutes).

- Carries food, i.e. glucose, amino acids.
- Carries enzymes, hormones and antibodies.
- Carries urea from liver to kidneys.
- Contains soluble fibrinogen involved in blood clotting.

(Solid part) Cells

Platelets (cell fragments) help the blood to clot when the skin is cut.

Fibrin threads

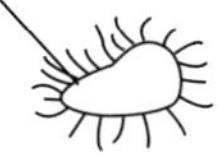

Platelet
Red blood cells get trapped in the threads. This forms a blood clot.
Blood clots stop:
- blood loss
- entry of germs.

Red blood cells carry oxygen

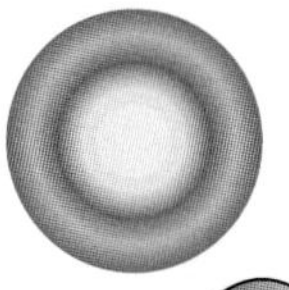

No nucleus

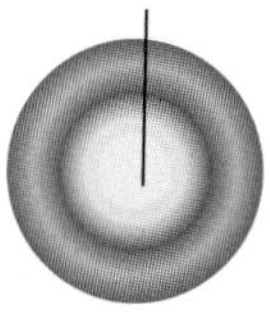

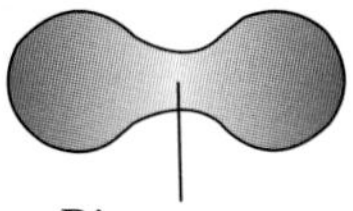

Biconcave

Red colour due to **haemoglobin** present.

White blood cells fight infections

Phagocytes

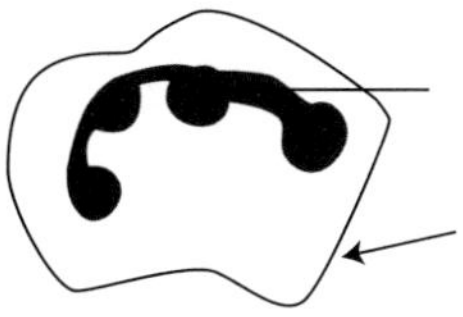

Lobed nucleus

Cell can change shape

These surround and digest bacteria.

Lymphocytes

Large nucleus

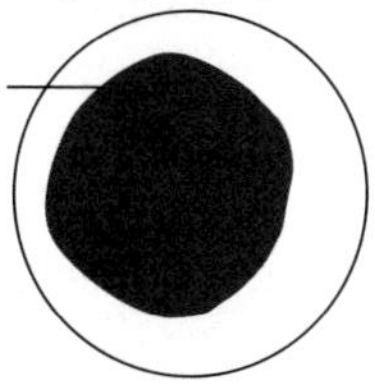

These produce antibodies which destroy bacteria and viruses.

In lungs
Oxygen enters red cells and joins with haemoglobin.

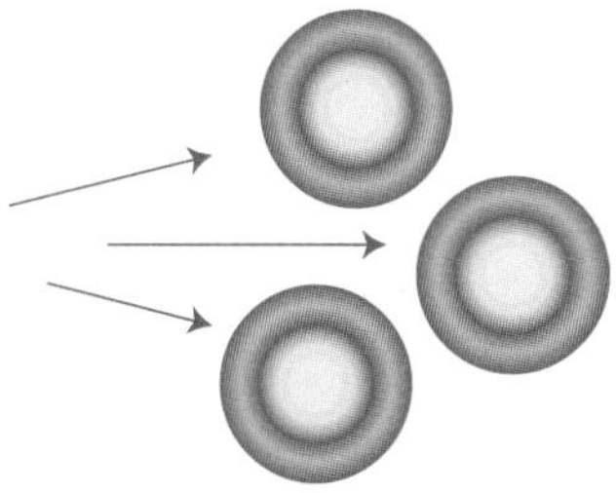

O_2 + Haemoglobin → Oxyhaemoglobin

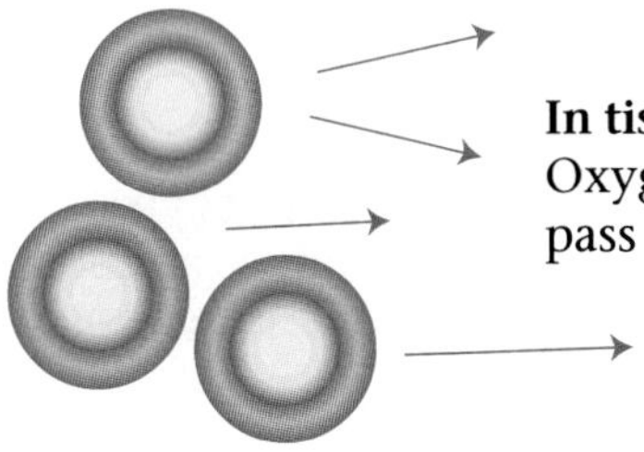

In tissues
Oxygen leaves red cells to pass to respiring cells.

Oxyhaemoglobin → Haemoglobin + Oxygen

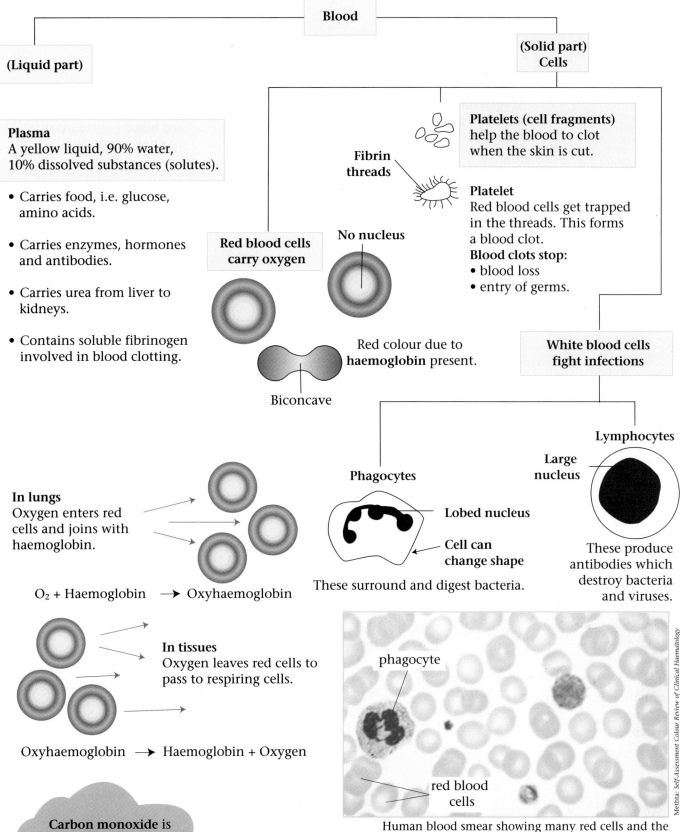

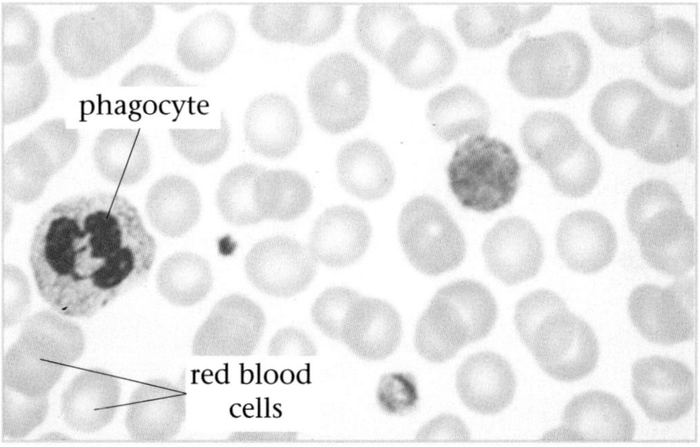

Mettha: Self-Assessment Colour Review of Clinical Haematology

Human blood smear showing many red cells and the phagocyte with its lobed nucleus.

Carbon monoxide is emitted from car exhausts and from burning fossil fuels. It is extremely dangerous as it combines with haemoglobin permanently, so preventing the pick-up of oxygen.

Questions:

1. Which blood cells have no nucleus?
2. What is the job of the white blood cells?
3. Name three substances carried dissolved in the plasma.
4. Why is it important that our blood clots when we cut ourselves?
5. How do phagocytes kill bacteria?
6. Which blood cells carry oxygen?

BLOOD CELLS

Red blood cells (carry oxygen)

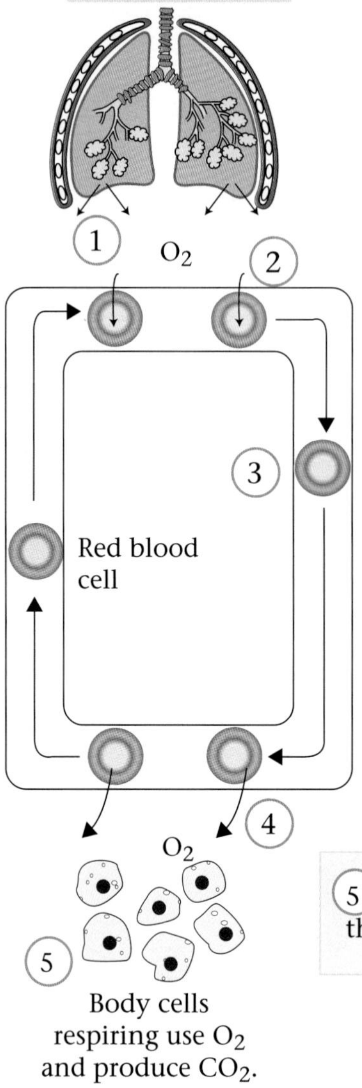

(1) In the lungs, **oxygen** diffuses into the blood and enters the **red blood cells.**

(2) Oxygen joins with haemoglobin in the red blood cells to make **oxyhaemoglobin.**

(3) Red cells with oxygen are pumped round the body in the blood.

(4) **Oxyhaemoglobin** breaks down releasing the **oxygen**.

(5) **Oxygen** diffuses to the cells for respiration.

(6) Red cells with haemoglobin return to the lungs to collect more oxygen and to get rid of carbon dioxide.

Body cells respiring use O_2 and produce CO_2.

White blood cells (fight disease)
There are two main types:

1. Phagocytes
2. Lymphocytes

(1) **Phagocytes**

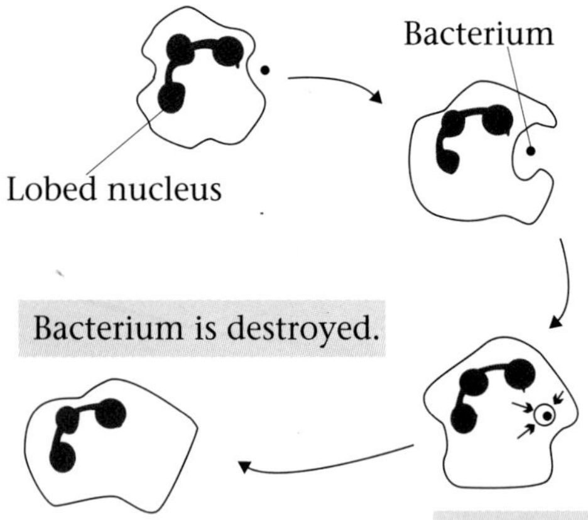

Phagocyte starts to surround the bacterium.

Bacterium is surrounded and digested.

Bacterium is destroyed.

(2) **Lymphocytes**

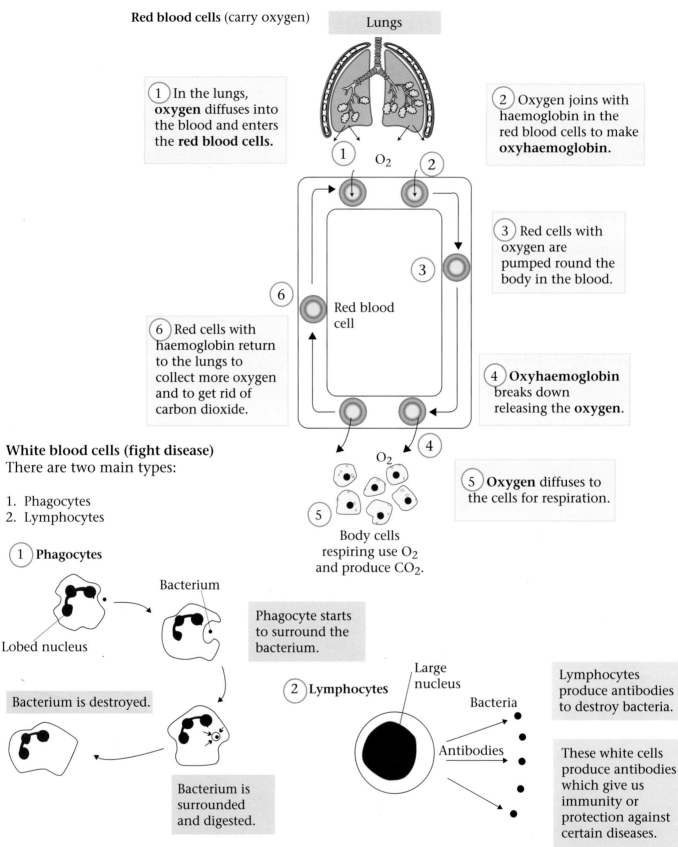

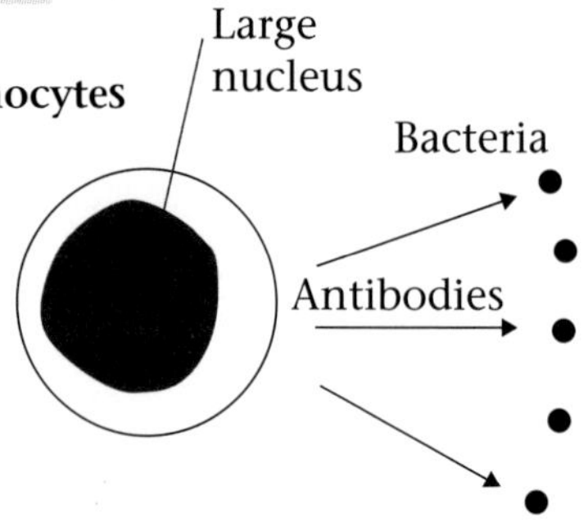

Lymphocytes produce antibodies to destroy bacteria.

These white cells produce antibodies which give us immunity or protection against certain diseases.

Questions:

1. Where does oxygen pass into the blood?
2. Which substance does oxygen join with in the red blood cell and what is formed?
3. What pumps the red cells round the body?
4. How does oxygen reach respiring cells?
5. Which white cell has a lobed nucleus?
6. How do lymphocytes destroy bacteria?
7. Carbon monoxide, in cigarette smoke, can also join with haemoglobin. This joining is permanent. How will this affect a smoker?

MOVEMENT OF BLOOD AROUND THE BODY

Blood is carried in tubes called **blood vessels.**

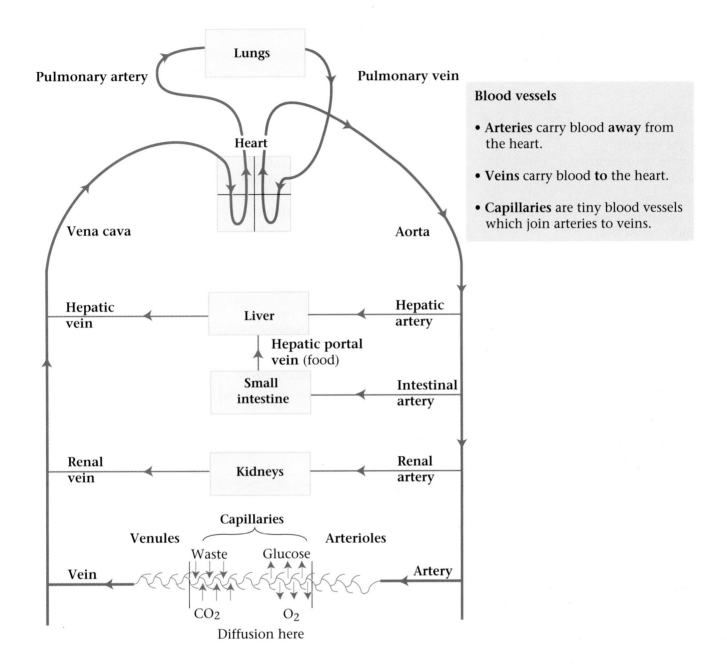

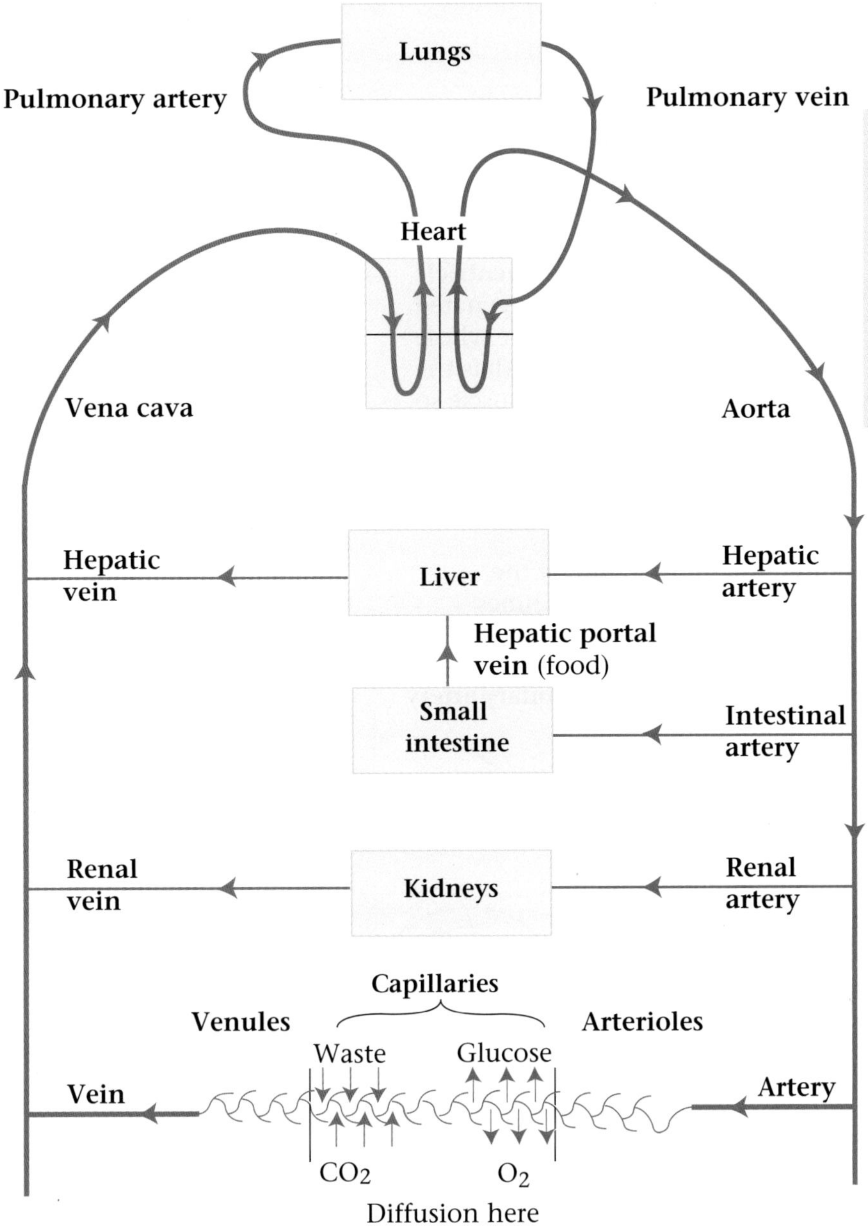

Blood vessels

- **Arteries** carry blood **away** from the heart.
- **Veins** carry blood **to** the heart.
- **Capillaries** are tiny blood vessels which join arteries to veins.

Capillaries are very thin, tiny blood vessels, with walls only one cell thick.

Here oxygen and glucose diffuse from the blood out to the cells. Carbon dioxide and waste, such as urea, diffuse from the body cells back into the blood. This is possible as capillaries are permeable.

Questions:

1. Which blood vessel enters the lungs?
2. What is added to blood in the lungs?
3. Which blood vessel carries oxygenated blood all round the body from the heart?
4. Why is the blood vessel leaving the kidneys deoxygenated?
5. What is carried in the hepatic portal vein?
6. Why do large arteries have to be split into tiny capillaries in body organs?
7. Which blood vessel leaves the liver, and what vessel does it join?

HEART (I)

Arteries carry blood **away** from the heart. **Veins** carry blood **to** the heart.

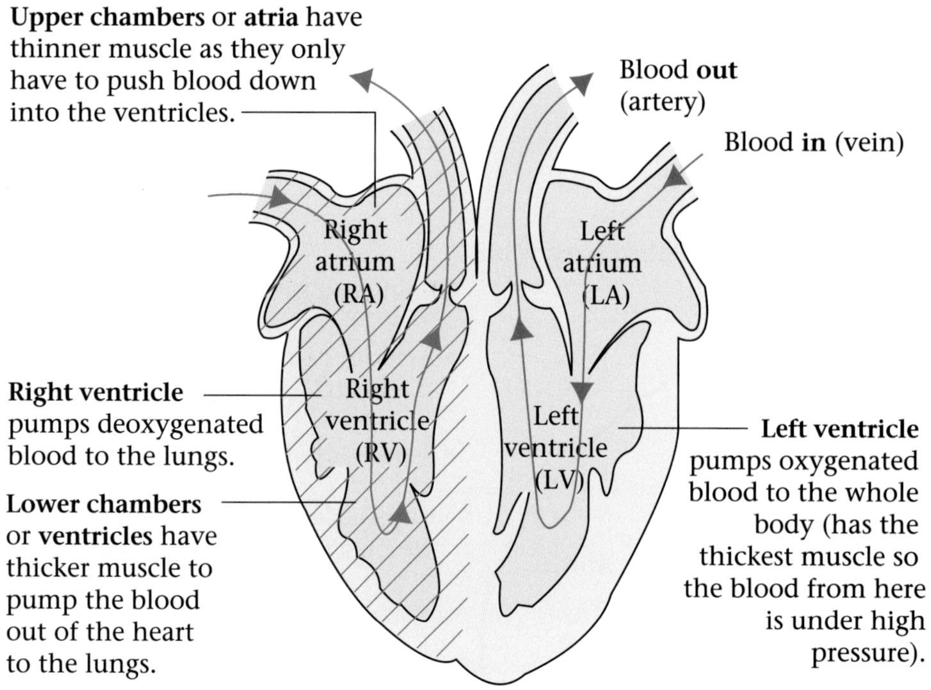

Upper chambers or **atria** have thinner muscle as they only have to push blood down into the ventricles.
Blood **out** (artery)
Blood **in** (vein)
Right atrium (RA)
Left atrium (LA)
Right ventricle (RV)
Left ventricle (LV)
Right ventricle pumps deoxygenated blood to the lungs.
Lower chambers or **ventricles** have thicker muscle to pump the blood out of the heart to the lungs.
Left ventricle pumps oxygenated blood to the whole body (has the thickest muscle so the blood from here is under high pressure).

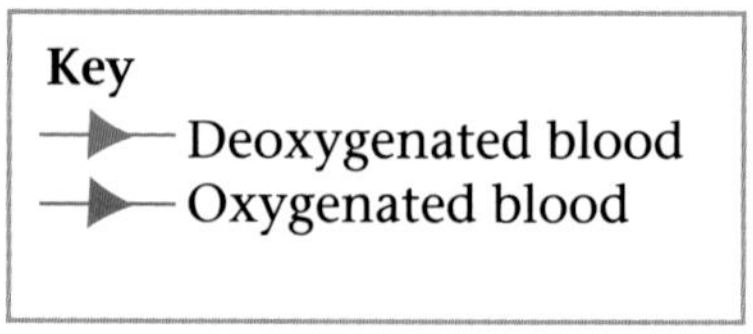

Key
Deoxygenated blood
Oxygenated blood

The heart is made of **cardiac muscle**. Beating starts in the wall of the right atrium at the **pacemaker**. The human heart beats about 70 times in a minute and causes the pulse which can be felt in the wrist.

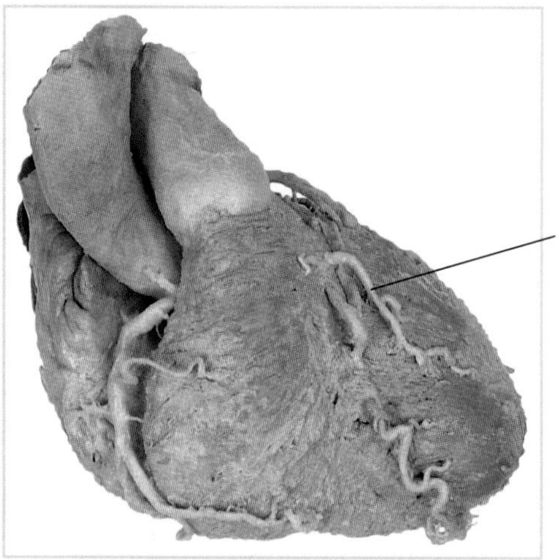

Courtesy of Ralph Hutchings.

Coronary artery brings food and oxygen to the heart muscle cells for **respiration**. This releases **energy** for **beating**.

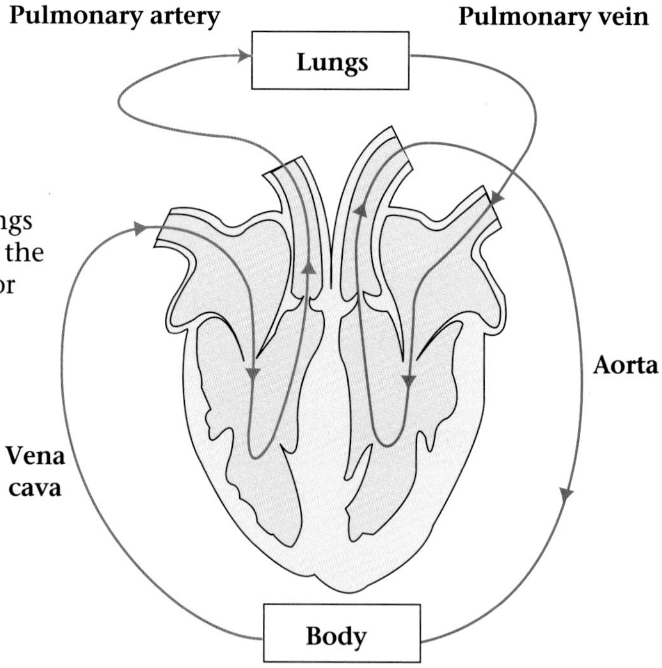

Pulmonary artery
Pulmonary vein
Lungs
Aorta
Vena cava
Body

1. **Healthy coronary artery.**

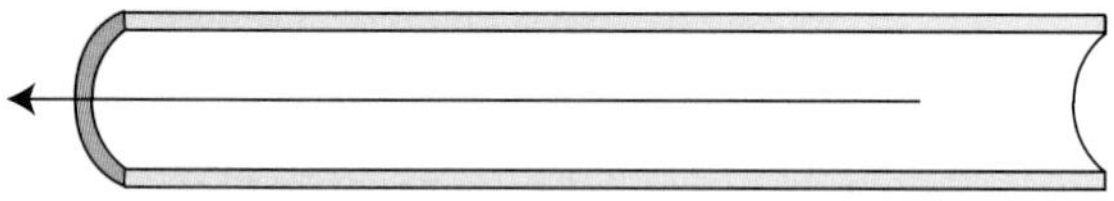

Blood flows freely – to the heart

Cholesterol is a substance made by the liver and found in the blood. The level of cholesterol produced depends upon diet and genetic factors. High levels of cholesterol increase the risk of disease of the heart and blood vessels.

2. **Unhealthy coronary artery** may be narrowed by smoking or fatty diet.

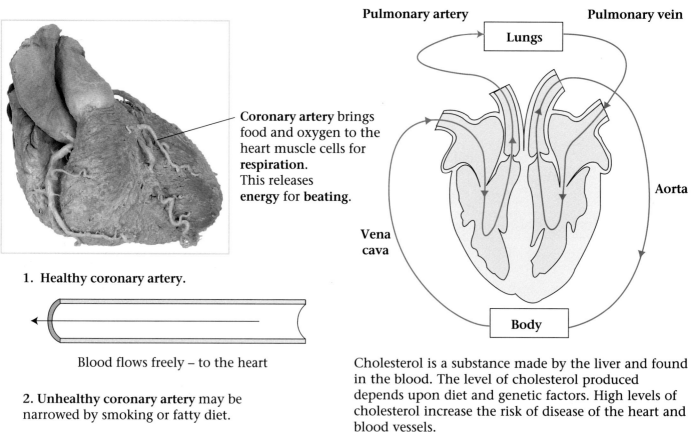

Blood clot

Artery narrowed

Blood clot can cause a blockage if the artery is narrow. If blood does not reach the heart, the muscle stops beating. This causes a **heart attack.**

Questions:
1. Which chamber of the heart has the thickest muscle and why?
2. What type of muscle makes up the heart?
3. Which blood vessel supplies the heart muscle itself with food and oxygen?

HEART (II)

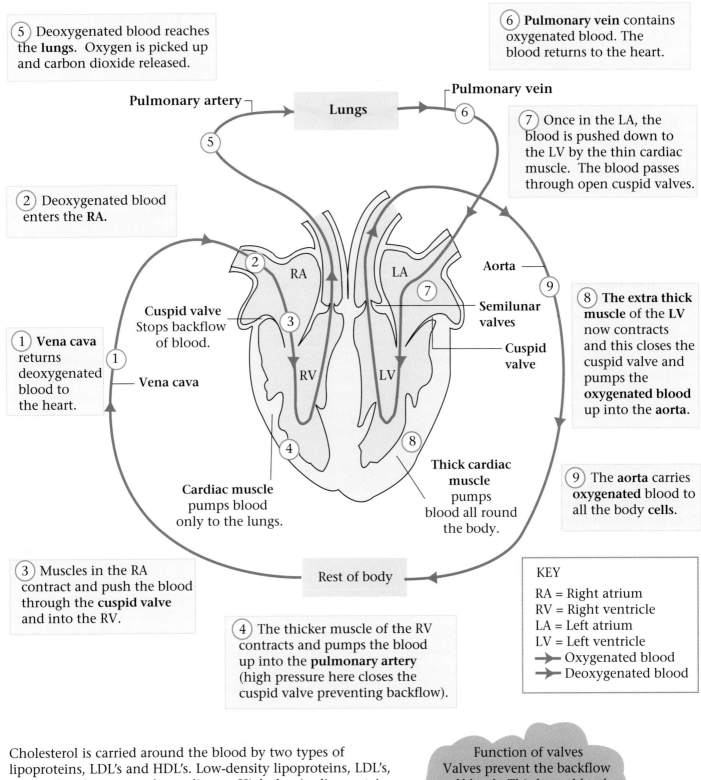

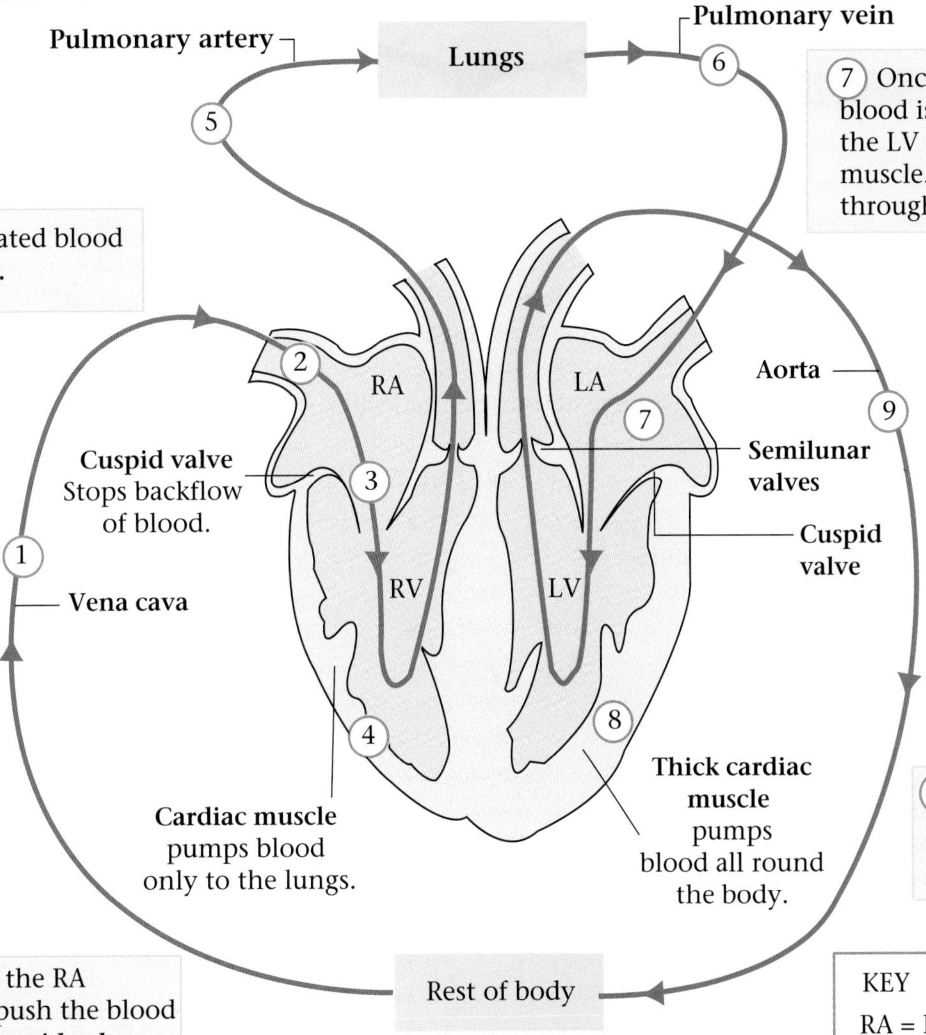

(1) **Vena cava** returns deoxygenated blood to the heart.

(2) Deoxygenated blood enters the **RA.**

(3) Muscles in the RA contract and push the blood through the **cuspid valve** and into the RV.

(4) The thicker muscle of the RV contracts and pumps the blood up into the **pulmonary artery** (high pressure here closes the cuspid valve preventing backflow).

(5) Deoxygenated blood reaches the **lungs.** Oxygen is picked up and carbon dioxide released.

(6) **Pulmonary vein** contains oxygenated blood. The blood returns to the heart.

(7) Once in the LA, the blood is pushed down to the LV by the thin cardiac muscle. The blood passes through open cuspid valves.

(8) **The extra thick muscle of the LV** now contracts and this closes the cuspid valve and pumps the **oxygenated blood** up into the **aorta.**

(9) The **aorta** carries **oxygenated** blood to all the body **cells.**

KEY

RA = Right atrium
RV = Right ventricle
LA = Left atrium
LV = Left ventricle
→ Oxygenated blood
→ Deoxygenated blood

Cholesterol is carried around the blood by two types of lipoproteins, LDL's and HDL's. Low-density lipoproteins, LDL's, are 'bad' and can cause heart disease. High-density lipoproteins, HDL's, are 'good'. The balance of these is vital to good health.

Function of valves
Valves prevent the backflow of blood. This keeps blood flowing in one direction.

The heart is a double pump, a **right** and **left** pump

Side of heart	Pressure	Blood	Sent to
Right Ventricle	Lower	Deoxygenated	Lungs
Left Ventricle	Higher	Oxygenated	Body

Questions:

1. What causes the higher pressure on the left side of the heart?
2. What happens to blood when it reaches the lungs?
3. What is the job of the valves in the heart?

BLOOD VESSELS

Arteries (from heart) | **Capillaries** | **Veins (to heart)**

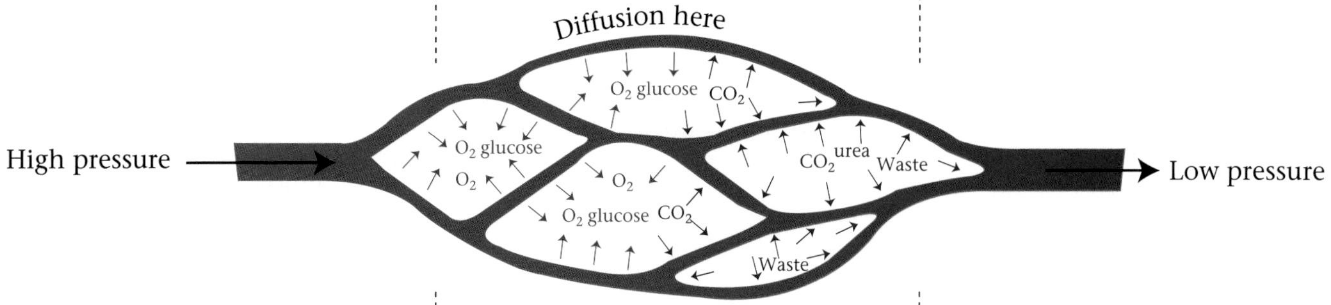

Arteries

(Take blood away from the heart.)

Blood under **pressure,** in spurts due to heartbeat. **Thick elastic wall** with muscle to withstand pressure.

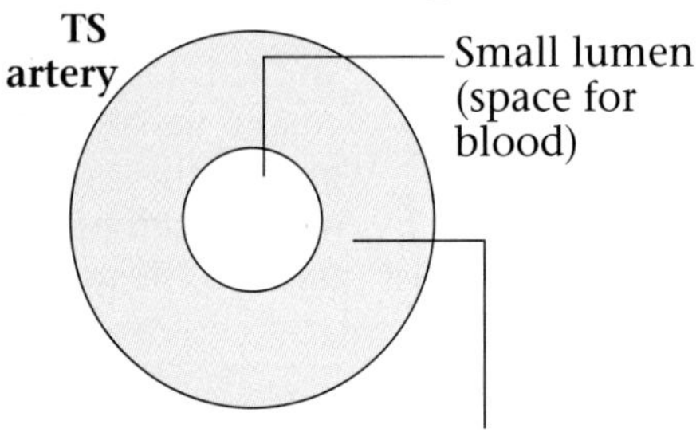

Thick wall of artery due to:

- **Elastic fibres** (withstand pressure changes).
- **Muscle** (can narrow or widen the lumen to decrease or increase blood flow through it). Constriction here leads to **vasoconstriction**. Widening here leads to **vasodilation**.
- No valves (blood is moved under pressure).
- **Impermeable**.

Capillaries

(Join arteries to veins.)

Low pressure.
Thin wall only one cell thick.

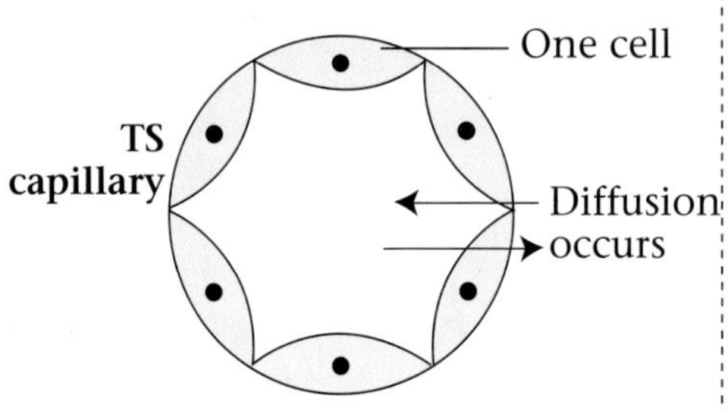

- No valves.
- Capillaries allow exchange of materials between blood and cells by diffusion.
- **Permeable**.

Veins

(Take blood to the heart.)

Low pressure.
Thin wall with little elasticity and muscle.

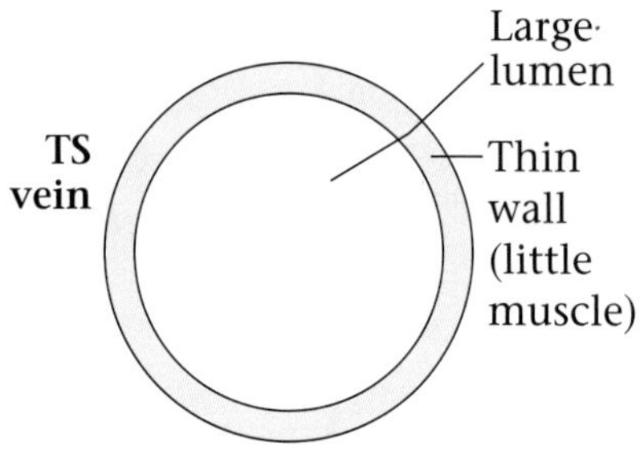

- Valves present.

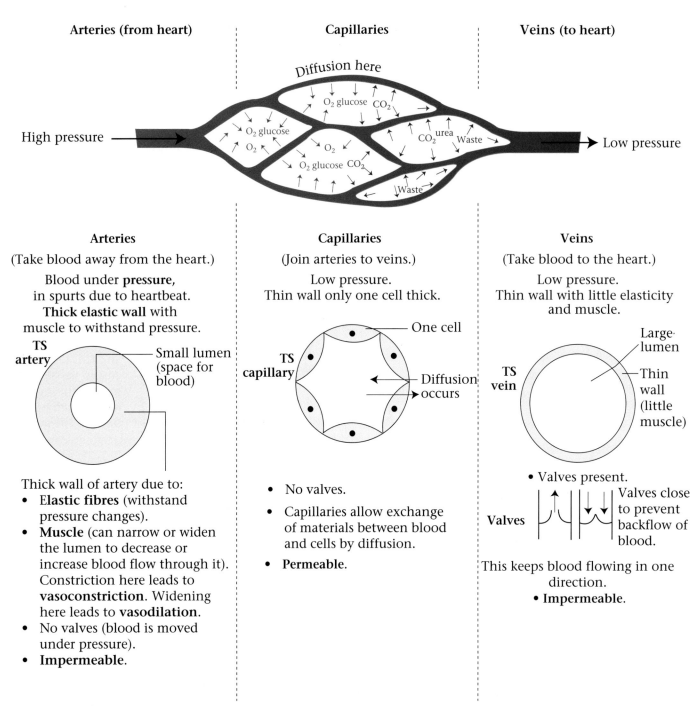

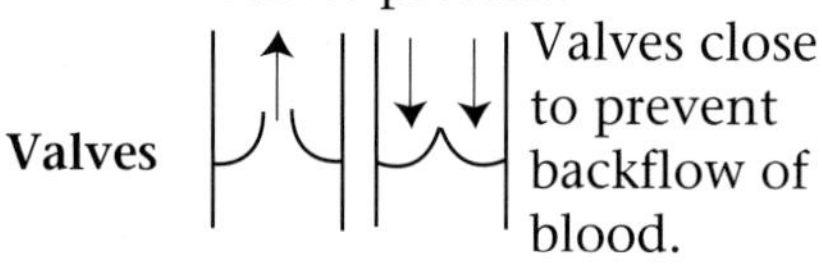

This keeps blood flowing in one direction.

- **Impermeable**.

Oxygen and **glucose** diffuse from the blood to the body cells. **Carbon dioxide** and **urea** diffuse from the cells back into the blood. This exchange of material takes place through the permeable capillaries.

Questions:

1. Which blood vessels take blood to the heart?
2. How do the valves work in veins, and why are they necessary?
3. Which blood vessels are permeable?
4. Name two substances which might diffuse out of blood and into the body cells.
5. What causes the high pressure in arteries?
6. Why must arteries have thick, elastic walls?

CAPILLARIES Exchange of materials.

Capillaries – tiny blood vessels.

Thin
Permeable
Large surface area

} Allow diffusion of materials between blood and cells.

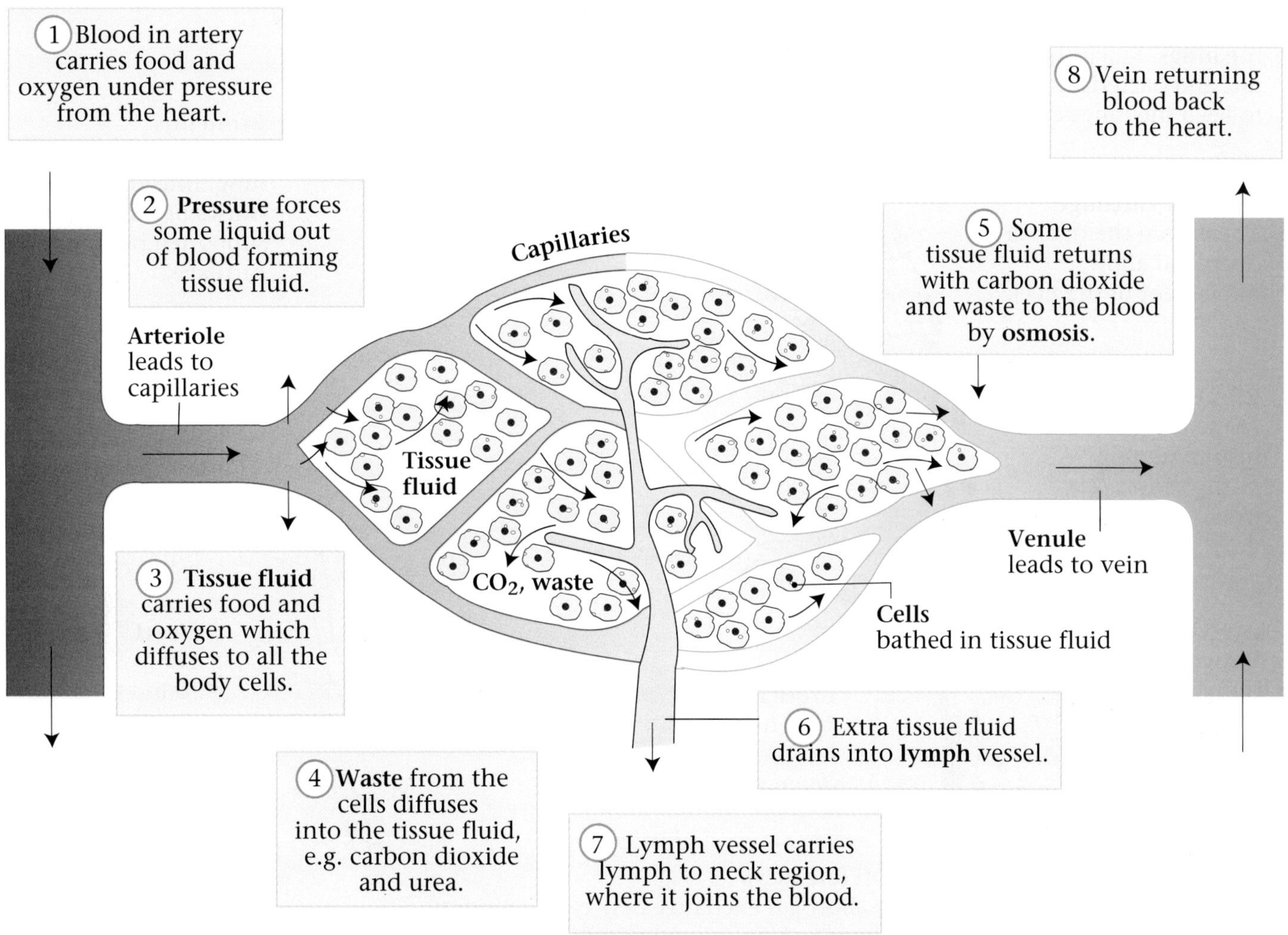

1 Blood in artery carries food and oxygen under pressure from the heart.
2 **Pressure** forces some liquid out of blood forming tissue fluid.
Arteriole leads to capillaries
3 **Tissue fluid** carries food and oxygen which diffuses to all the body cells.
4 **Waste** from the cells diffuses into the tissue fluid, e.g. carbon dioxide and urea.
Capillaries
Tissue fluid
CO_2, waste
5 Some tissue fluid returns with carbon dioxide and waste to the blood by **osmosis.**
6 Extra tissue fluid drains into **lymph** vessel.
7 Lymph vessel carries lymph to neck region, where it joins the blood.
8 Vein returning blood back to the heart.
Venule leads to vein
Cells bathed in tissue fluid

- Blood in the artery is under pressure.
- The artery divides into smaller arterioles and capillaries.
- Some of the liquid part of the blood, plasma is pushed through the permeable capillary wall to form **tissue fluid**. (Blood cells and plasma proteins are too large to filter through.)
- Oxygen and glucose in the tissue fluid diffuse to the body cells.
- Carbon dioxide and urea diffuse from the body cells into the tissue fluid.
- At the venule end of the capillaries, some tissue fluid returns to the blood by **osmosis**.
- The remaining tissue fluid drains into lymph vessels to form **lymph**.

Questions:

1. What is tissue fluid and how is it formed?
2. How do body cells get their oxygen and food?
3. Why must capillaries be permeable?
4. Body cells produce carbon dioxide and waste, how does this return to the blood?
5. What happens to the tissue fluid that does not return to the blood immediately?

GAS EXCHANGE AND RESPIRATION

THE HUMAN THORAX

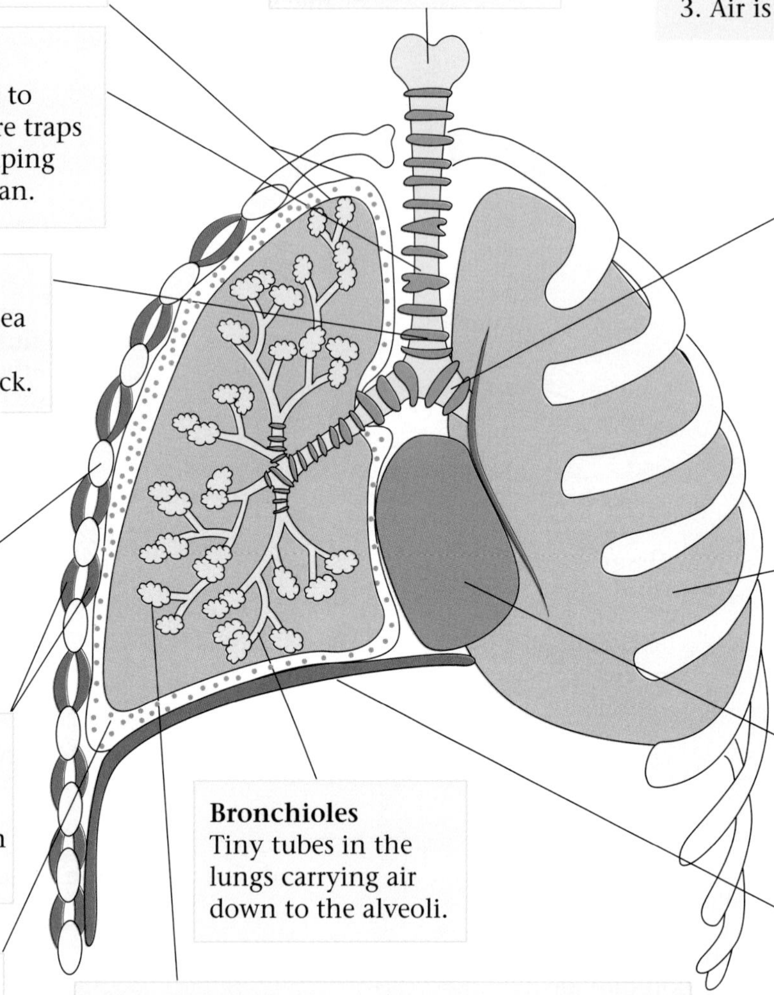
Pleural membranes
Contain a fluid which cushions the lungs.
Larynx (voice box)
Infection here causes laryngitis.
In the nose
1. Air is cleaned by cilia hairs.
2. Air is warmed by blood vessels.
3. Air is moistened by mucus.
Trachea (windpipe)
This carries air down to the lungs. Mucus here traps dirt and bacteria, helping to keep the lungs clean.
Bronchus
This tube enters the lung. Infection here causes bronchitis. It is supported by rings of cartilage.
Rings of cartilage
These keep the trachea open and allow movement of the neck.
Ribs (cut)
These protect the heart and lungs in the thorax from damage.
Left lung
Intercostal muscles (between the ribs)
These move the rib cage up and down during breathing.
Heart
Pumps blood round the body.
Bronchioles
Tiny tubes in the lungs carrying air down to the alveoli.
Pleural fluid
Allows lungs to move freely.
Alveoli (air sacs)
These allow gas exchange between the lungs and the blood. Each is surrounded by a network of blood capillaries taking oxygen away.
Diaphragm
This muscle moves up and down causing air to rush into or out of the lungs (**ventilation**).

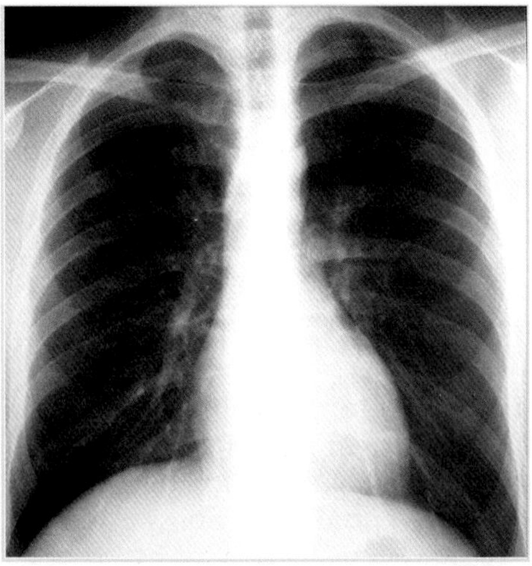

From McMinn *et al*: *Concise Handbook of Human Anatomy*

X-ray of human thorax clearly showing the ribs and lungs (dark shadow). The position of the heart can be seen by the displaced lungs on the right side.

Questions:
1. What protects the lungs and heart?
2. What is the role of mucus in the trachea?
3. What happens to air in the nose?
4. Describe the function of the alveoli.
5. What is the job of the rings of cartilage?

GAS EXCHANGE IN THE ALVEOLI

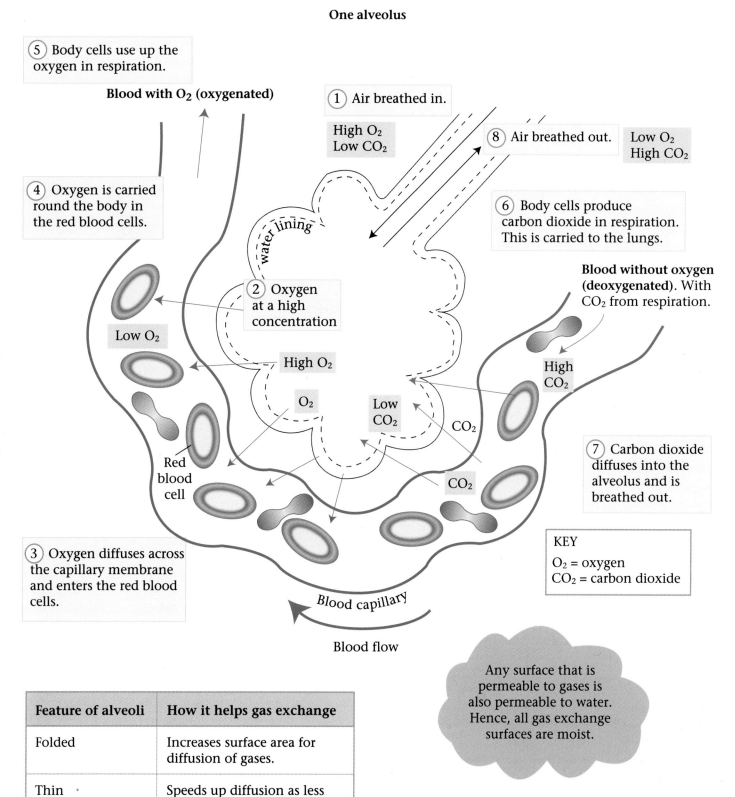

One alveolus

⑤ Body cells use up the oxygen in respiration.

Blood with O₂ (oxygenated)

① Air breathed in.

High O₂
Low CO₂

⑧ Air breathed out.

Low O₂
High CO₂

④ Oxygen is carried round the body in the red blood cells.

⑥ Body cells produce carbon dioxide in respiration. This is carried to the lungs.

Blood without oxygen (deoxygenated). With CO₂ from respiration.

water lining

② Oxygen at a high concentration

Low O₂

High O₂

O₂

Low CO₂

CO₂

High CO₂

Red blood cell

⑦ Carbon dioxide diffuses into the alveolus and is breathed out.

③ Oxygen diffuses across the capillary membrane and enters the red blood cells.

CO₂

KEY
O₂ = oxygen
CO₂ = carbon dioxide

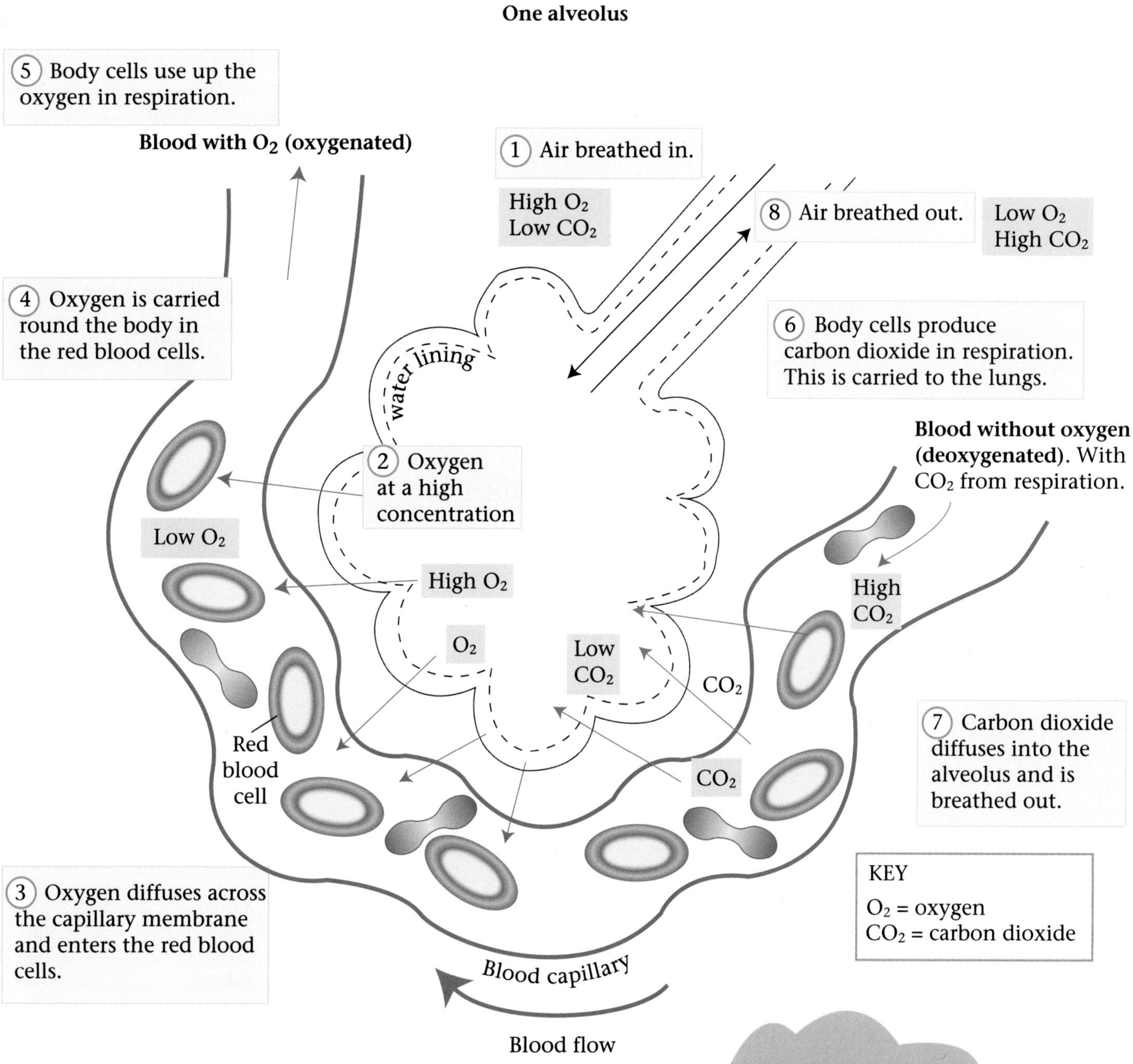

Any surface that is permeable to gases is also permeable to water. Hence, all gas exchange surfaces are moist.

Feature of alveoli	How it helps gas exchange
Folded	Increases surface area for diffusion of gases.
Thin	Speeds up diffusion as less distance to travel.
Large number of alveoli	Increases the surface area for diffusion.
Close to blood capillary	Reduces the diffusion distance.

Questions:
1. Why must the alveoli be only one cell thick?
2. Which blood cells carry oxygen round the body?
3. What is blood with oxygen called?
4. Alveoli are folded, how does this help in the exchange of gases?
5. Why is it important that body cells get oxygen?
6. Carbon dioxide diffuses from the blood, into the alveoli. Explain this statement.

BREATHING (I)

Changes in the position of the ribs and diaphragm when breathing.

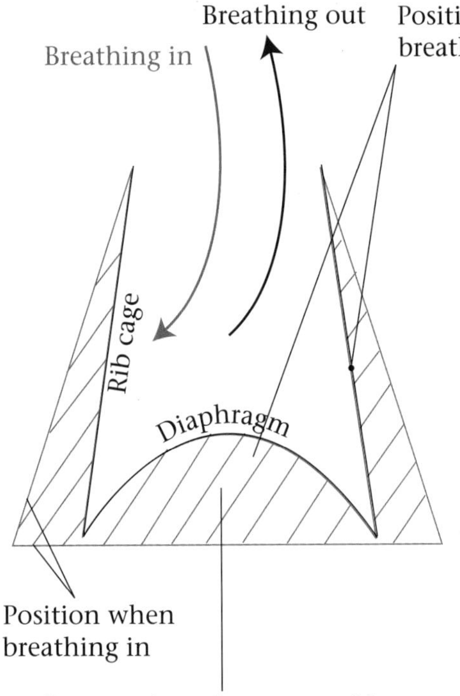

Gas	Air breathed in	Air breathed out	Changes
Nitrogen	79%	79%	None.
Oxygen	20%	16%	Oxygen enters blood so less is breathed out.
Carbon dioxide	0.04%	4%	Carbon dioxide is produced by respiring cells – so more carbon dioxide is breathed out.
Water vapour	Varies	Saturated	Water is produced by moist cells lining lungs and in respiration.

Air always moves from high to low pressure.

Breathing in:	Breathing out:
Diaphragm contracts and flattens ↓ Intercostal muscles pull ribs up and out ↓ Both increase volume of thorax ↓ This lowers the pressure in the thorax – air outside has higher pressure ↓ Air rushes into lungs	Diaphragm relaxes and rises to a dome shape ↓ Intercostal muscles pull ribs down and in ↓ Both decrease volume of thorax ↓ This increases the pressure in the thorax – air outside has lower pressure. ↓ Air is forced out of lungs

Questions:

1. What two actions increase the volume of the thorax?
2. How does an increase in volume affect the pressure?
3. What causes air to rush into the lungs?
4. Which gas is taken into the body during breathing?
5. Give three differences between air breathed in and air breathed out?
6. Why is air forced out of the body when the diaphragm rises into a dome shape?

BREATHING (II)

Control of breathing

Exercise requires energy

↓

Respiration rate increases to provide this energy

↓

Respiration produces high levels of carbon dioxide (oxygen is used up)

↓

High level of carbon dioxide in the blood is detected in the brain

↓

Brain sends impulses to diaphragm and intercostal muscles

↓

Breathing rate increases

↓

Extra carbon dioxide is removed, extra oxygen is taken in

Breathing is controlled by the brain. The brain responds to carbon dioxide levels in the blood (not lack of oxygen).

Graph showing effect of carbon dioxide on breathing rate

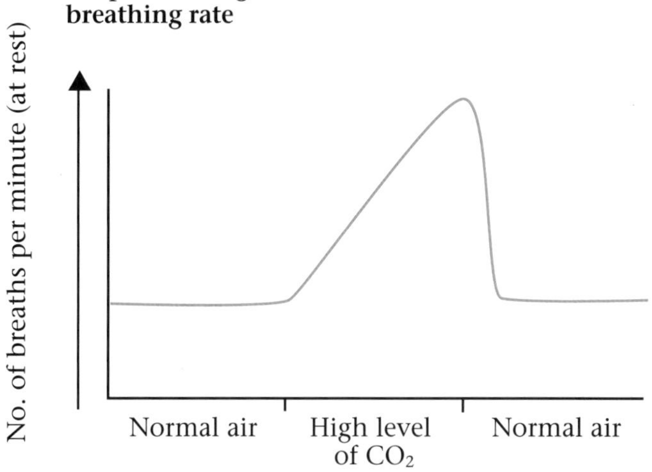

Trachea and cleaning (in a non-smoker):

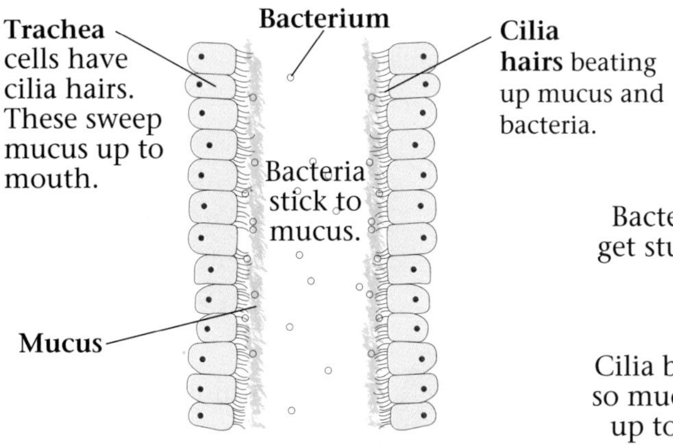

Bacteria and dirt get stuck in mucus

↓

Cilia beat upwards so mucus is carried up to the mouth

↓

Mucus and bacteria are swallowed

↓

In stomach, acid kills most of bacteria – lungs remain clean

This shows that an increase in carbon dioxide increases the breathing rate.

Bacteria and mucus are swallowed.

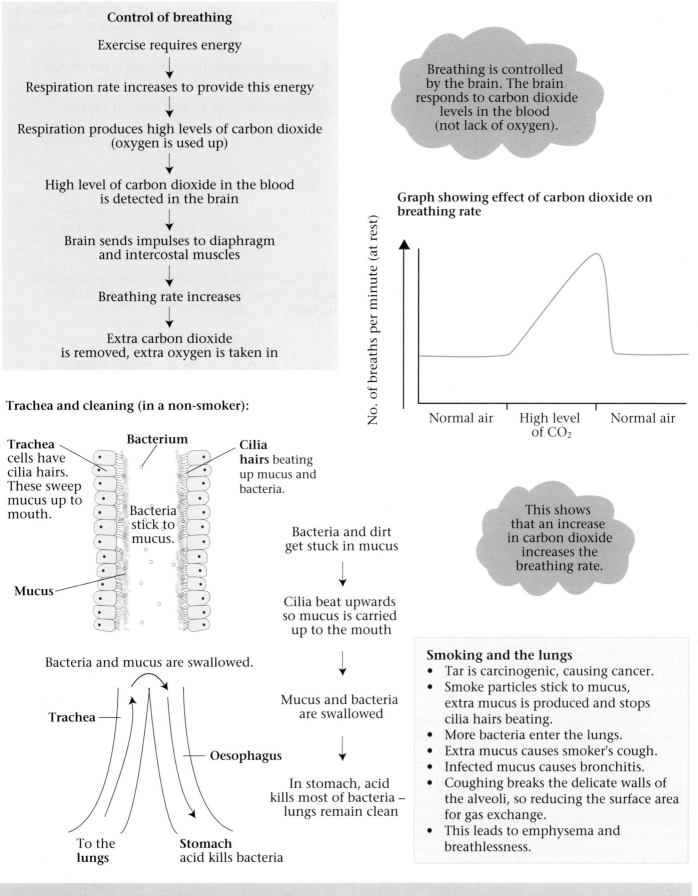

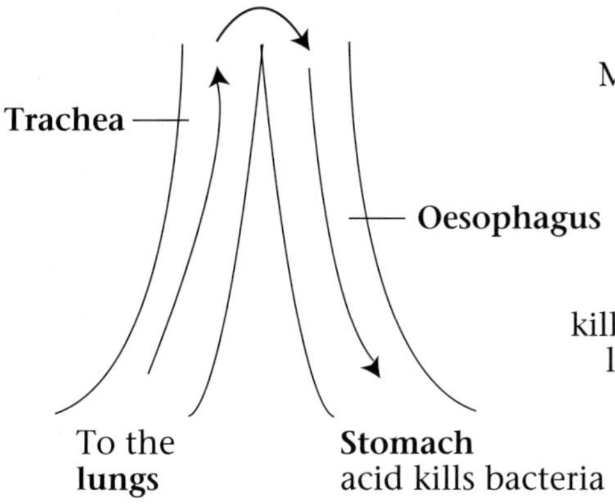

Smoking and the lungs

- Tar is carcinogenic, causing cancer.
- Smoke particles stick to mucus, extra mucus is produced and stops cilia hairs beating.
- More bacteria enter the lungs.
- Extra mucus causes smoker's cough.
- Infected mucus causes bronchitis.
- Coughing breaks the delicate walls of the alveoli, so reducing the surface area for gas exchange.
- This leads to emphysema and breathlessness.

Questions:

1. What controls the breathing rate?
2. What increases the breathing rate?
3. How do trachea cells clean the air breathed in?
4. Why is it better to swallow bacteria than breathe them in?
5. Which part of cigarette smoke causes cancer?
6. Why do smokers suffer from more lung infections?

RESPIRATION Aerobic and anaerobic.

All cells respire to release the energy from food (usually sugar).

The energy released during respiration is used for muscle contraction, to build up amino acids into protein and to maintain a constant temperature in mammals and birds.

1. Aerobic respiration – oxygen is needed.

Aerobic respiration takes place in the **mitochondria** inside cells.

A cell

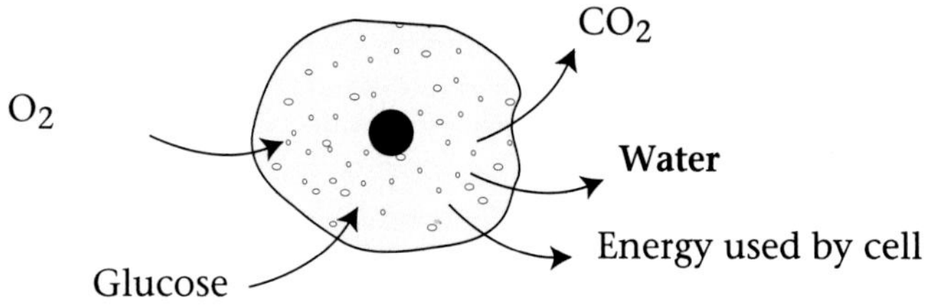

Energy → For growth
→ For movement
→ For all cellular activities

Oxygen + Glucose → **Energy** + carbon dioxide + water

This is the equation for **aerobic respiration**

$6O_2 + C_6H_{12}O_6 \longrightarrow \text{Energy} + 6CO_2 + 6H_2O$

Glucose is completely broken down in **aerobic respiration.** This releases **lots of energy.**

2. Anaerobic respiration – oxygen is not needed

Cells can respire for a while without oxygen.

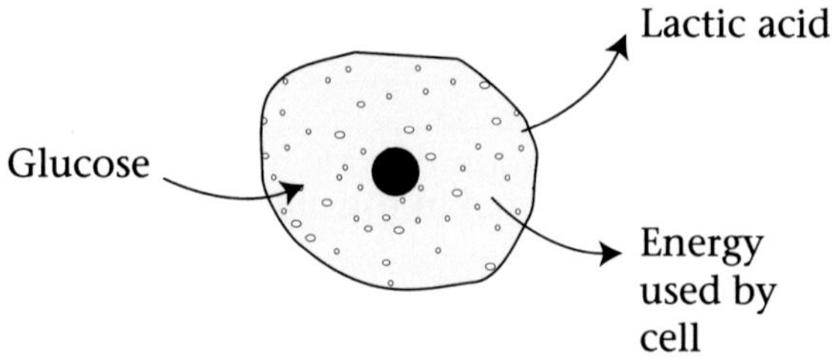

This occurs when the body cells run out of oxygen, such as during vigorous exercise. **Lactic acid** builds up in the muscles.

This is the equation for **anaerobic respiration**

Glucose → Energy + lactic acid

The build-up of lactic acid causes **cramp**. To remove the **lactic acid**, you breathe rapidly.

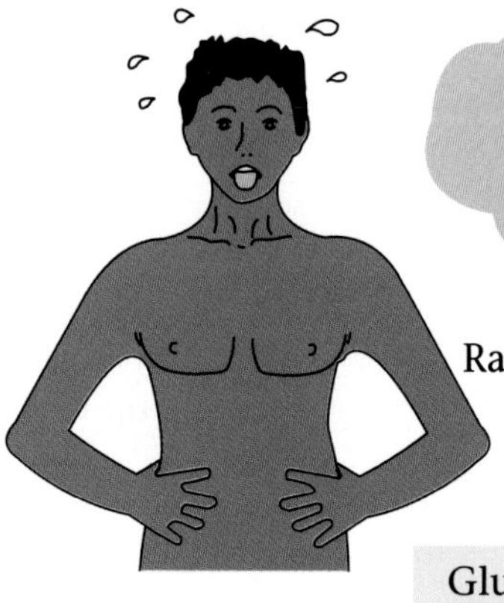

The amount of oxygen needed to remove the lactic acid is called the **oxygen debt.**

Rapid breathing

Glucose is **not** completely broken down. **Little energy** is released (energy is trapped in lactic acid).

Small animals like ***Amoeba*** have a **large surface area** compared to their volume, so oxygen can reach every part of the cell by **diffusion**. Aerobic respiration therefore takes place.

Amoeba

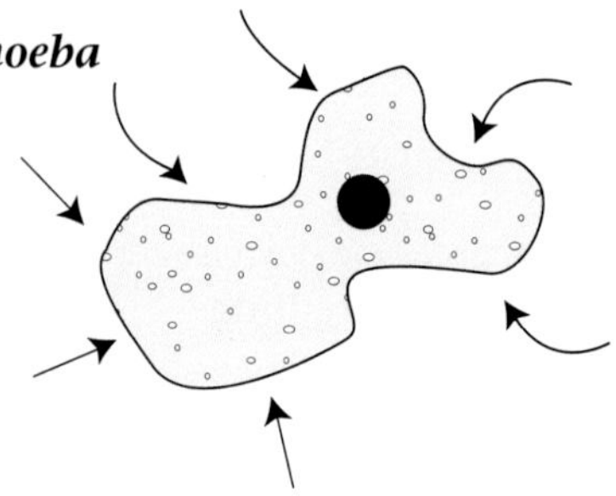

O_2 diffuses in

Yeast can respire anaerobically and releases alcohol. This is used to make wine.

Yeast cell

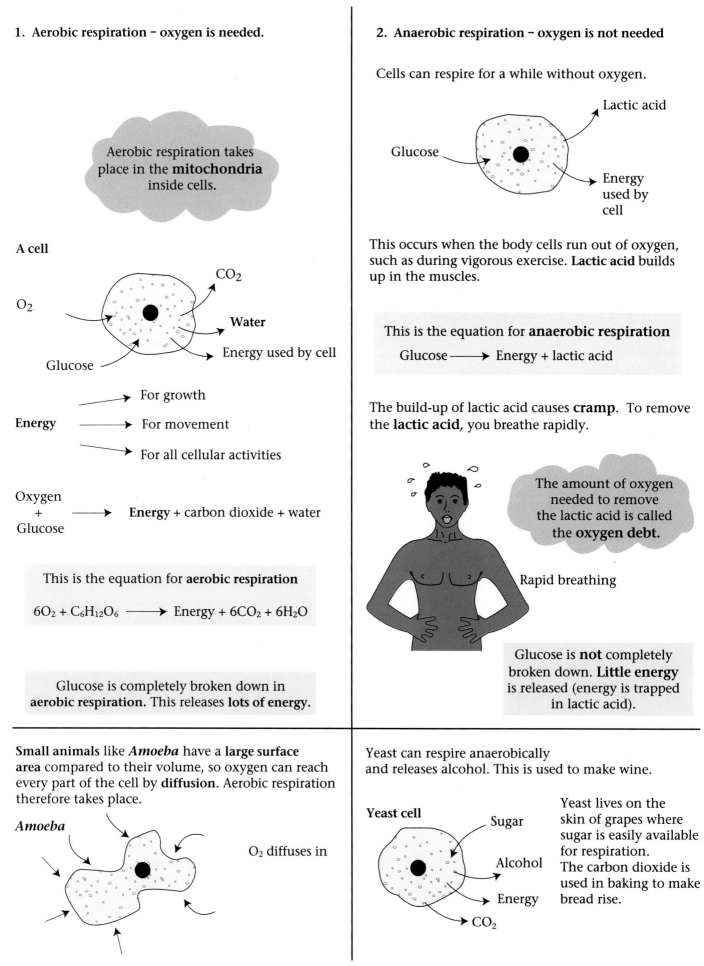

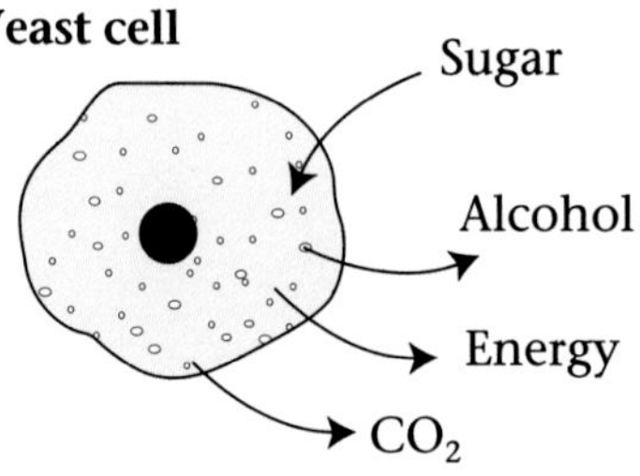

Yeast lives on the skin of grapes where sugar is easily available for respiration.
The carbon dioxide is used in baking to make bread rise.

DRUGS Drugs are substances that alter the way the body works.

Depressants (slow down transmission of nerves at the synapse)

1. Alcohol

Alcohol
- Causes **stomach** ulcers.
- **Addictive** Body is dependent on it and withdrawal symptoms may appear if drug not taken.
- **Slows reactions.**
- Damages **brain.**
- Damages **liver** causing cirrhosis.
- Damages **neurones.**
- Causes **clumsiness** and poor judgement.
- Causes **aggressive** behaviour.
- Leads to **obesity.**

Liver can break down one unit of alcohol per hour, e.g. one glass of wine, half a pint of beer, or one measure of spirit. More alcohol can damage the liver.

2. Solvents

Solvents*
- **Reduce oxygen intake** causing tiredness.
- Stop the ability to **concentrate.**
- Slow **breathing rate.**
- **Hallucinogenic.**
- Slow **heart rate.**
- Damage **brain.**
- Damage **neurones.**
- Loss of **coordination.**
- **Confusion** and loss of control
- Damage **liver** and **kidneys.**

*In glue, aerosol paints, and lighter fuel.

Amount of alcohol drunk	Effect on body
One pint of beer	Likely to have accident.
↓	Reaction time slows.
	Person stumbles and is clumsy.
10 pints of beer	Person could pass out.

Stimulants (speed up transmission of nerves at the synapse)

1. Nicotine in tobacco

Effects of smoking
- Causes **heart disease.**
- Tar causes **cancer.**
- Speeds up **heart beat.**
- Nicotine is **addictive** and there are withdrawal symptoms if drug not taken.
- Produces excess mucus in **lung** stopping action of cilia hairs sweeping bacteria out.
- **Reduces** oxygen carrying capacity, leading to tiredness.
- Causes **coughing** (smokers' cough).
- Causes **bronchitis.**
- Causes **emphysema.**
- Causes **low birth weight** babies if mother smokes.

2. Caffeine

Caffeine*
- Speeds up **heart beat.**
- Causes **trembling.**
- Speeds up **breathing rate.**
- Damages **stomach lining.**

*In coffee, tea and cola.

Attitudes to smoking have changed as scientific evidence of its harmful effects has built up. Many restaurants and all public buildings now ban smoking.

Useful drugs

- Antibiotics e.g. penicillin → Kills bacteria causing infection.
- Painkillers e.g. aspirin → Reduces sensation of pain in brain.

HOMEOSTASIS

THE ENDOCRINE SYSTEM Endocrine glands produce hormones.

Endocrine gland → Hormone (a chemical messenger) —Blood→ Target organs

Some hormones are used illegally in sport, to build up muscles and strength.

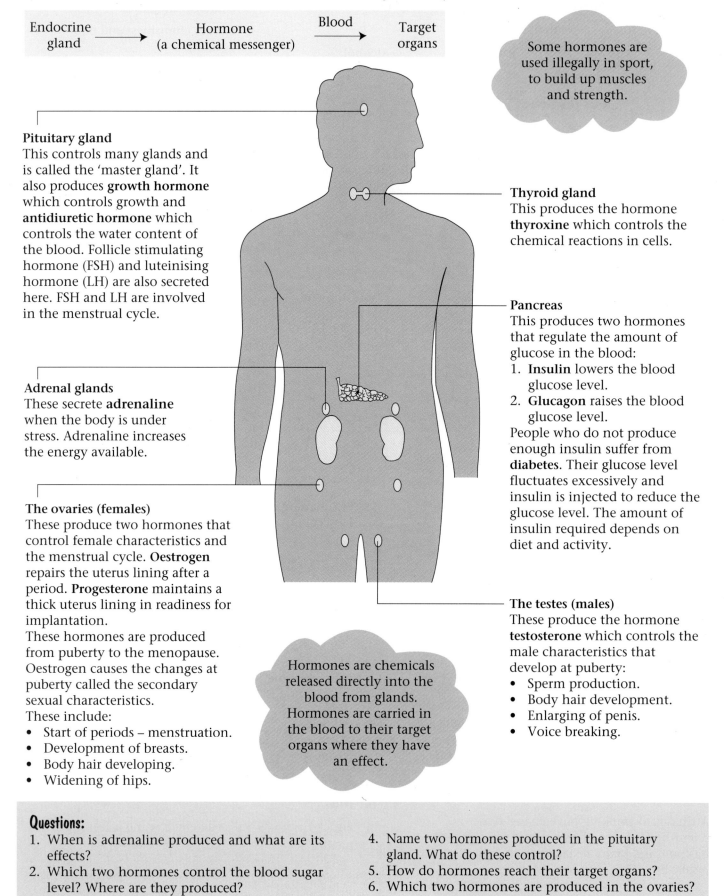

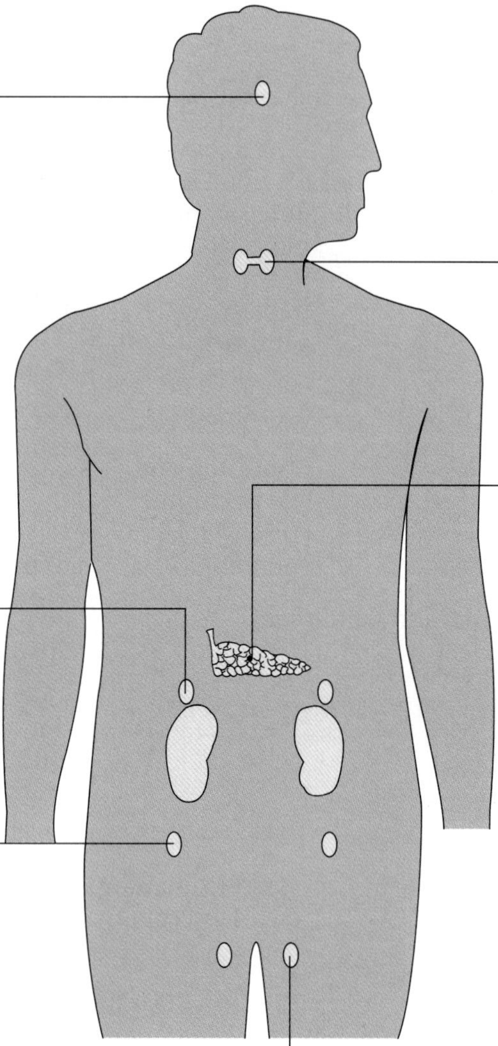

Pituitary gland
This controls many glands and is called the 'master gland'. It also produces **growth hormone** which controls growth and **antidiuretic hormone** which controls the water content of the blood. Follicle stimulating hormone (FSH) and luteinising hormone (LH) are also secreted here. FSH and LH are involved in the menstrual cycle.

Adrenal glands
These secrete **adrenaline** when the body is under stress. Adrenaline increases the energy available.

The ovaries (females)
These produce two hormones that control female characteristics and the menstrual cycle. **Oestrogen** repairs the uterus lining after a period. **Progesterone** maintains a thick uterus lining in readiness for implantation.
These hormones are produced from puberty to the menopause. Oestrogen causes the changes at puberty called the secondary sexual characteristics.
These include:
- Start of periods – menstruation.
- Development of breasts.
- Body hair developing.
- Widening of hips.

Thyroid gland
This produces the hormone **thyroxine** which controls the chemical reactions in cells.

Pancreas
This produces two hormones that regulate the amount of glucose in the blood:
1. **Insulin** lowers the blood glucose level.
2. **Glucagon** raises the blood glucose level.

People who do not produce enough insulin suffer from **diabetes.** Their glucose level fluctuates excessively and insulin is injected to reduce the glucose level. The amount of insulin required depends on diet and activity.

The testes (males)
These produce the hormone **testosterone** which controls the male characteristics that develop at puberty:
- Sperm production.
- Body hair development.
- Enlarging of penis.
- Voice breaking.

Hormones are chemicals released directly into the blood from glands. Hormones are carried in the blood to their target organs where they have an effect.

Questions:
1. When is adrenaline produced and what are its effects?
2. Which two hormones control the blood sugar level? Where are they produced?
3. When does testosterone start to be produced?
4. Name two hormones produced in the pituitary gland. What do these control?
5. How do hormones reach their target organs?
6. Which two hormones are produced in the ovaries?

ADRENAL GLANDS

These produce the hormone **adrenaline**. Adrenaline prepares the body for emergencies, e.g. fighting or running away, by making more energy available. (It is known as the 'flight or fight' hormone.)

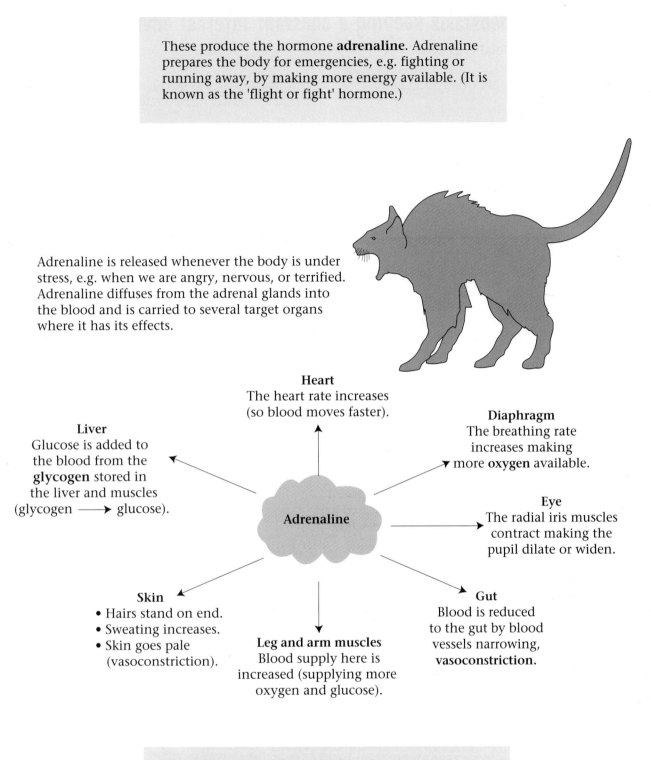

Adrenaline is released whenever the body is under stress, e.g. when we are angry, nervous, or terrified. Adrenaline diffuses from the adrenal glands into the blood and is carried to several target organs where it has its effects.

Adrenaline

Heart
The heart rate increases (so blood moves faster).

Diaphragm
The breathing rate increases making more **oxygen** available.

Eye
The radial iris muscles contract making the pupil dilate or widen.

Gut
Blood is reduced to the gut by blood vessels narrowing, **vasoconstriction.**

Leg and arm muscles
Blood supply here is increased (supplying more oxygen and glucose).

Skin
- Hairs stand on end.
- Sweating increases.
- Skin goes pale (vasoconstriction).

Liver
Glucose is added to the blood from the **glycogen** stored in the liver and muscles (glycogen ⟶ glucose).

The blood contains more oxygen (from increased breathing) and more glucose (from the liver). Therefore more glucose and oxygen reaches the muscles enabling the respiration rate there to increase. Increased respiration results in more energy available.

Questions:
1. Before a race athletes are nervous. Why is this an advantage for them?
2. Name four target organs affected by adrenaline.
3. How is the blood in the gut decreased?

CONTROL OF BLOOD SUGAR LEVEL

An example of homeostasis keeping a constant internal environment.

The sugar carried in the blood is glucose.

Digestion and absorption of carbohydrates **adds** glucose to the blood.

During exercise lots of glucose is **removed** from blood by muscle cells to release energy in respiration.

1. The role of insulin

2. The role of glucagon

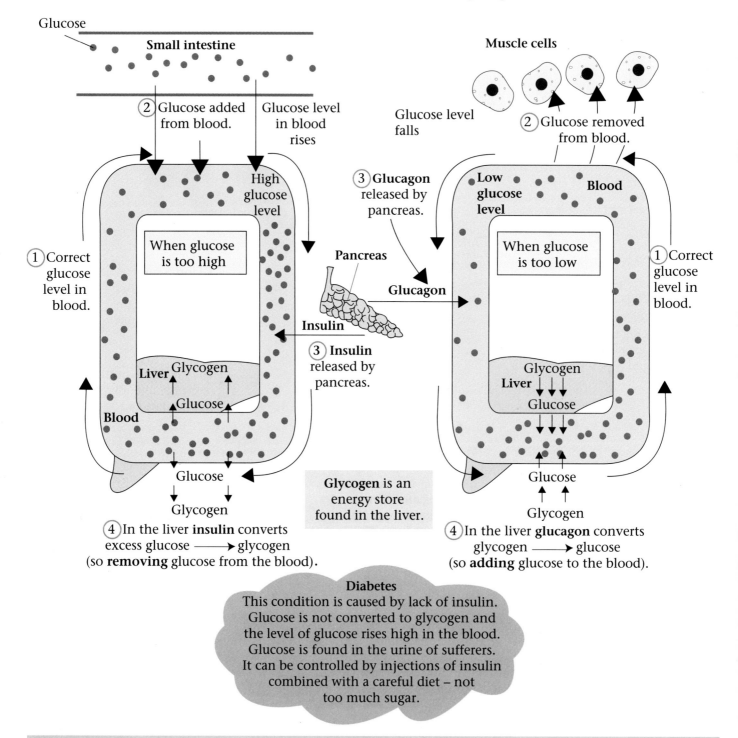

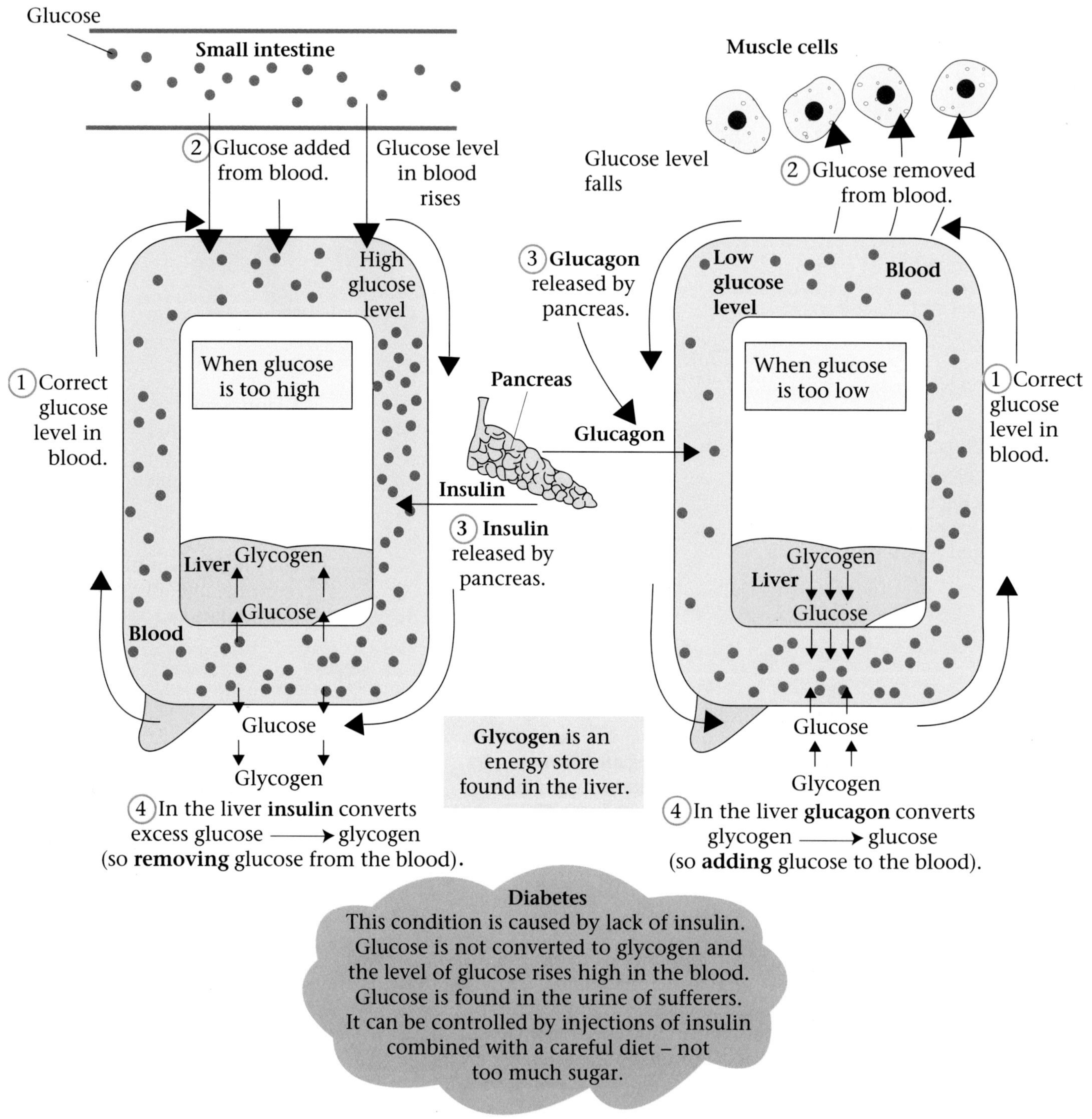

Glucose
Small intestine
2 Glucose added from blood.
Glucose level in blood rises
High glucose level
When glucose is too high
1 Correct glucose level in blood.
Liver
Glycogen
Glucose
Blood
Glucose
Glycogen
4 In the liver insulin converts excess glucose → glycogen (so removing glucose from the blood).
Pancreas
Insulin
3 Insulin released by pancreas.
3 Glucagon released by pancreas.
Glucagon
Glycogen is an energy store found in the liver.
Muscle cells
Glucose level falls
2 Glucose removed from blood.
Low glucose level
Blood
When glucose is too low
1 Correct glucose level in blood.
Glycogen
Liver
Glucose
Glucose
Glycogen
4 In the liver glucagon converts glycogen → glucose (so adding glucose to the blood).

Diabetes

This condition is caused by lack of insulin. Glucose is not converted to glycogen and the level of glucose rises high in the blood. Glucose is found in the urine of sufferers. It can be controlled by injections of insulin combined with a careful diet – not too much sugar.

Questions:

1. Which hormone reduces the glucose level and how does it do this?
2. Where is glucose added to the blood?
3. How is glucose used up in the body?
4. Which hormone increases the glucose level in the blood?
5. Where are the hormones produced that control glucose levels?
6. Excess glucose is stored as glycogen in which organ?

HOMEOSTASIS AND THE LIVER

The liver keeps the level of amino acids and glucose at a constant level, i.e. homeostasis.

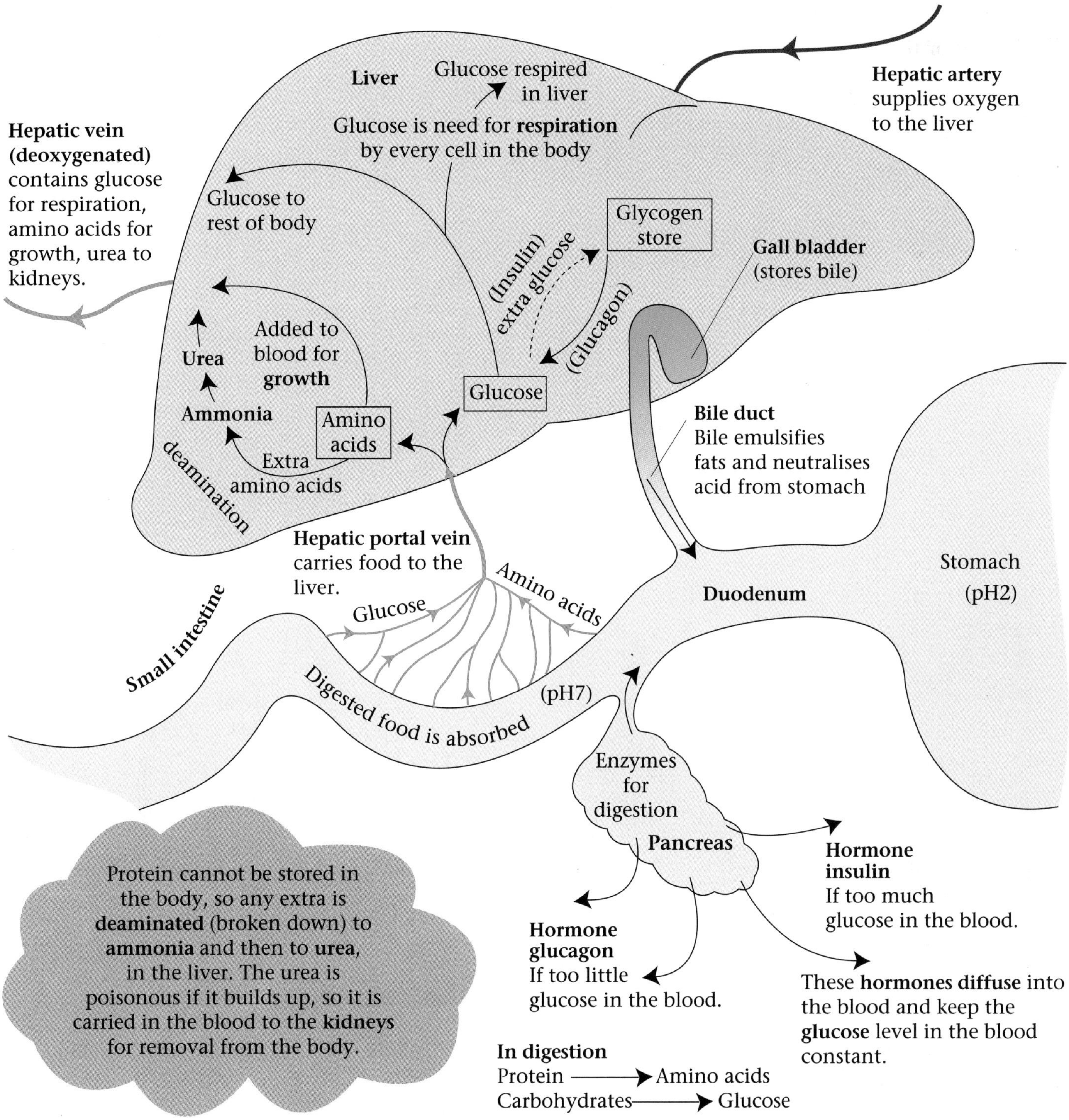

Liver
Glucose respired in liver
Glucose is need for respiration by every cell in the body
Hepatic artery supplies oxygen to the liver
Hepatic vein (deoxygenated) contains glucose for respiration, amino acids for growth, urea to kidneys.
Glucose to rest of body
Glycogen store
(Insulin) extra glucose
(Glucagon)
Gall bladder (stores bile)
Added to blood for growth
Urea
Ammonia
Amino acids
Glucose
Extra amino acids
deamination
Bile duct
Bile emulsifies fats and neutralises acid from stomach
Hepatic portal vein carries food to the liver.
Amino acids
Glucose
Duodenum
Stomach
(pH2)
Small intestine
Digested food is absorbed
(pH7)
Enzymes for digestion
Pancreas
Hormone insulin
If too much glucose in the blood.
Hormone glucagon
If too little glucose in the blood.
These hormones diffuse into the blood and keep the glucose level in the blood constant.

Protein cannot be stored in the body, so any extra is **deaminated** (broken down) to **ammonia** and then to **urea**, in the liver. The urea is poisonous if it builds up, so it is carried in the blood to the **kidneys** for removal from the body.

In digestion
Protein ⟶ Amino acids
Carbohydrates ⟶ Glucose

Glucose and amino acids are carried in the hepatic portal vein to the liver.

Fate of glucose

- Some glucose is respired by the liver.
- Some glucose is added to the hepatic vein and carried to other body cells, for respiration.
- Excess glucose is converted to **glycogen** in the liver and stored.
- The hormone **insulin**, secreted by the pancreas, converts glucose to glycogen.
- If glucose levels fall in the blood, the hormone **glucagon** from the pancreas, causes the conversion of glycogen back to glucose.

Fate of amino acids

- Some amino acids are used by the liver cells for growth.
- Some amino acids are added to the hepatic vein and carried to other body cells for growth.
- Excess amino acids are deaminated, broken down to ammonia and then converted to urea which is less poisonous (in the liver).
- Urea is added to the hepatic vein and carried to the kidneys for removal in the urine.

THE SKIN

Functions of the skin
- It is waterproof and stops us drying up.
- It stops the entry of germs.
- It helps to control our body temperature.
- It has sense receptors which keep us aware of danger.

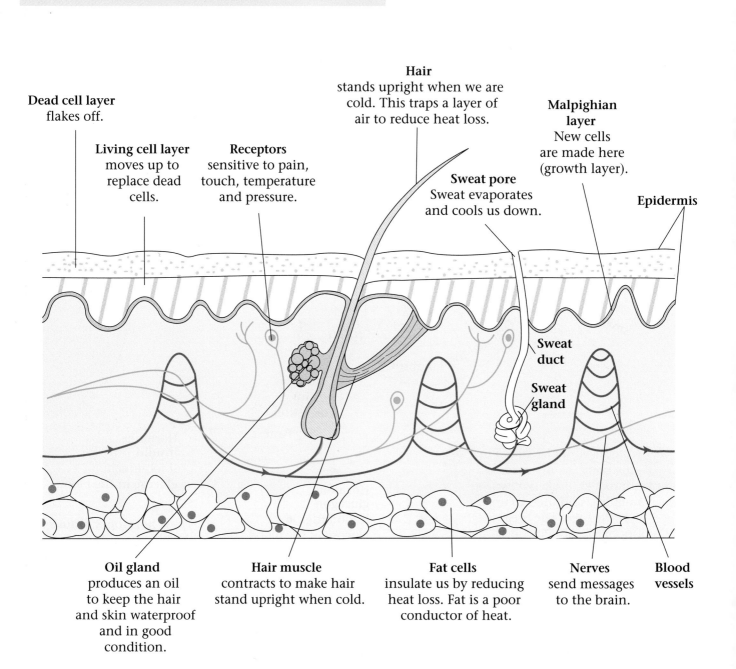

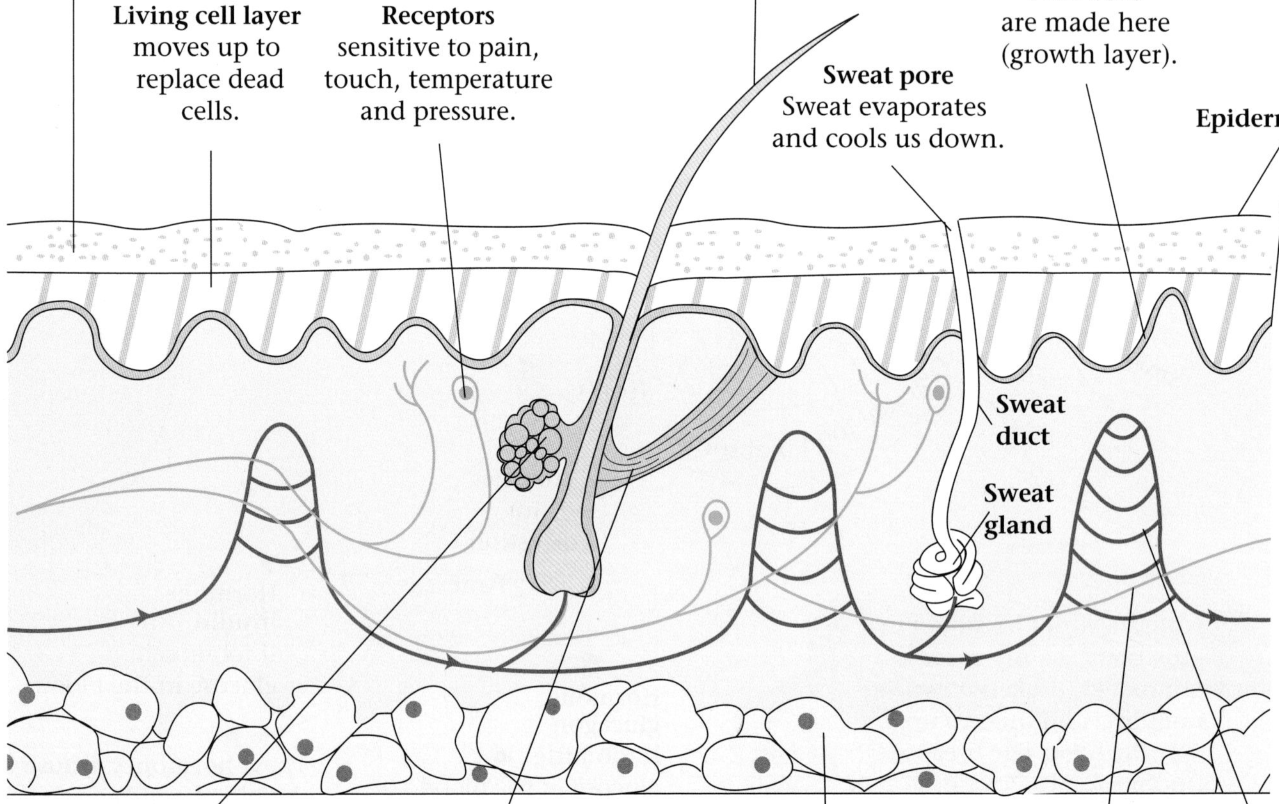

Hair
stands upright when we are cold. This traps a layer of air to reduce heat loss.
Dead cell layer
flakes off.
Malpighian layer
New cells are made here (growth layer).
Living cell layer
moves up to replace dead cells.
Receptors
sensitive to pain, touch, temperature and pressure.
Sweat pore
Sweat evaporates and cools us down.
Epidermis
Sweat duct
Sweat gland
Oil gland
produces an oil to keep the hair and skin waterproof and in good condition.
Hair muscle
contracts to make hair stand upright when cold.
Fat cells
insulate us by reducing heat loss. Fat is a poor conductor of heat.
Nerves
send messages to the brain.
Blood vessels

Excess fat from food is stored as fat under the skin.

Questions:
1. Where are new cells made in the skin?
2. Where is oil produced in the skin and why?
3. How does the hair and fat in the skin help to keep us warm?
4. What is produced when we are too hot, to cool us down?
5. Give two functions of the skin.
6. What is the skin sensitive to?

TEMPERATURE REGULATION This keeps the body temperature constant.

Sweat glands

Structure in skin: Sweat glands

Too hot (sun, exercise, fever)

Hot

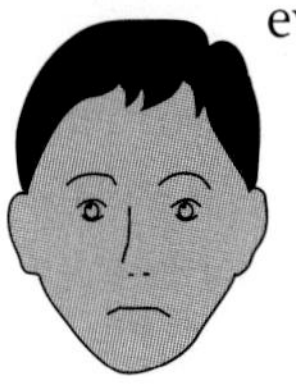

More sweat evaporates.

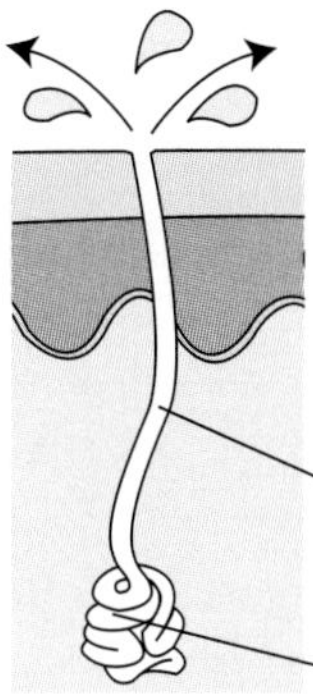

Evaporation produces cooling.

Sweat duct

Sweat gland

Increased sweating cools us down.

Too cold (cold weather, few clothes)

Cold

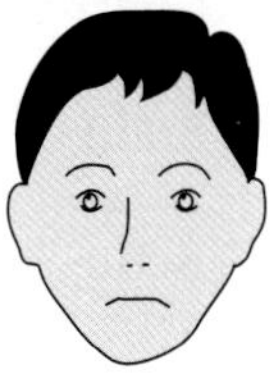

Less sweat.

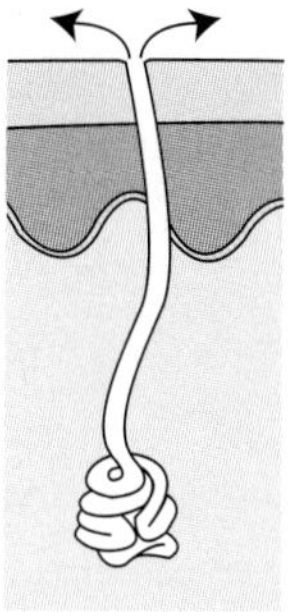

Less sweating, less evaporation, less cooling.

Blood vessels

Structure in skin: Blood vessels

Increased heat lost by radiation.

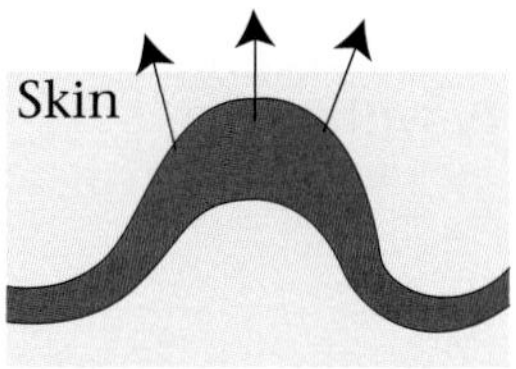

Vasodilation
Blood vessels **widen** allowing more warm blood close to the surface. This increases heat loss.

Little heat lost by radiation.

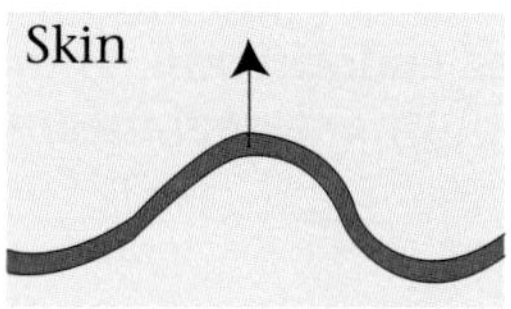

Vasoconstriction
Blood vessels **narrow** so less blood flows near the surface, reducing heat loss.

Hairs

Structure in skin: Hairs

Hairs lie flat.

Heat loss

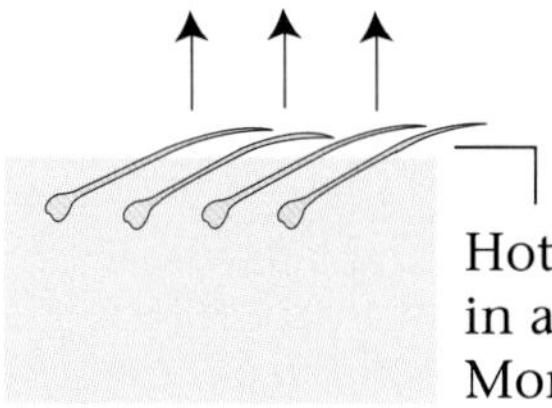

Hot air not trapped in air space.
More heat is lost by radiation.

Heat loss increases

Temperature falls → Back to normal

Hairs stand upright.

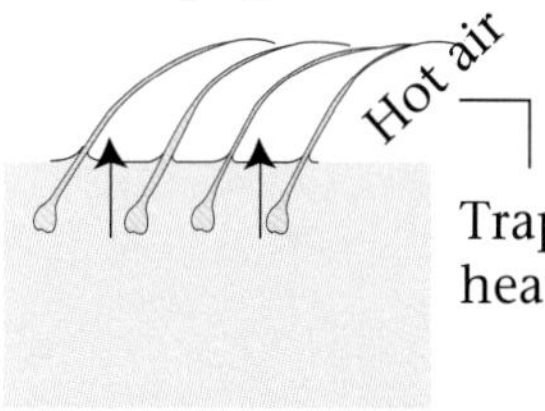

Trapped air is heated by body.

Hair muscle contracts causing goose pimples.
This traps air which acts as an **insulator**.
This reduces heat loss by radiation.

Heat loss decreases

Temperature rises → Back to normal

The temperature of the human body is approximately 37°C.

This is an example of **homeostasis** – keeping a **constant internal environment**.

Shivering

This muscular activity warms us up as it increases the rate of respiration.

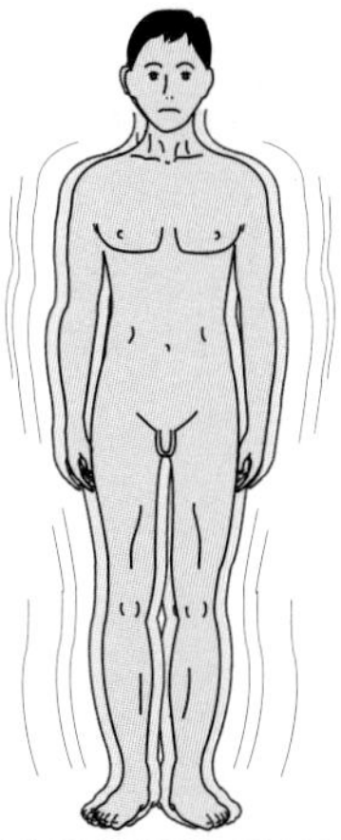

Questions:

1. What makes us too hot?
2. What is vasodilation and how does it increase heat loss?
3. When does the hair muscle contract and what does this achieve?
4. How does vasoconstriction reduce heat loss?

HOMEOSTASIS e.g. control of body temperature.

This means keeping a **constant internal environment**. Cells can then function most efficiently and an organism stays healthy. In humans, it is important that temperature, sugar level and water levels stay constant.

Condition	Controller	Organ involved	Hormone
Temperature	Brain	Skin	None
Water level	Brain	Kidneys	Anti-diuretic hormone (ADH)
Sugar level	Pancreas	Liver	Insulin and glucagon

Changes in the above conditions are detected and brought back to their optimum level.

It is important that temperature remains constant for enzyme action in cells. Mammals and birds must maintain a constant temperature to survive. They are called **endothermic** as they control their temperature from within the body, not by external means.

Normal body temperature (normal level).

Body temperature **rises** (exercise, illness, hot weather).

High body temperature is detected by brain. Impulses sent to skin to **increase** heat loss.

Corrective mechanisms:
- Vasodilation.
- Increased sweating.
- Hairs lie flat.
- Increased heat loss by radiation.

Body temperature **falls**.

Normal body temperature.

Body temperature falls (cold weather).

Low body temperature is **detected** by brain. Impulses sent to skin to **reduce** heat loss.

Corrective mechanisms:
- Vasoconstriction.
- Decreased sweating.
- Shivering.
- Hairs stand on end.
- Decreased heat loss by radiation.

Body temperature **rises**.

This is maintained by **negative feedback mechanisms**.

Definition of a negative feedback mechanism.
Any change in the normal level is detected, corrected, and returned to the normal level.

Questions:
1. Why is it important that the temperature in mammals stays constant?
2. Which two groups of vertebrates have a constant body temperature?
3. What controls the temperature of the body?
4. What is vasoconstriction and when does it occur?

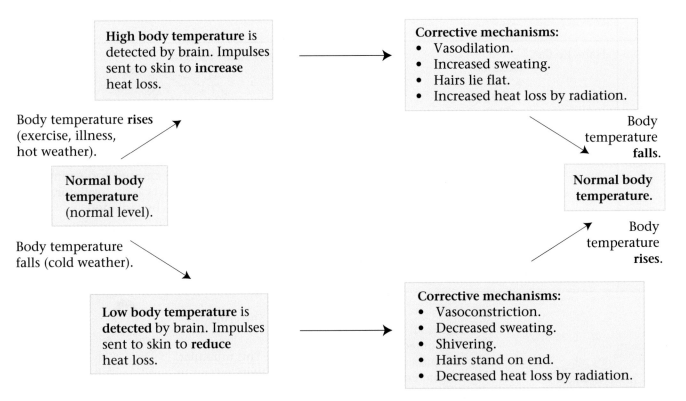

EXCRETION IN HUMANS

Excretion is the removal of waste produced by cells.

Excretory wastes

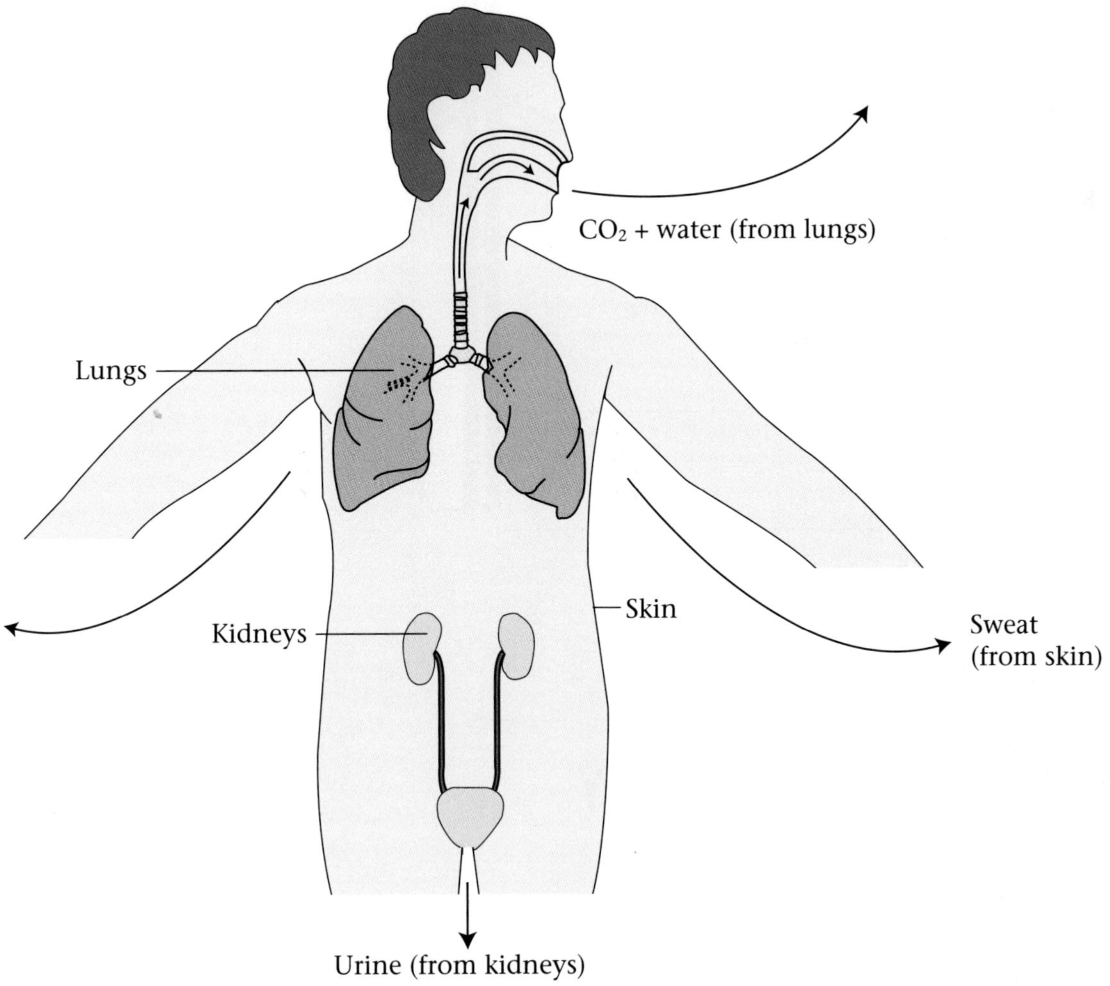

Waste is produced by the skin, kidneys, and lungs		
Excretory organ	**What it excretes**	**Purpose**
Skin	Sweat	Cools the body.
Kidneys	Urine	Controls the water content of the body. Removes harmful urea.
Lungs	Carbon dioxide + water	These waste substances are produced by respiring cells.

Questions:

1. Which three organs produce excretory waste?
2. Where is sweat made?
3. Where is the excess water in our body removed?
4. What excretory wastes are produced in respiration?
5. When do we sweat a lot and why?

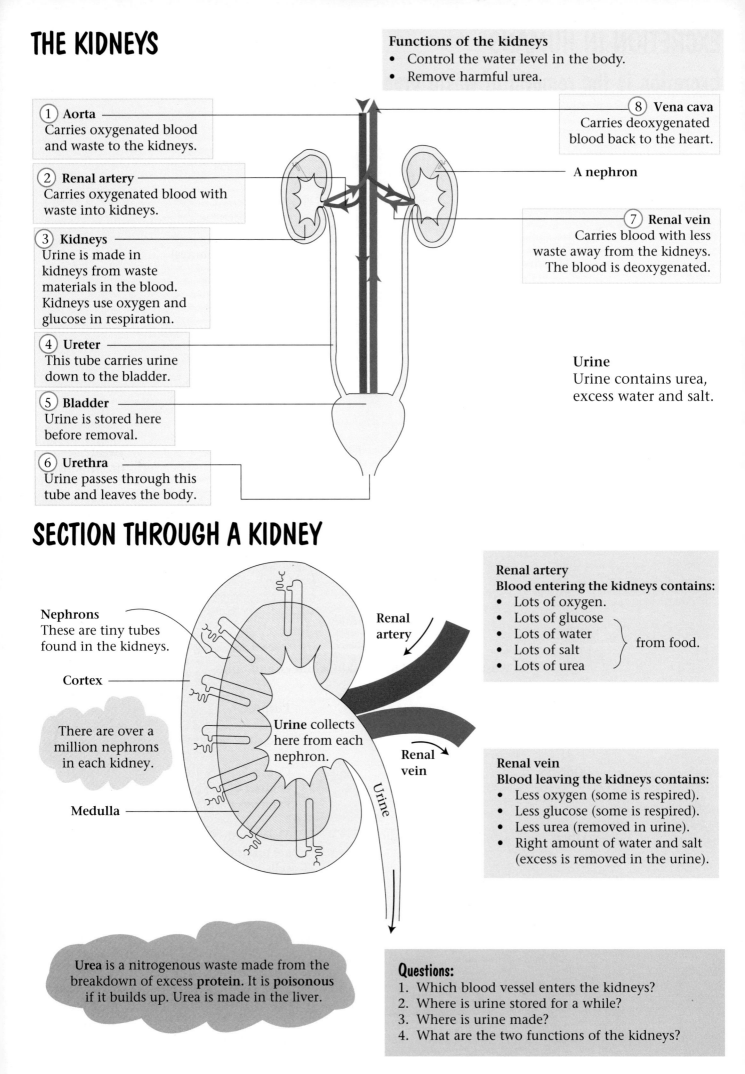

THE KIDNEYS

Functions of the kidneys
- Control the water level in the body.
- Remove harmful urea.

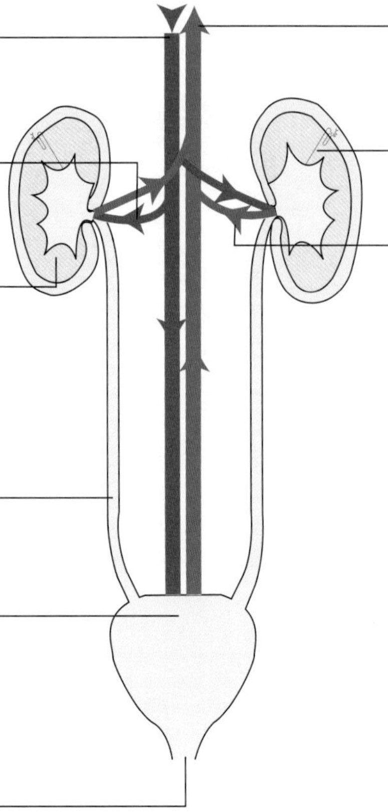

1 **Aorta**
Carries oxygenated blood and waste to the kidneys.
2 **Renal artery**
Carries oxygenated blood with waste into kidneys.
3 **Kidneys**
Urine is made in kidneys from waste materials in the blood. Kidneys use oxygen and glucose in respiration.
4 **Ureter**
This tube carries urine down to the bladder.
5 **Bladder**
Urine is stored here before removal.
6 **Urethra**
Urine passes through this tube and leaves the body.
8 **Vena cava**
Carries deoxygenated blood back to the heart.
A nephron
7 **Renal vein**
Carries blood with less waste away from the kidneys. The blood is deoxygenated.

Urine
Urine contains urea, excess water and salt.

SECTION THROUGH A KIDNEY

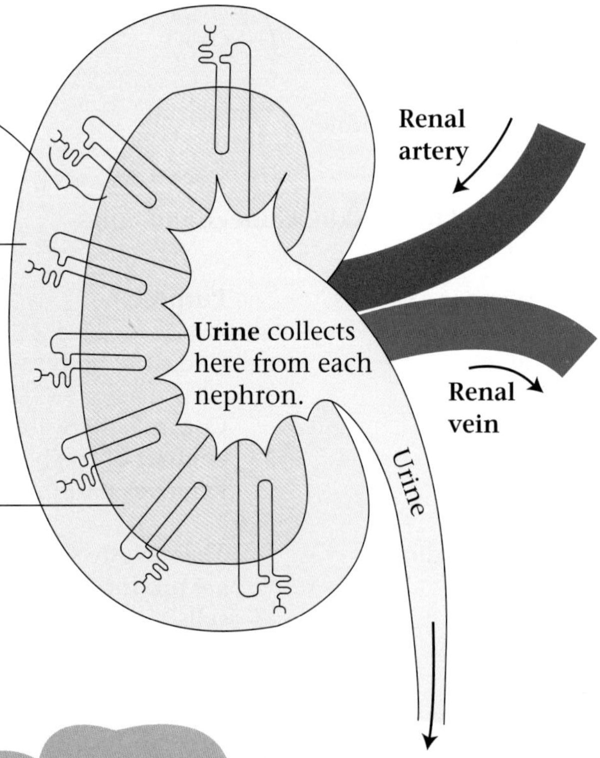

Nephrons
These are tiny tubes found in the kidneys.
Cortex
Medulla
Urine collects here from each nephron.
Renal artery
Renal vein
Urine

There are over a million nephrons in each kidney.

Renal artery
Blood entering the kidneys contains:
- Lots of oxygen.
- Lots of glucose
- Lots of water
- Lots of salt
- Lots of urea

(glucose, water, salt and urea: from food.)

Renal vein
Blood leaving the kidneys contains:
- Less oxygen (some is respired).
- Less glucose (some is respired).
- Less urea (removed in urine).
- Right amount of water and salt (excess is removed in the urine).

Urea is a nitrogenous waste made from the breakdown of excess **protein.** It is **poisonous** if it builds up. Urea is made in the liver.

Questions:
1. Which blood vessel enters the kidneys?
2. Where is urine stored for a while?
3. Where is urine made?
4. What are the two functions of the kidneys?

A NEPHRON

What happens in the nephron?
- Blood is filtered.
- Useful materials are reabsorbed back into the blood.
- Waste is removed from the body as urine.

The renal artery divides into many smaller arterioles. One arteriole enters each nephron.

(2) Narrow blood vessel comes out of glomerulus. This causes **pressure** to build up.

(6) The blood here contains all the blood cells but little plasma.

(1) Wide blood vessel (arteriole) enters renal capsule.

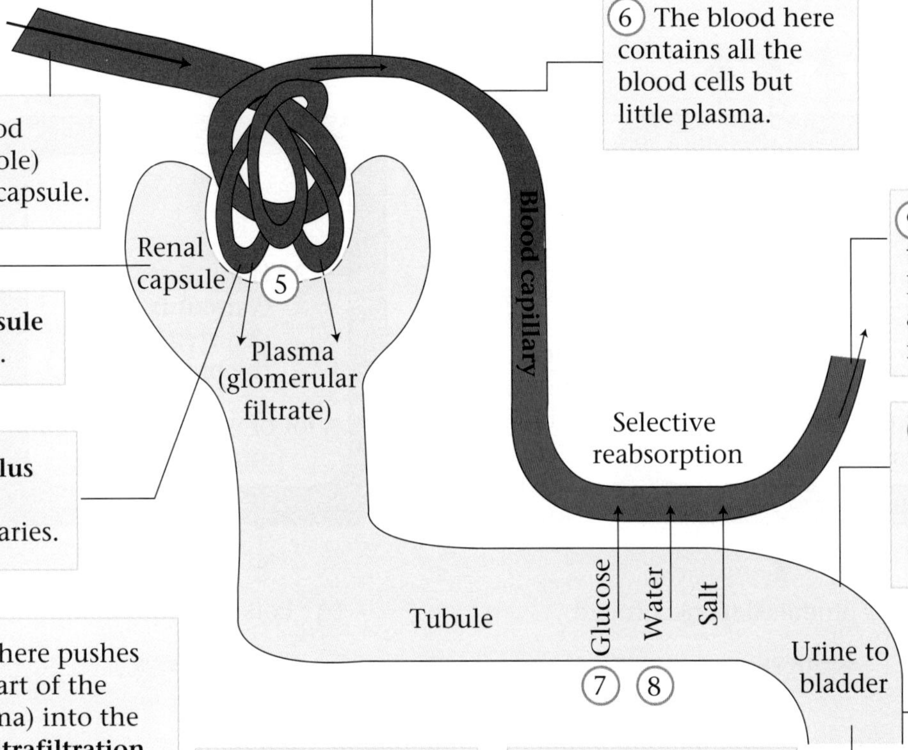

(9) The blood leaving the kidneys contains little urea and the right amount of water. This is **osmoregulation**.

(3) **Renal capsule** is like a sieve.

(4) **Glomerulus** is a knot of blood capillaries.

(10) The liquid that remains in the tubule contains lots of **urea**, excess **water** and **salt**. This is **urine**.

(5) **Pressure** here pushes the liquid part of the blood (plasma) into the **tubule** by **ultrafiltration** (red and white cells and proteins are too big to get through). This forms glomerular filtrate.

(7) All the **glucose** passes back to the blood. It is needed for respiration.

(8) The right amount of **water** passes back to the blood.

This is **selective reabsorption**.

(11) **Collecting duct. Urine** collects here from many nephrons. It passes to the bladder and out of the body.

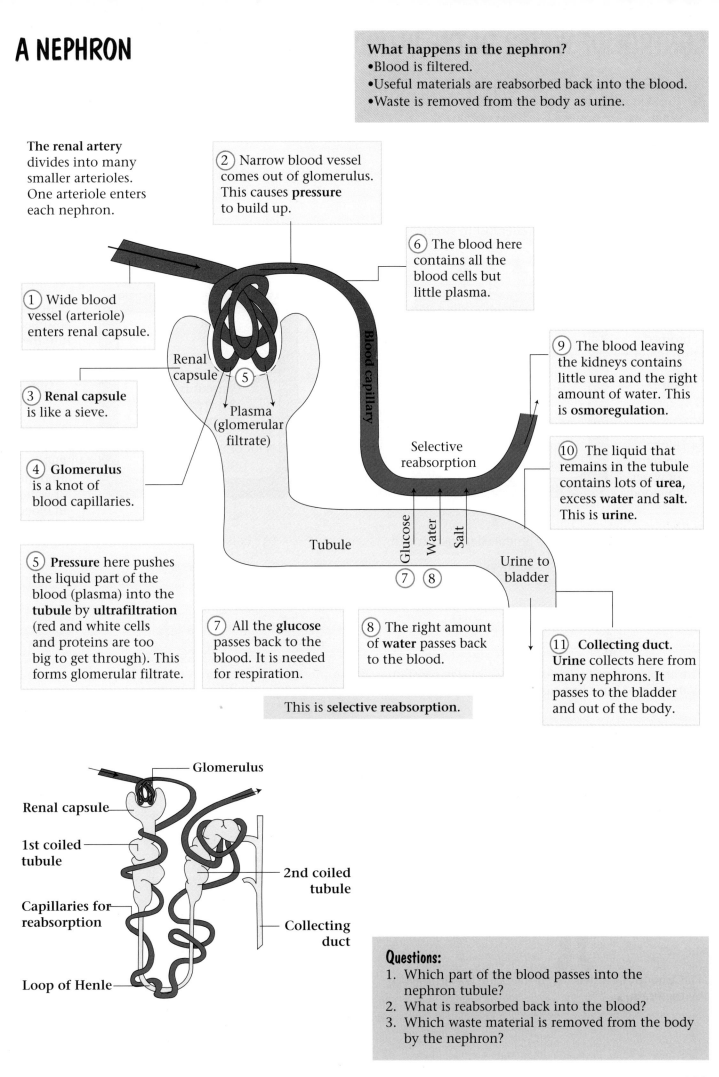

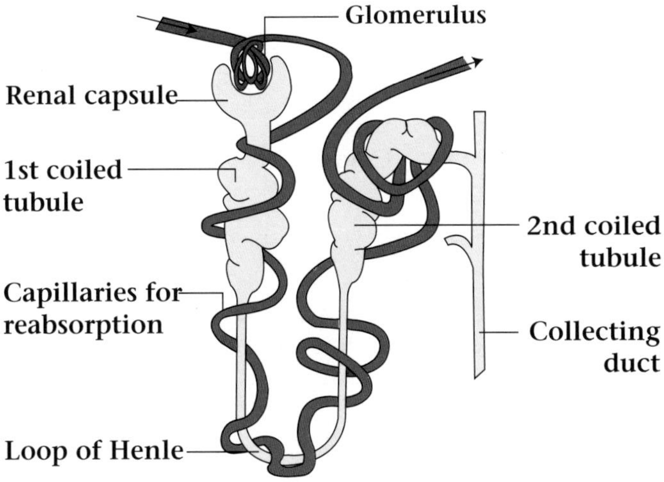

Questions:
1. Which part of the blood passes into the nephron tubule?
2. What is reabsorbed back into the blood?
3. Which waste material is removed from the body by the nephron?

CONTROL OF WATER IN THE BLOOD An example of homeostasis.

How sweating affects the urine produced.
- The skin helps to control body temperature.
- When we are hot, we sweat to cool down.
- This reduces the water content of blood.
- Less water is lost in urine to conserve water.

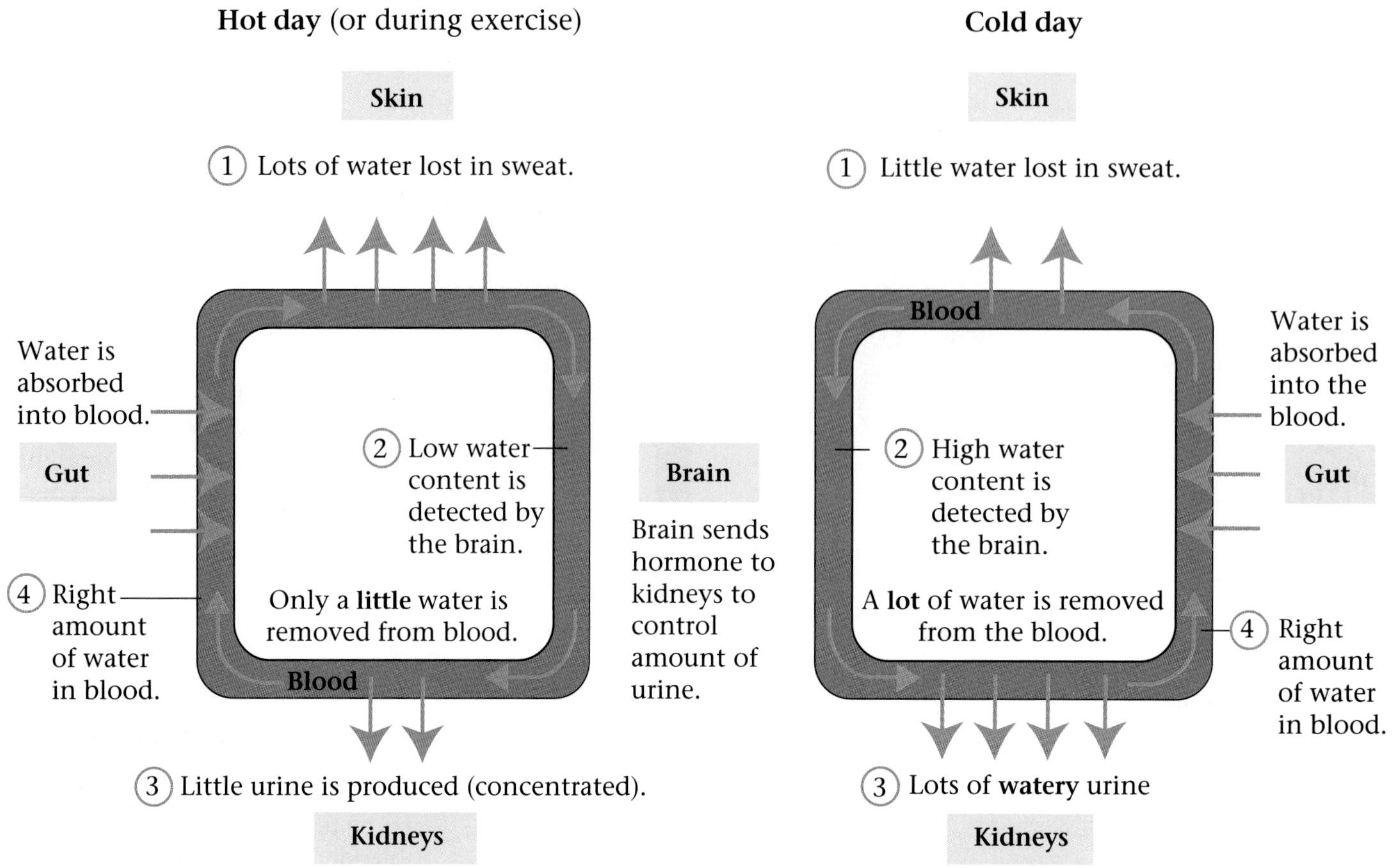

Hot day (or during exercise)
Skin
1 Lots of water lost in sweat.
Water is absorbed into blood.
Gut
2 Low water content is detected by the brain.
Only a little water is removed from blood.
4 Right amount of water in blood.
Blood
3 Little urine is produced (concentrated).
Kidneys
Brain
Brain sends hormone to kidneys to control amount of urine.
Cold day
Skin
1 Little water lost in sweat.
Blood
Water is absorbed into the blood.
Gut
2 High water content is detected by the brain.
A lot of water is removed from the blood.
4 Right amount of water in blood.
3 Lots of watery urine
Kidneys

The amount of water in our body must stay constant. Water gained must be equal to water lost.

Only water lost in the **urine** can be controlled. We have no control over the amount of water lost through sweating or breathing out.

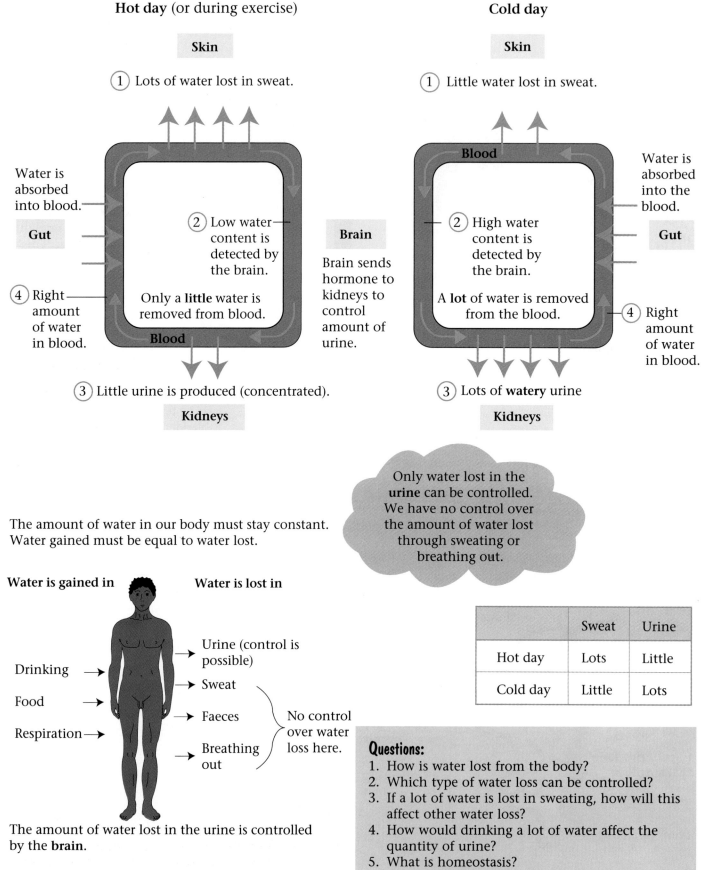

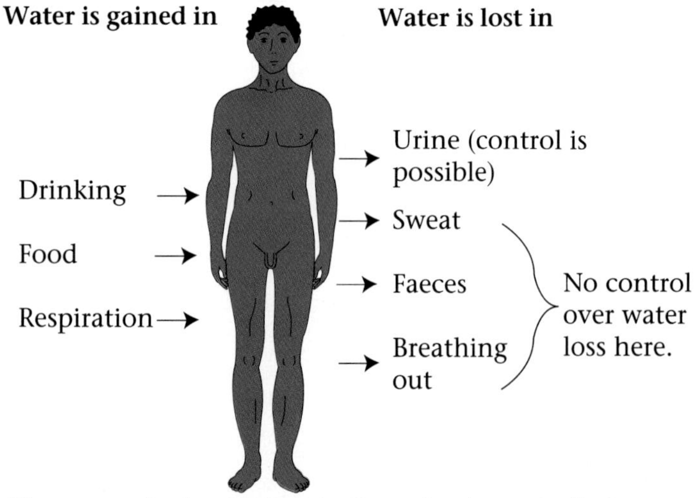

Water is gained in
Water is lost in
Drinking
Food
Respiration
Urine (control is possible)
Sweat
Faeces
Breathing out
No control over water loss here.

	Sweat	Urine
Hot day	Lots	Little
Cold day	Little	Lots

The amount of water lost in the urine is controlled by the **brain**.

Questions:
1. How is water lost from the body?
2. Which type of water loss can be controlled?
3. If a lot of water is lost in sweating, how will this affect other water loss?
4. How would drinking a lot of water affect the quantity of urine?
5. What is homeostasis?
6. What controls the level of water in the blood?

HORMONAL CONTROL OF WATER LEVEL

The **brain** releases **ADH** which controls the amount of **urine** produced (ADH = anti-diuretic hormone).

Less ADH means less water reabsorbed back into blood.

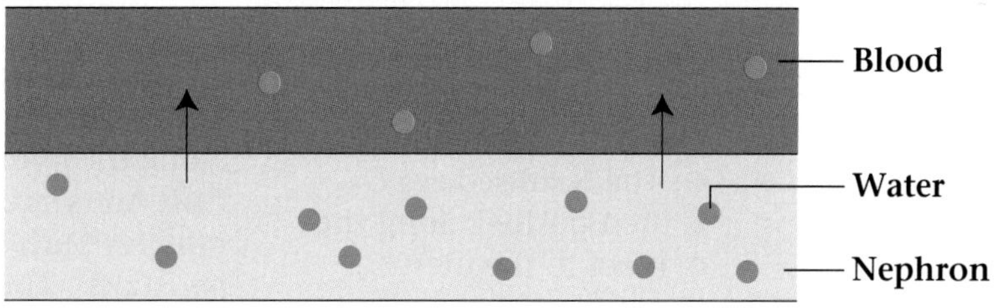

Water levels **rises** due to drinking

High water level detected by brain which releases **less ADH** hormone.

Carried to kidneys

Less ADH means:

- Less water reabsorbed back into the blood.
- This lowers water level of blood.
- A large quantity of **dilute** urine is produced.

Water level of blood falls

Normal water level in blood

In kidneys.

Normal water level in blood

Water level **falls** due to exercise (sweating) or little drinking

Low water level is detected by brain which releases **more ADH** hormone.

Carried to kidneys

More ADH means

- More water is reabsorbed back into the blood.
- This raises water level of blood.
- A small quantity of **concentrated** urine is produced.

Water level of blood rises

More ADH means more water reabsorbed back into blood.

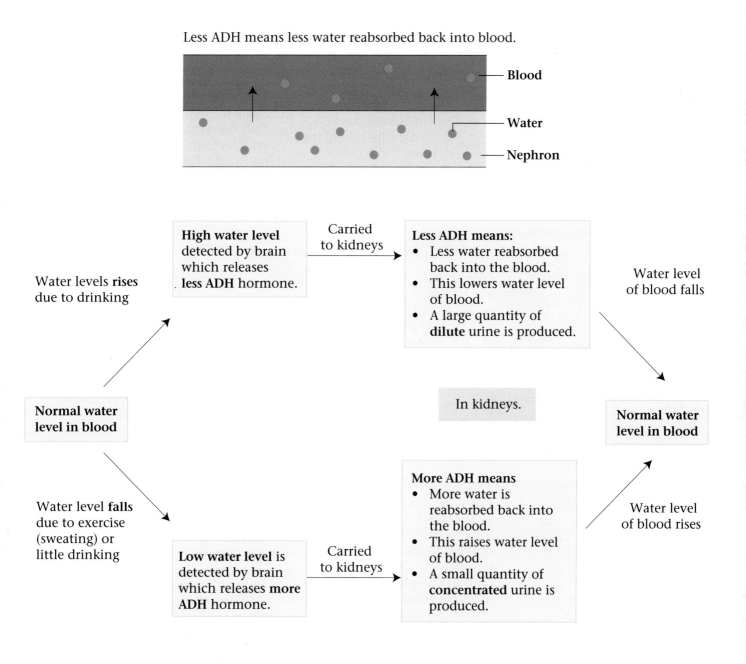

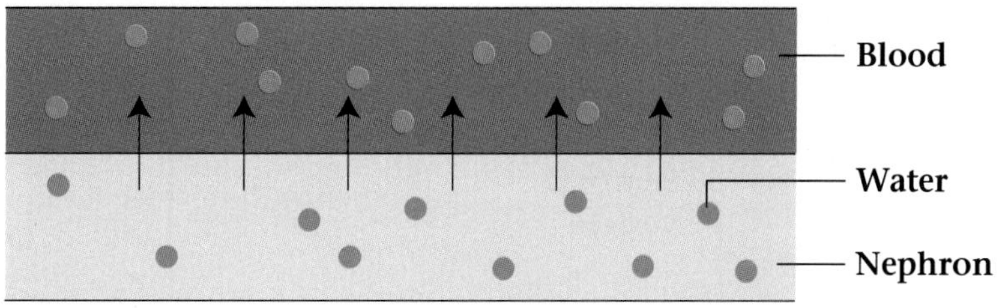

Questions:

1. Where is the water level detected in the body?
2. Which hormone is produced to control the water level?
3. Where is the water level altered by this hormone?
4. On hot days do we produce more or less urine and why?

REPRODUCTION

HUMAN REPRODUCTIVE SYSTEMS

1. Female reproductive system

(1) **Ovary**
Here the hormones **oestrogen** and **progesterone** are produced. These cause the menstrual cycle and the secondary sexual characteristics. One ovum is released very 28 days from alternate ovaries. The release of the ovum is called **ovulation.**

(6) **The fertilised egg**
is then pushed along the oviduct to the uterus to continue developing.

(2) **The oviduct**
The ovum from the ovary is pushed along the oviduct to the uterus. Cilia hairs sweep it along and muscles push the ovum along by peristalsis. The ovum lives for about one day.

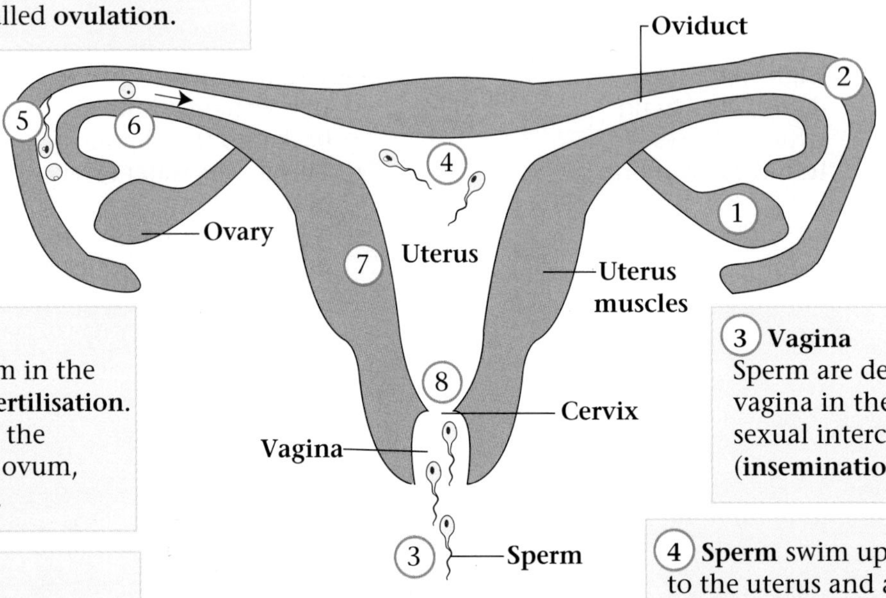

(5) **The ovum**
joins with a sperm in the oviduct. This is **fertilisation.** Only the head of the sperm enters the ovum, the tail drops off.

(3) **Vagina**
Sperm are deposited in the vagina in the process called sexual intercourse (**insemination**).

(7) **The uterus**
Here the fertilised egg attaches to the uterus wall. This is **implantation**. The fetus grows, stretching the muscular uterus wall.

(8) **The cervix**
is the narrow (1 cm wide) opening to the uterus. It has to widen to 9 cm diameter for birth to take place. Uterus muscles contract and widen the cervix. Then strong uterus muscles push the baby out at **birth.**

(4) **Sperm** swim up the vagina to the uterus and along the oviducts. They can live for about two days. Only the strong, healthy sperm complete the journey; the less fit die. This is survival of the fittest and ensures that only a healthy sperm fertilises the ovum.

2. Male reproductive system
Sperm passes from the penis into the female vagina during sexual intercourse. This is called **insemination**.

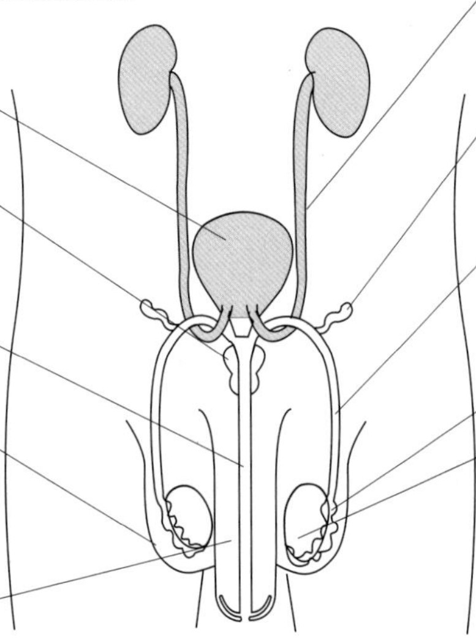

Fertilisation

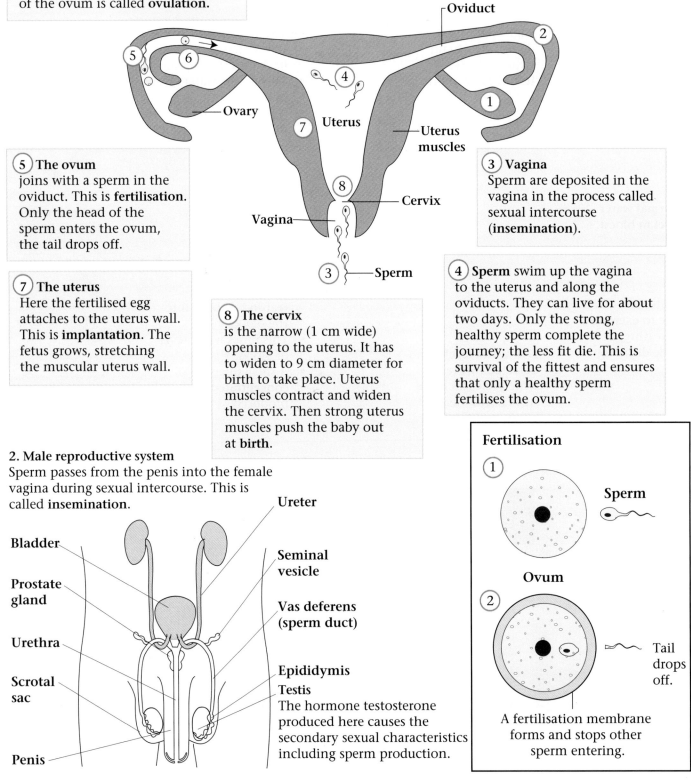

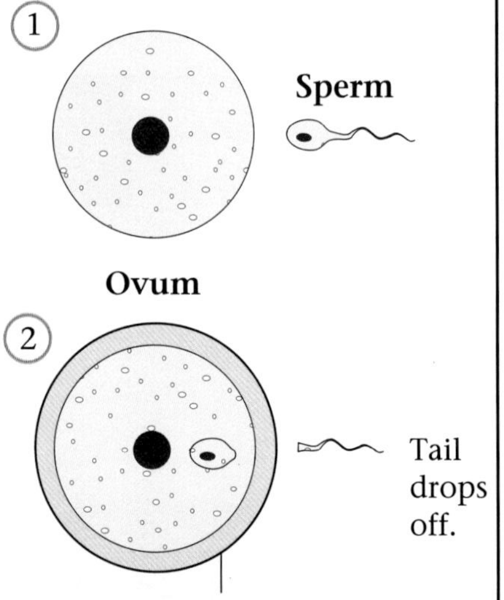

A fertilisation membrane forms and stops other sperm entering.

THE MENSTRUAL CYCLE (28-day cycle).

This 28-day cycle is controlled by hormones. **One egg** is produced each cycle from alternate ovaries. The egg lives for only about one day and fertilisation is possible then. The uterus lining **(endometrium)** thickens up in preparation for a possible pregnancy. If fertilisation does not occur, this is shed from the body in the monthly period.

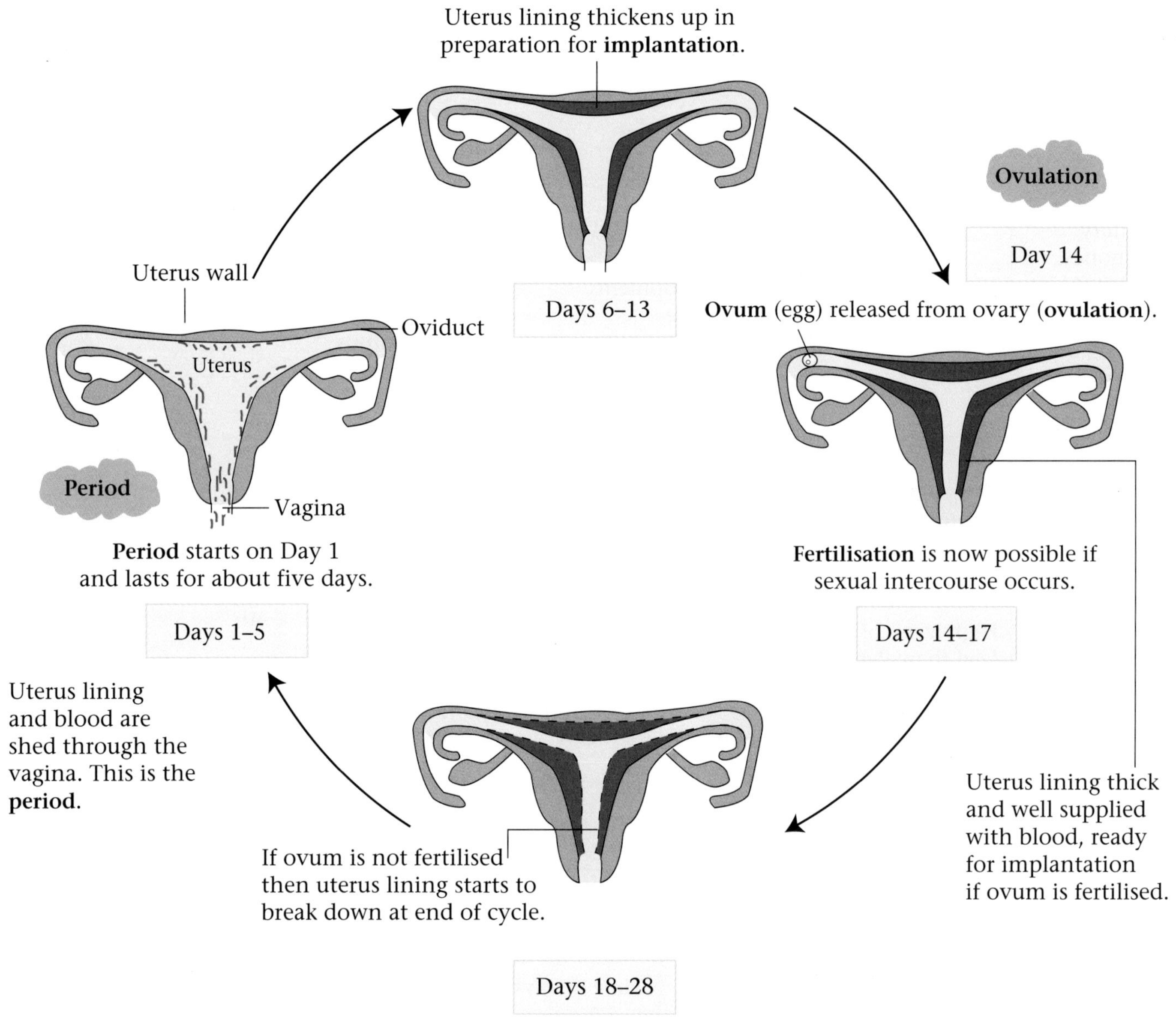
Uterus lining thickens up in preparation for **implantation**.
Days 6–13
Ovulation
Day 14
Ovum (egg) released from ovary (**ovulation**).
Fertilisation is now possible if sexual intercourse occurs.
Days 14–17
Uterus lining thick and well supplied with blood, ready for implantation if ovum is fertilised.
If ovum is not fertilised then uterus lining starts to break down at end of cycle.
Days 18–28
Uterus lining and blood are shed through the vagina. This is the **period**.
Days 1–5
Period starts on Day 1 and lasts for about five days.
Period
Uterus wall
Oviduct
Uterus
Vagina

Menstrual cycle starts at **puberty**, ends at the **menopause**.

Questions:

1. Where is the egg produced?
2. Where might fertilisation take place?
3. What is the purpose of the uterus lining thickening up?
4. What is ovulation?

HORMONAL CONTROL OF THE MENSTRUAL CYCLE

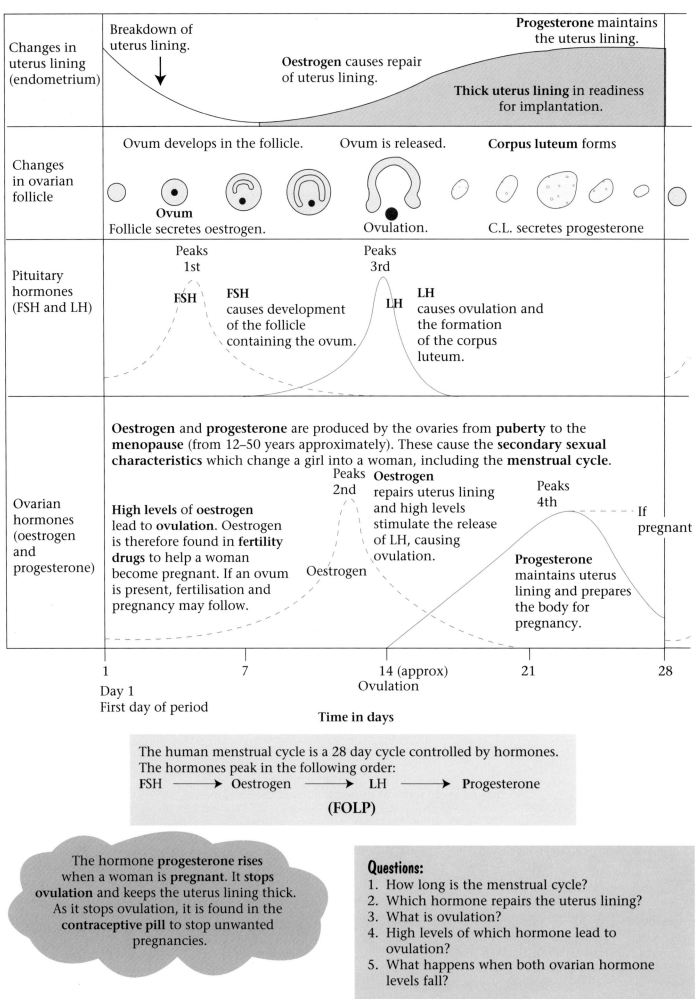

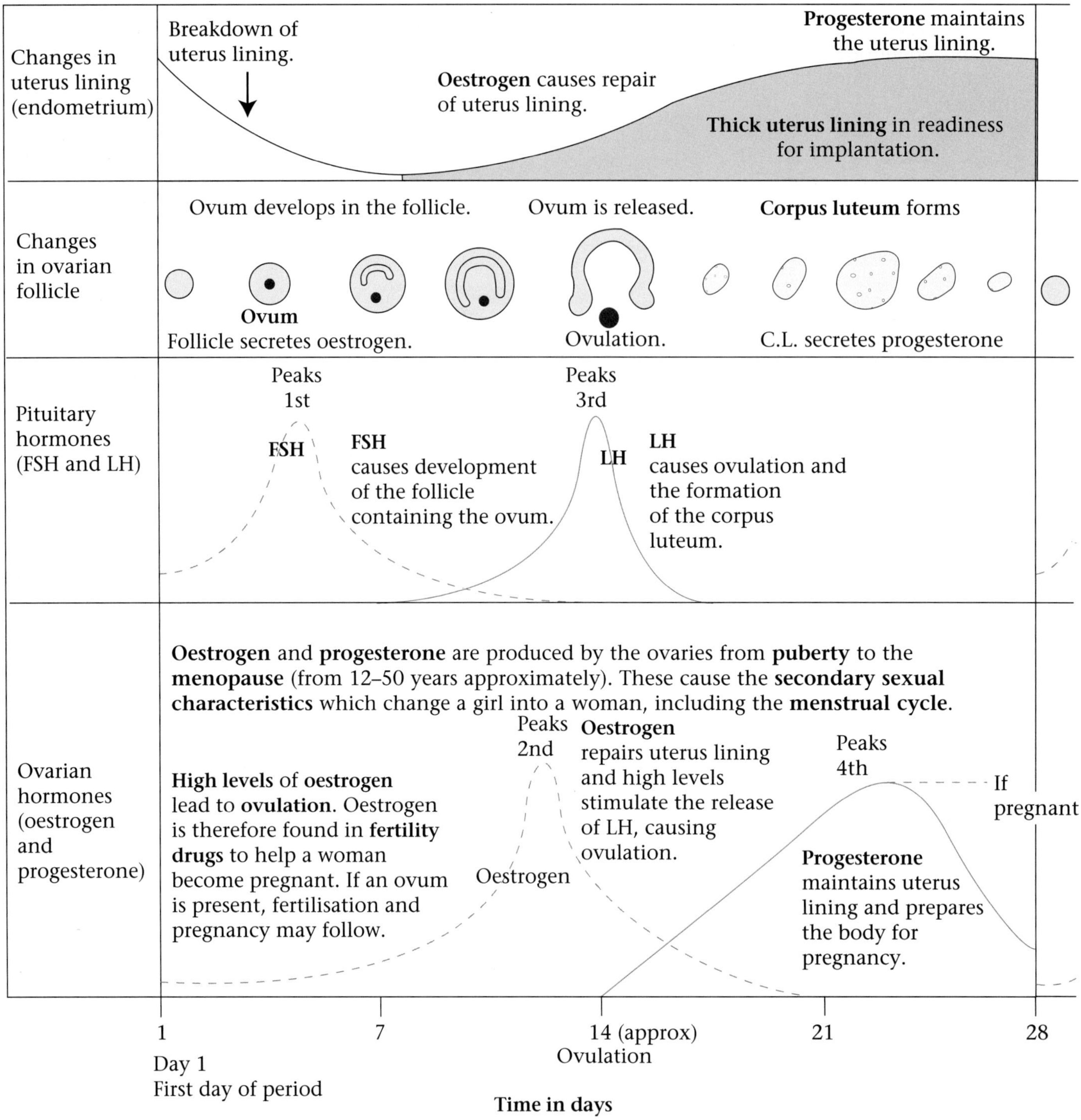

Changes in uterus lining (endometrium)
Breakdown of uterus lining.
Oestrogen causes repair of uterus lining.
Progesterone maintains the uterus lining.
Thick uterus lining in readiness for implantation.
Changes in ovarian follicle
Ovum develops in the follicle.
Ovum is released.
Corpus luteum forms
Ovum
Follicle secretes oestrogen.
Ovulation.
C.L. secretes progesterone
Pituitary hormones (FSH and LH)
Peaks 1st
FSH
FSH causes development of the follicle containing the ovum.
Peaks 3rd
LH
LH causes ovulation and the formation of the corpus luteum.
Ovarian hormones (oestrogen and progesterone)
Oestrogen and progesterone are produced by the ovaries from puberty to the menopause (from 12–50 years approximately). These cause the secondary sexual characteristics which change a girl into a woman, including the menstrual cycle.
High levels of oestrogen lead to ovulation. Oestrogen is therefore found in fertility drugs to help a woman become pregnant. If an ovum is present, fertilisation and pregnancy may follow.
Peaks 2nd
Oestrogen
Oestrogen repairs uterus lining and high levels stimulate the release of LH, causing ovulation.
Peaks 4th
If pregnant
Progesterone maintains uterus lining and prepares the body for pregnancy.
1
7
14 (approx)
Ovulation
21
28
Day 1
First day of period
Time in days

The human menstrual cycle is a 28 day cycle controlled by hormones.
The hormones peak in the following order:
FSH ⟶ Oestrogen ⟶ LH ⟶ Progesterone

(FOLP)

The hormone **progesterone rises** when a woman is **pregnant**. It **stops ovulation** and keeps the uterus lining thick. As it stops ovulation, it is found in the **contraceptive pill** to stop unwanted pregnancies.

Questions:

1. How long is the menstrual cycle?
2. Which hormone repairs the uterus lining?
3. What is ovulation?
4. High levels of which hormone lead to ovulation?
5. What happens when both ovarian hormone levels fall?

THE PLACENTA Exchange of material between the mother and fetus.

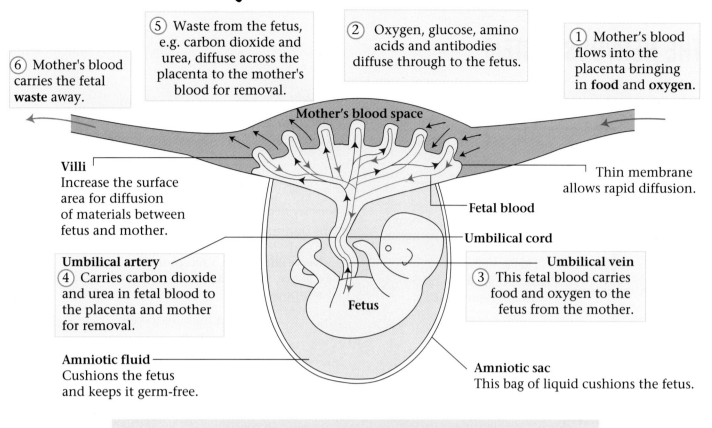

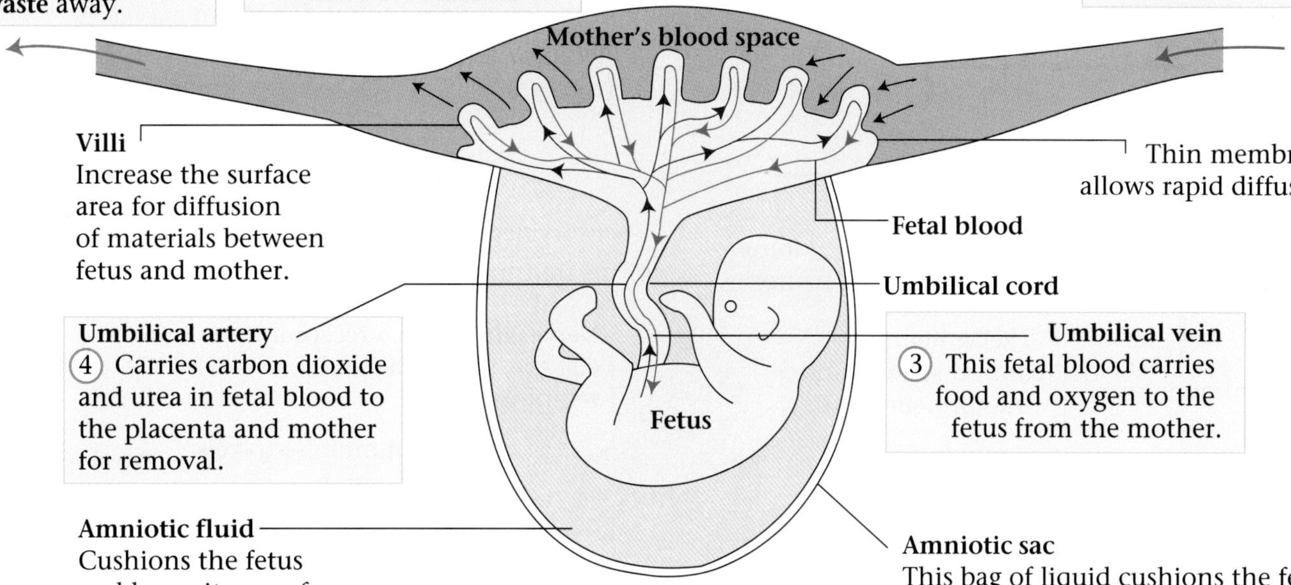

5 Waste from the fetus, e.g. carbon dioxide and urea, diffuse across the placenta to the mother's blood for removal.
2 Oxygen, glucose, amino acids and antibodies diffuse through to the fetus.
1 Mother's blood flows into the placenta bringing in **food** and **oxygen.**
6 Mother's blood carries the fetal **waste** away.
Mother's blood space
Villi
Increase the surface area for diffusion of materials between fetus and mother.
Thin membrane allows rapid diffusion.
Fetal blood
Umbilical cord
Umbilical artery
4 Carries carbon dioxide and urea in fetal blood to the placenta and mother for removal.
Umbilical vein
3 This fetal blood carries food and oxygen to the fetus from the mother.
Fetus
Amniotic fluid
Cushions the fetus and keeps it germ-free.
Amniotic sac
This bag of liquid cushions the fetus.

If the mother smokes, this leads to low birthweight babies and premature births. Smoking deprives the baby of oxygen.

The **placenta** keeps the mother's and fetal blood *apart* while allowing exchange of materials. This separation is essential to:
- Stop the high pressure of the mother's blood destroying the fetal blood vessels.
- Stop the mother's blood rejecting the fetus .

Other substances that can cross the placenta to the fetus	
Substance	**Effect on baby**
Antibodies	Gives immunity to new born baby.
HIV	Causes AIDS.
Rubella virus (German measles)	Causes blindness, deafness, and brain damage.
*Nicotine	Increases heart rate.
*Carbon monoxide	Reduces oxygen reaching fetus.
(*From smoking)	

Thalidomide is a drug that was developed as a sleeping pill, but was also found to be effective at reducing morning sickness in pregnant women. Unfortunately, it was not tested before use, and many babies were born with severe limb abnormalities to mothers who took the drug in pregnancy. Many babies had no arms or legs. Recently, thalidomide is being used successfully to treat leprosy.

Questions:
1. Why must the mother's and fetal blood be kept apart?
2. Name two substances that pass from the mother to the fetus.
3. How does the fetus get rid of waste?
4. What features of the placenta increase the rate of diffusion?

COORDINATION

THE NERVOUS SYSTEM

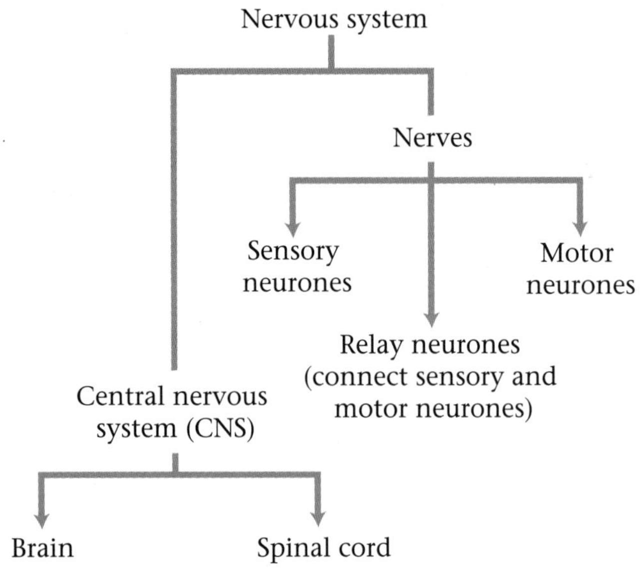

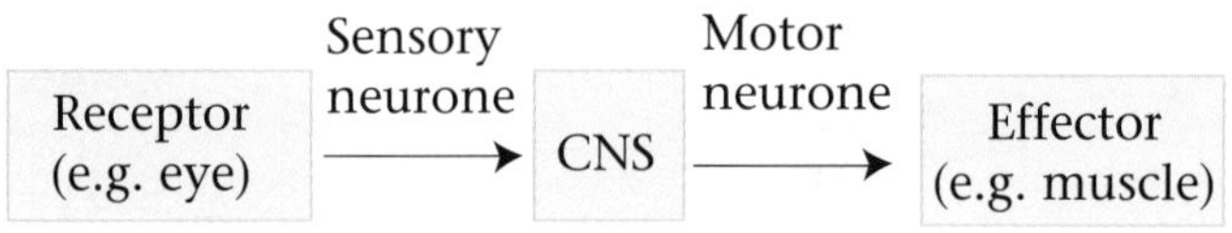

Sensory nerve cell (neurone)

This transmits an impulse from a **receptor** to the *CNS*.

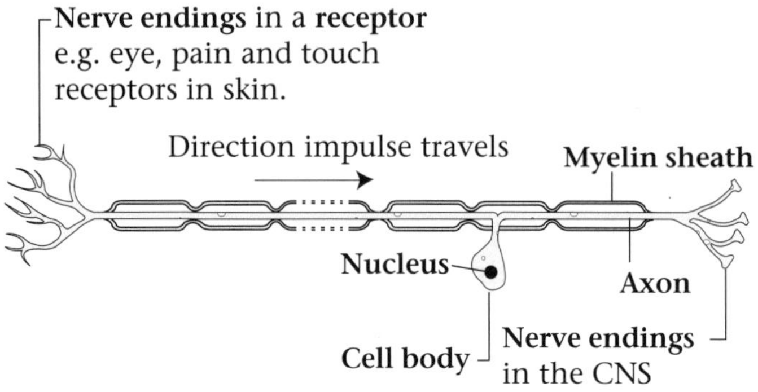

Central nervous system – CNS

Cerebral hemisphere
This controls all **voluntary** activities. Also thinking, speech, sight, learning and memory.

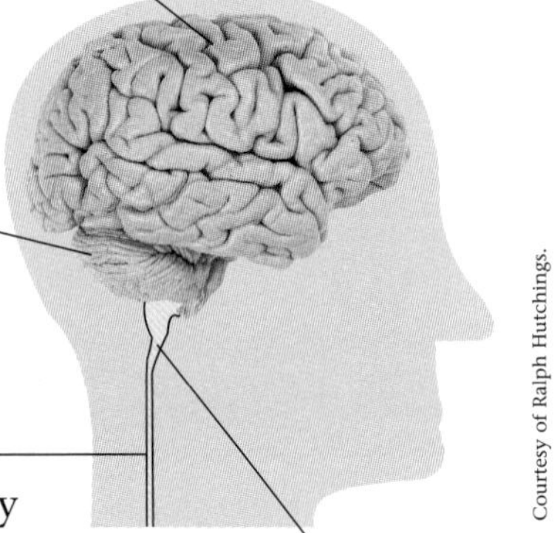

Cerebellum
This controls our sense of balance and lets us make precise and accurate movements.

Spinal cord
This controls involuntary reflex actions, e.g. removing hand from fire, knee jerk reflex.

Medulla oblongata
This controls our breathing rate, heart rate and swallowing.

Motor nerve cell (neurone)

This transmits an impulse from the *CNS* to an *effector* (muscle or gland).

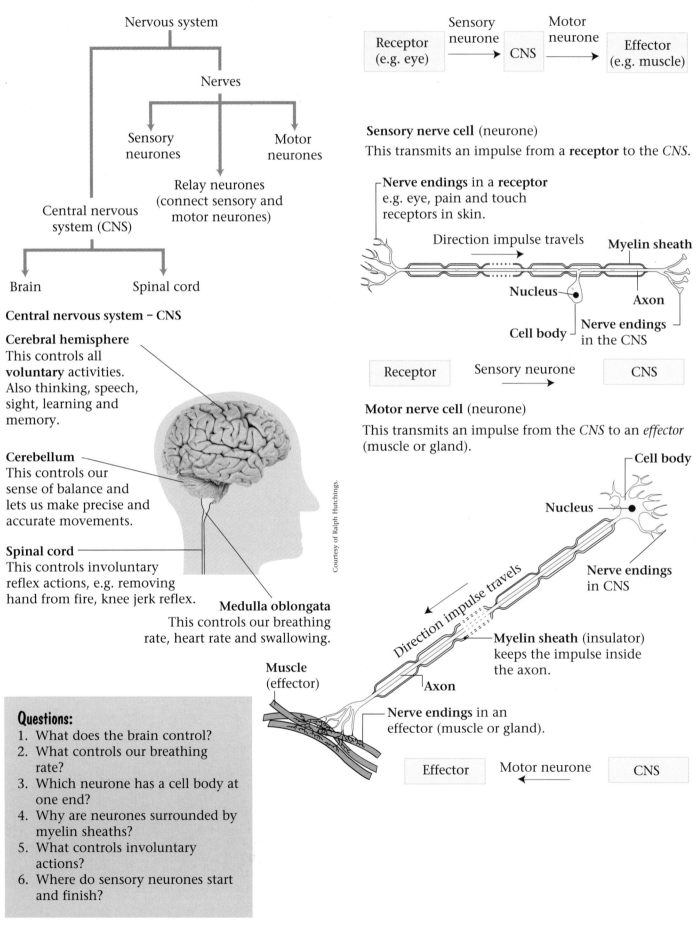

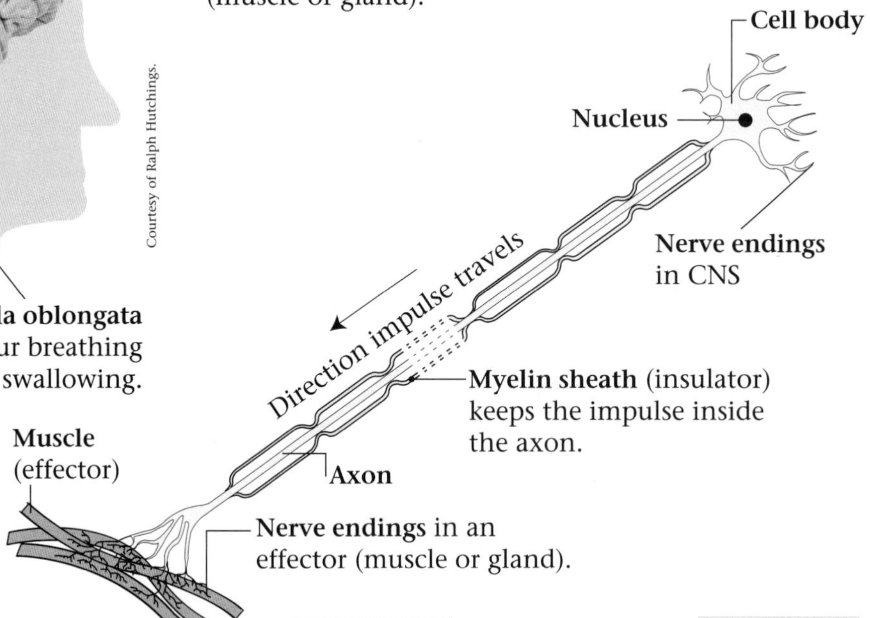

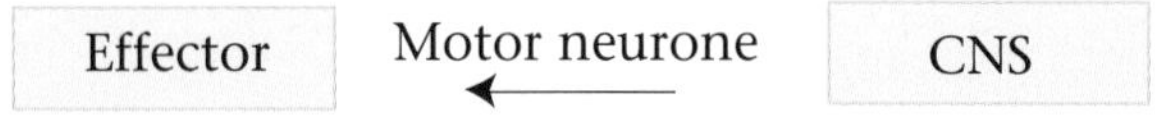

Questions:

1. What does the brain control?
2. What controls our breathing rate?
3. Which neurone has a cell body at one end?
4. Why are neurones surrounded by myelin sheaths?
5. What controls involuntary actions?
6. Where do sensory neurones start and finish?

THE SPINAL CORD This controls reflex actions, e.g. knee jerk, withdrawal of hand from fire.

(1) Hand in fire stimulates pain **receptor** in skin.

(2) Message is sent as an impulse along the **sensory neurone** to the spinal cord.

(3) Sensory neurone is found in the **dorsal root** of spinal nerve.

(4) Message passes by diffusion of chemicals across the **synapse**. This causes an impulse to pass to the **relay neurone**.

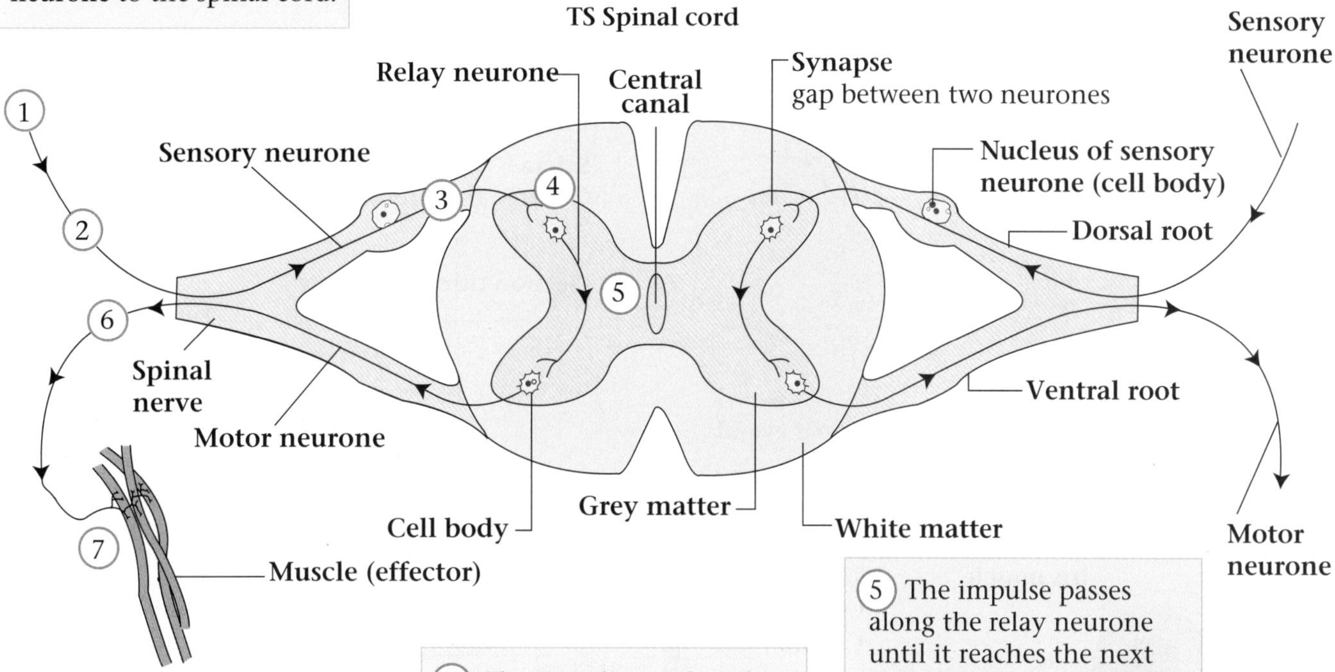

(6) Motor (effector) neurone carries impulse to the **effector** (muscle).

(7) The impulse reaches the muscle (effector), causing it to contract. The hand is removed from the fire.

(5) The impulse passes along the relay neurone until it reaches the next synapse. Here once again, a chemical transmitter passes across to the **motor neurone**.

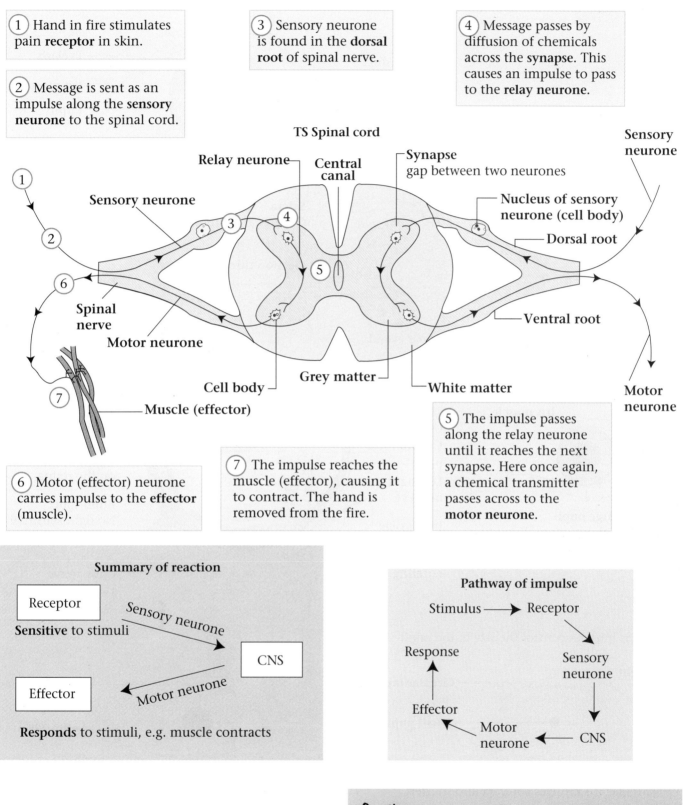

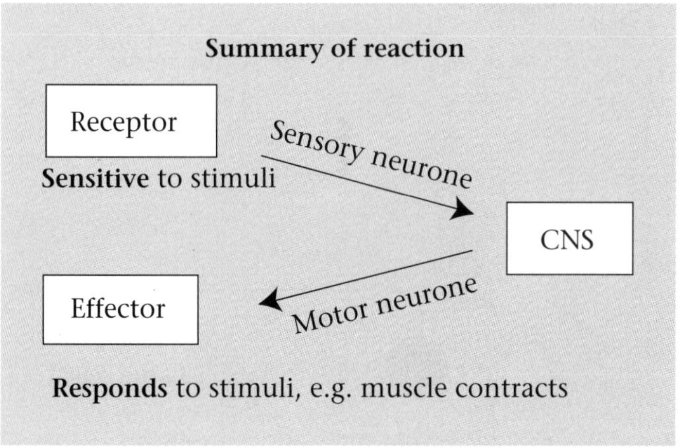

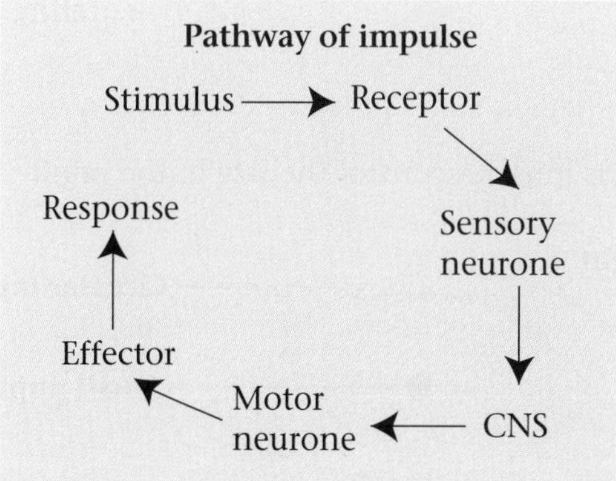

Questions:

1. What does the spinal cord control?
2. What is stimulated by pain?
3. How is a message sent along to the spinal cord?
4. What is a synapse and how do messages cross this?
5. Which neurone carries an impulse to the effector?
6. What is an effector?

Definition:
A reflex action is a rapid, automatic response to a stimulus. These involuntary actions are usually controlled by the spinal cord.

THE EYE

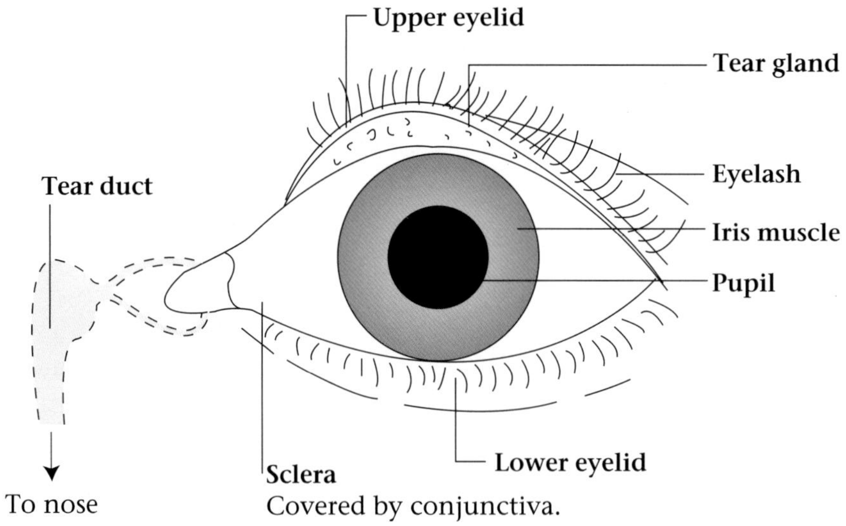

Response to bright and dim light

Bright light

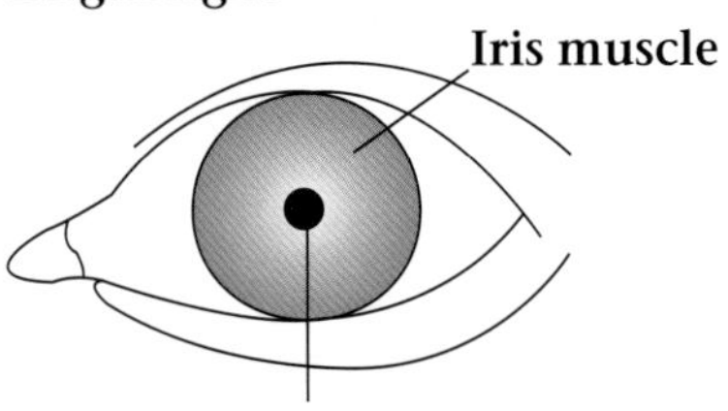

Small pupil
This stops too much light entering the eye.

Dim light

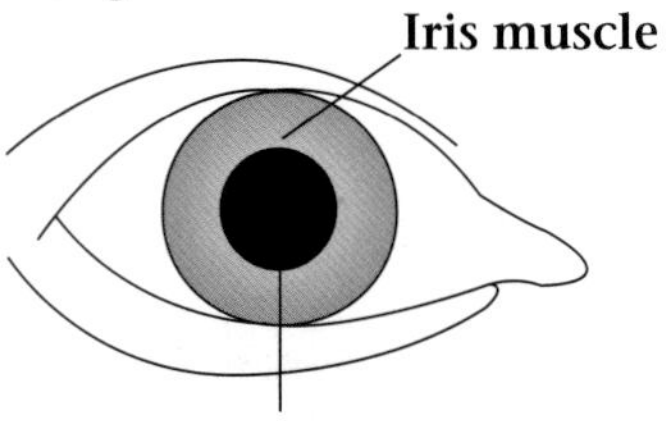

Large pupil
This allows more light into the eye.

Section through the eye

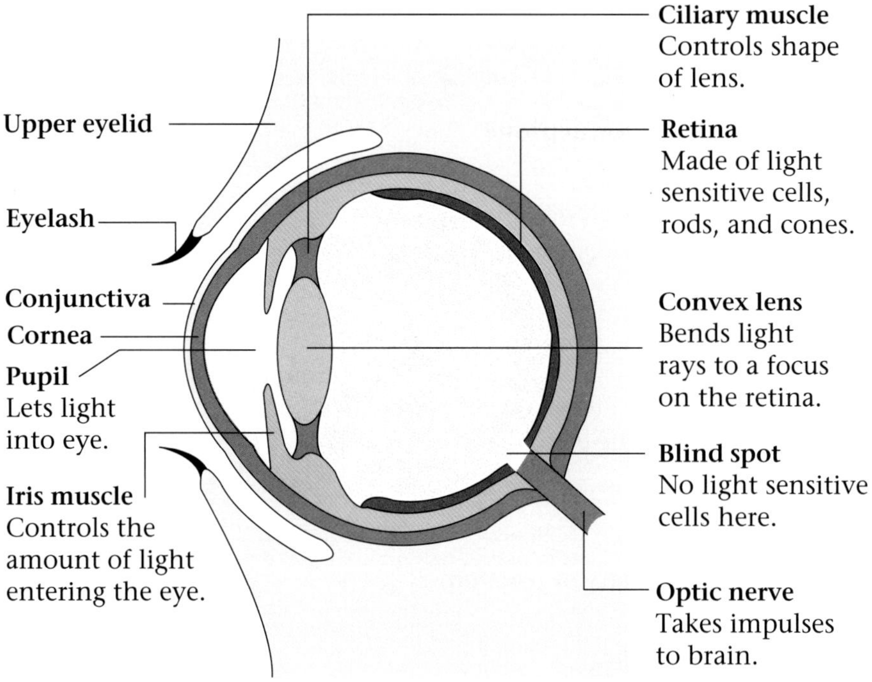

The iris muscles control the size of the pupil

Bright light

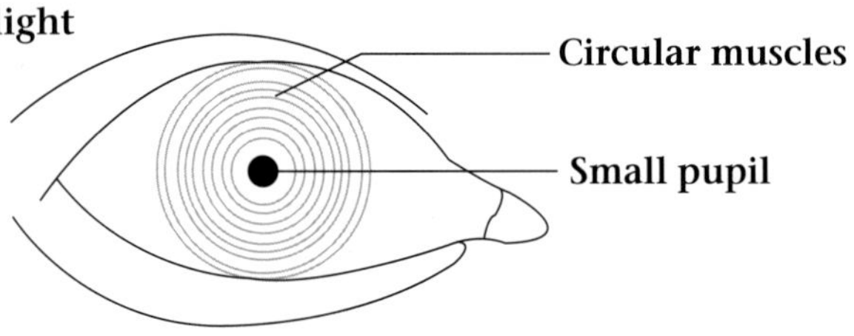

Circular iris muscles contract and shorten.

Dim light

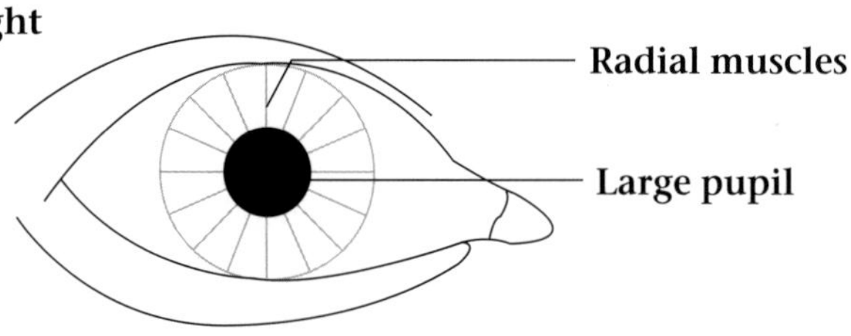

Radial iris muscles contract and shorten.

Questions:

1. Which part of the eye is made of light sensitive cells?
2. What type of lens is found in the eye and what does it do to light rays?
3. Which nerve takes impulses from the eye to the brain?
4. Which muscles control the size of the pupil?
5. What size is the pupil in bright conditions and why?
6. Which muscle can control the shape of the lens in the eye?

VISION — HOW WE SEE

The **retina** is made of **light sensitive cells**. These are stimulated by light. This information is sent to the brain which interprets the message.

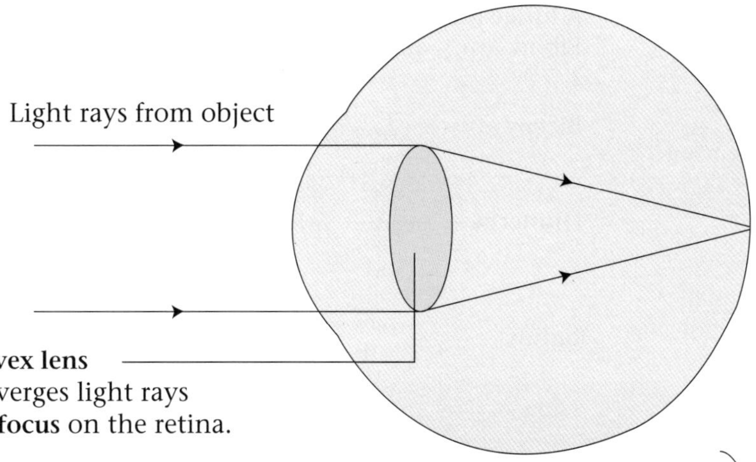

If light rays focus on the retina, the image is clear.

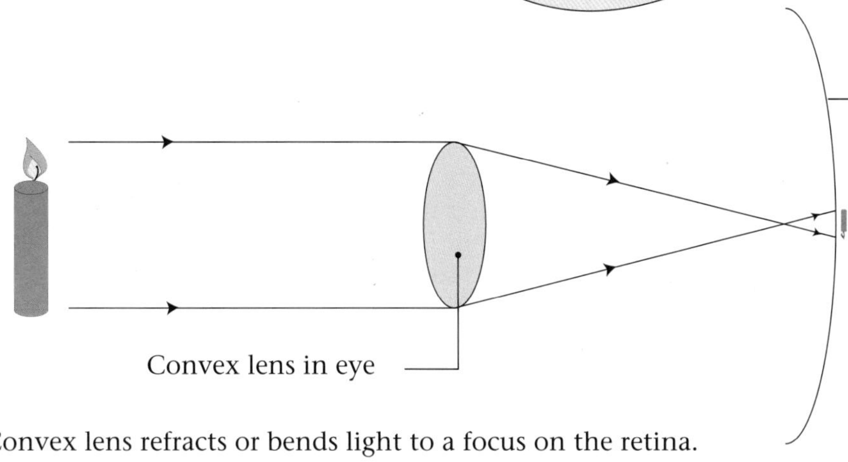

The image is:
1. **Inverted** (upside down).
2. **Diminished** (smaller).

Convex lens refracts or bends light to a focus on the retina.

Object	Light rays entering eye	Ciliary muscle	Lens shape	Effect
Near	Light rays are diverging 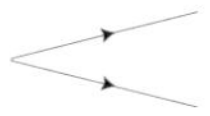	Contracts	Fatter 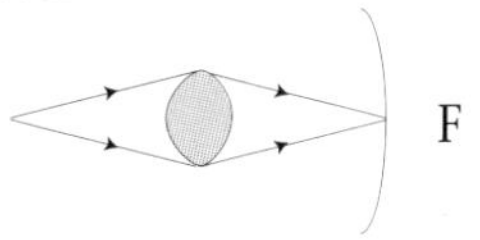	**Converges** the **diverging** light rays onto the **retina** to give a **clear image.**
Distant	Light rays are parallel 	Relaxes	Thinner 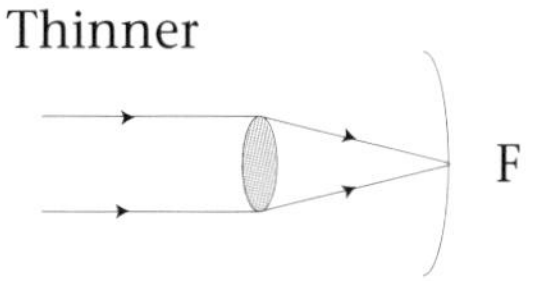	**Converges** the **parallel** light rays onto the **retina** to give a **clear image.**

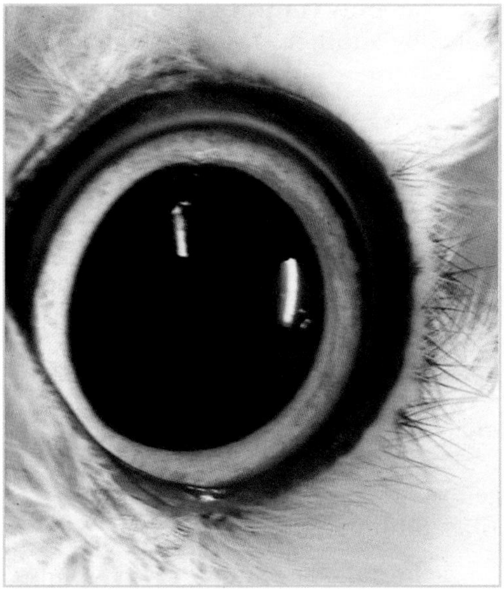
Courtesy of Neil Forbes

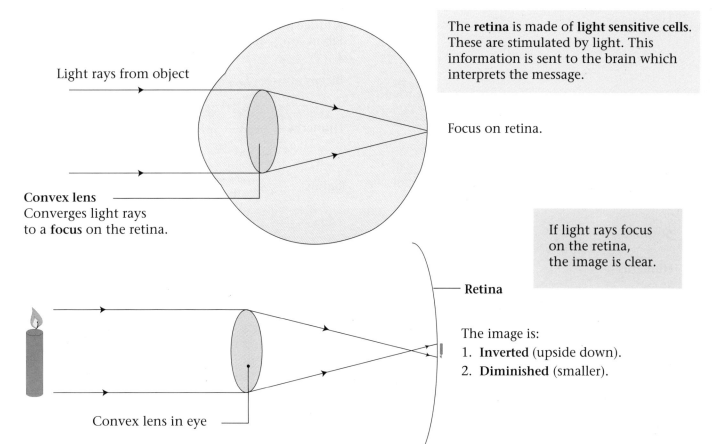

The large eyes of an owl provide excellent vision for finding and catching prey such as mice. Their forward facing eyes provide binocular vision enabling them to judge distance accurately for hunting.

Questions:
1. What shape lens is needed for looking at near objects?
2. What effect does the lens have on light rays entering the eye?
3. Which muscle controls the shape of the lens?

SKELETON AND MOVEMENT

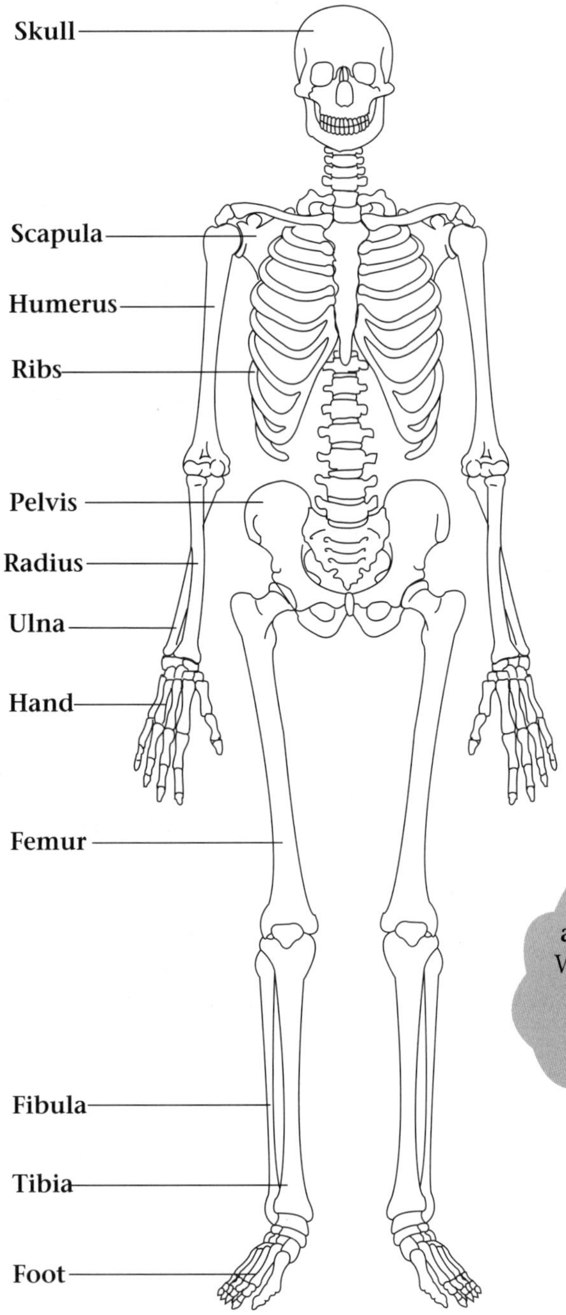

A hinge joint
Elbow joint, an example of a **synovial joint**.

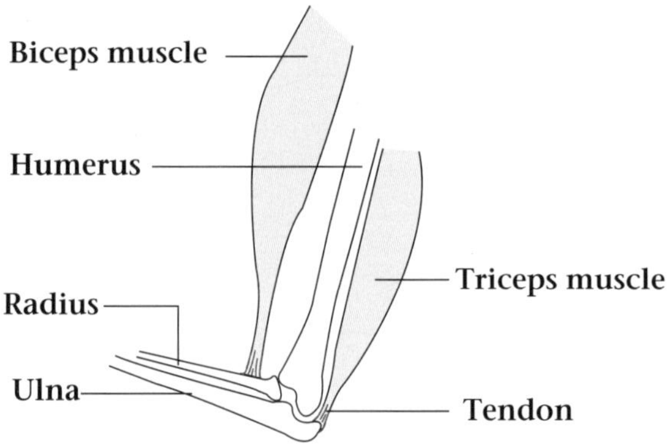

Biceps muscle contracted (short and fat).

When the biceps muscle contracts it gets shorter and fatter and pulls the lower arm up. It makes the **triceps relax** by pulling it long and thin.

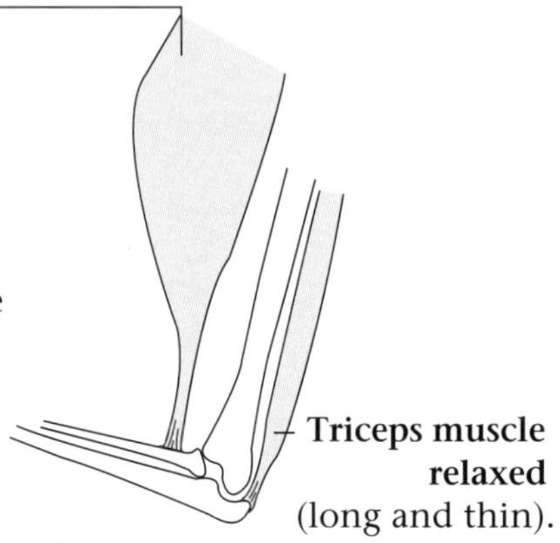

Biceps muscle relaxed.

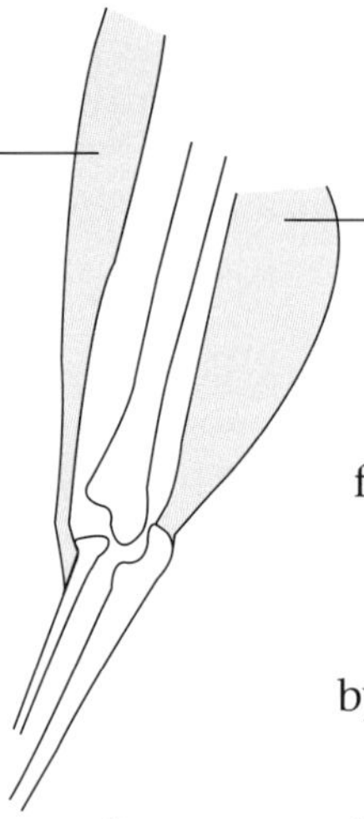

Triceps muscle contracted. When this happens, it shortens and fattens and pulls the lower arm down. It makes the **biceps relax** by pulling it long and thin.

Muscles work in **antagonistic pairs.** When one of a pair is contracted the other is relaxed.

Muscles can only contract, they are made to relax by another muscle contracting.

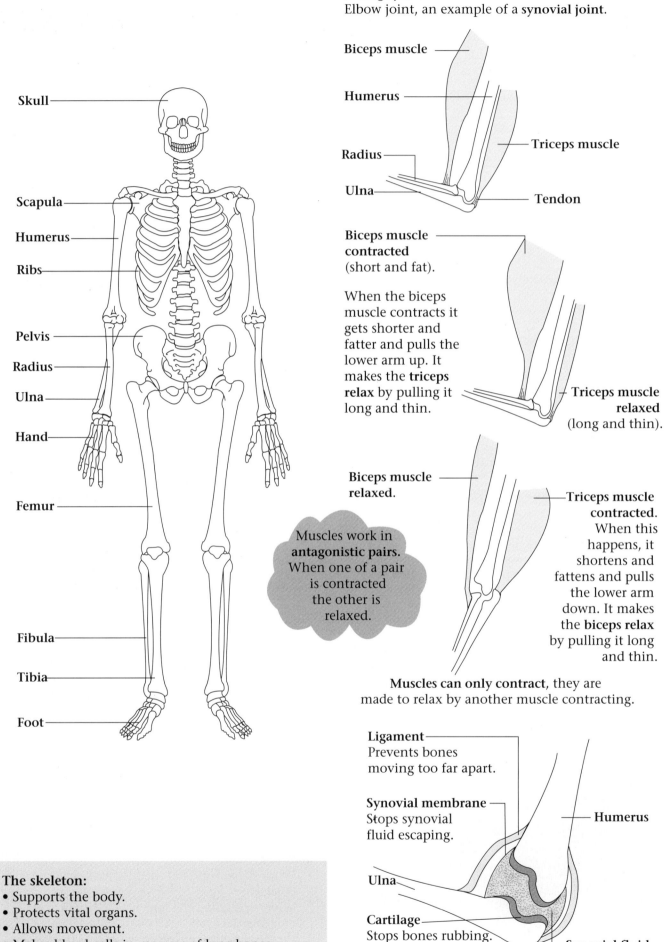

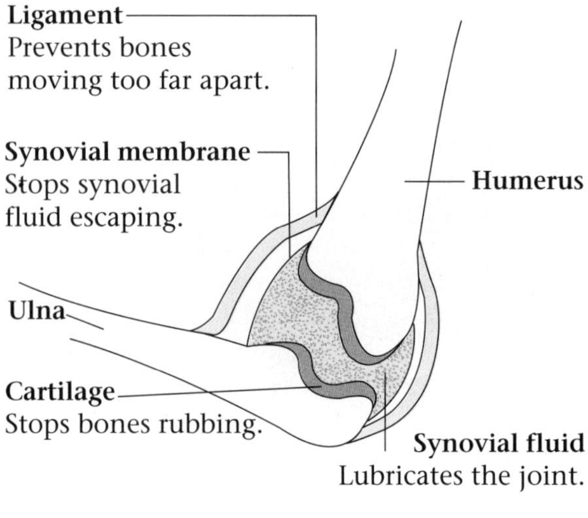

The skeleton:
- Supports the body.
- Protects vital organs.
- Allows movement.
- Makes blood cells in marrow of long bones.

PLANTS

LEAVES AND PHOTOSYNTHESIS Photosynthesis is the process by which plants make sugar.

Equation of photosynthesis

$$\text{Water} + \text{carbon dioxide} \xrightarrow[\text{Light}]{\text{Chlorophyll}} \text{Sugar} + \text{oxygen}$$

$$6H_2O + 6CO_2 \longrightarrow C_6H_{12}O_6 + 6O_2$$

Photosynthesis takes place only during the day as light is needed.

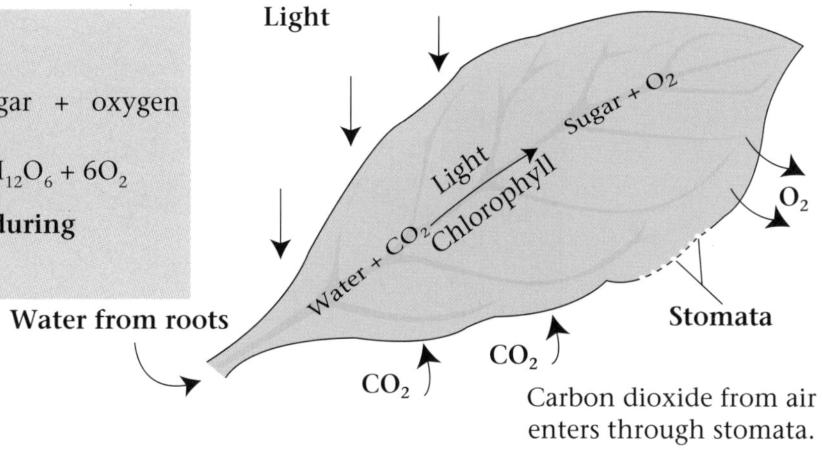

How leaves are adapted for photosynthesis

- Leaves are thin and flat to trap more light.
- Upper layers are transparent to let light through to palisade layer.
- Palisade layer has lots of chloroplasts to trap light.
- Palisade cells are arranged so that the maximum number receive light.
- Leaves can turn to face the light.
- Veins bring in water for photosynthesis.
- Stomata allow entry of carbon dioxide.
- Spongy mesophyll allows gases to circulate.

Leaf structure TS

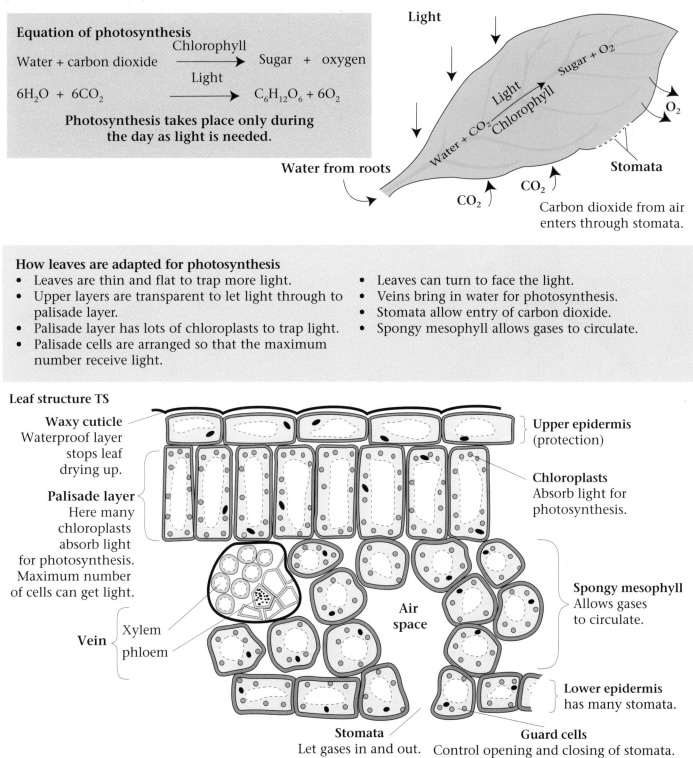

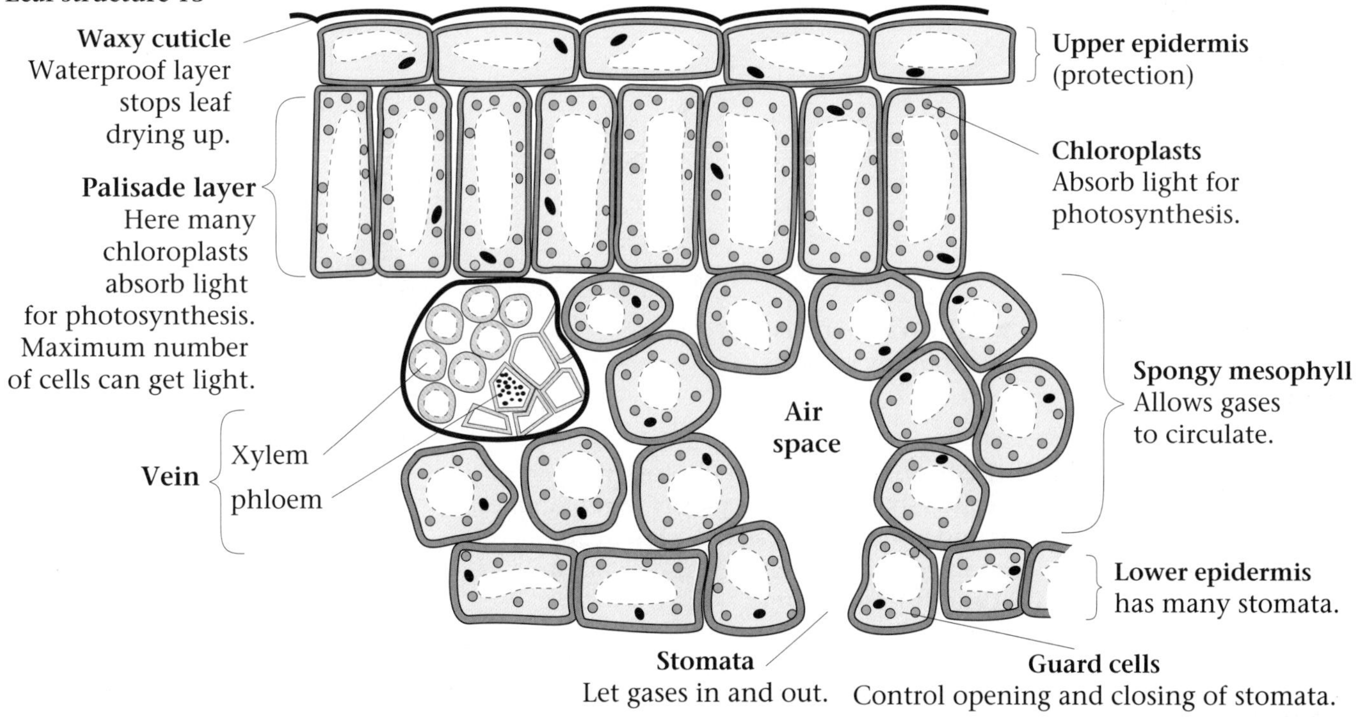

Questions:

1. What is photosynthesis?
2. In which part of the plant does photosynthesis occur?
3. Why does photosynthesis take place only during the day?
4. Carbon dioxide is needed for photosynthesis. How does it enter the plant?
5. In which layer of the leaf does most photosynthesis take place and why?
6. Describe three ways in which leaves are suited for photosynthesis.

TRANSPORT IN PLANTS

Photosynthesis takes place mainly in the **chloroplasts** of the **palisade layer** in the leaf.

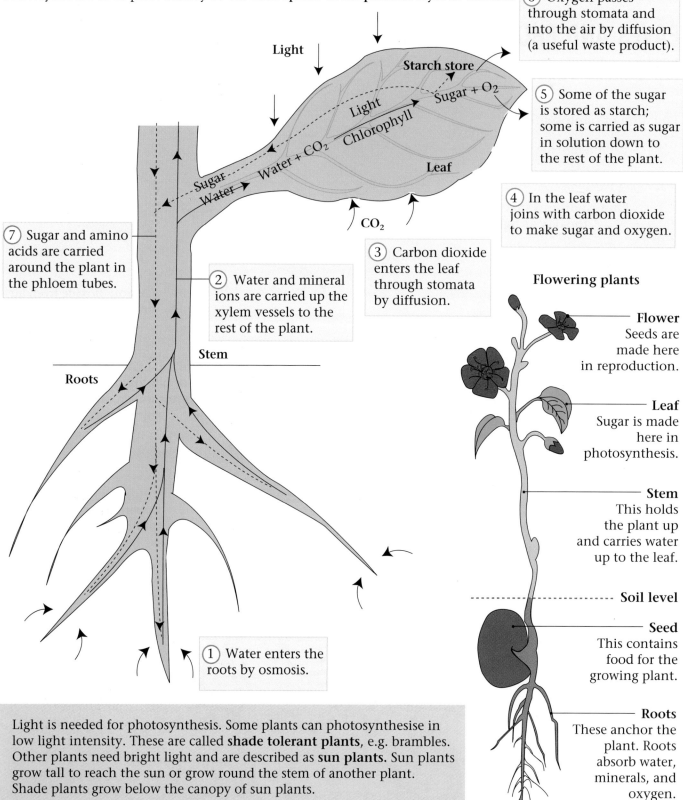

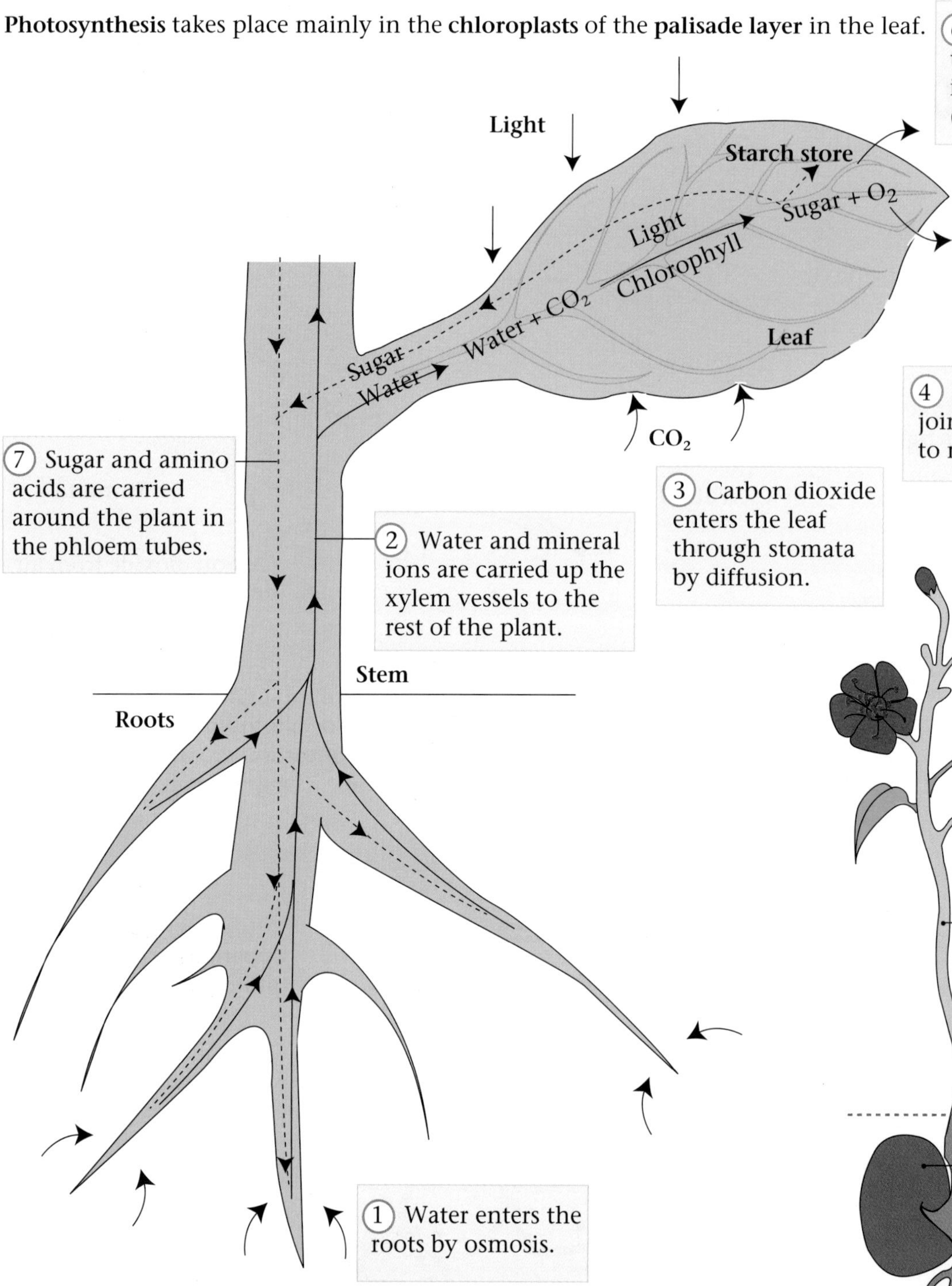

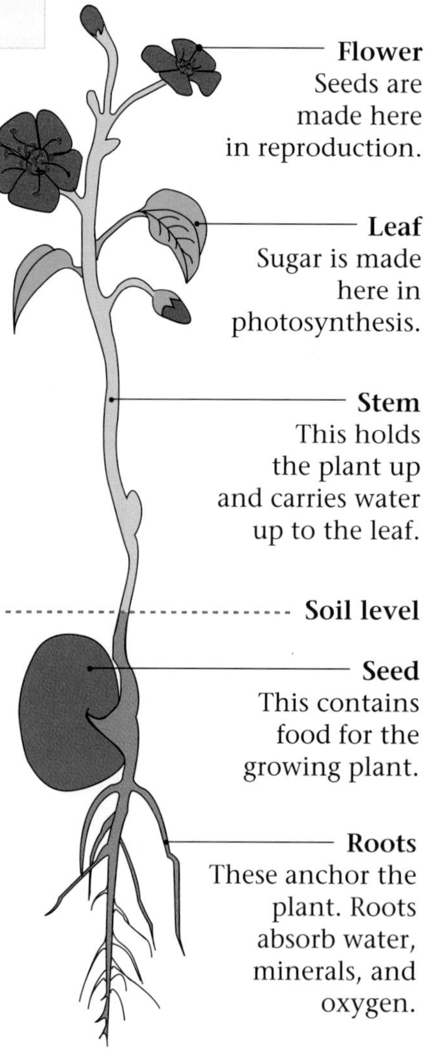

Light is needed for photosynthesis. Some plants can photosynthesise in low light intensity. These are called **shade tolerant plants**, e.g. brambles. Other plants need bright light and are described as **sun plants.** Sun plants grow tall to reach the sun or grow round the stem of another plant. Shade plants grow below the canopy of sun plants.

Questions:

1. Where is sugar made?
2. How is this sugar carried to other parts of the plant?
3. Water is needed for photosynthesis. How is it carried to the leaves?
4. Which gas is produced as a waste product in photosynthesis and what happens to it?
5. What happens to the sugar that stays in the leaf?
6. What two materials are taken into the leaf for photosynthesis?
7. In the stem, xylem and phloem are found. What do they each carry and in what direction?

USES OF SUGAR MADE IN PHOTOSYNTHESIS

Sugar made in the leaves is carried all over the plant in the phloem tubes. This is possible as sugar is **soluble**. This transport of sugar is called **translocation.**

Sugar can be used to form many useful substances in the plant.

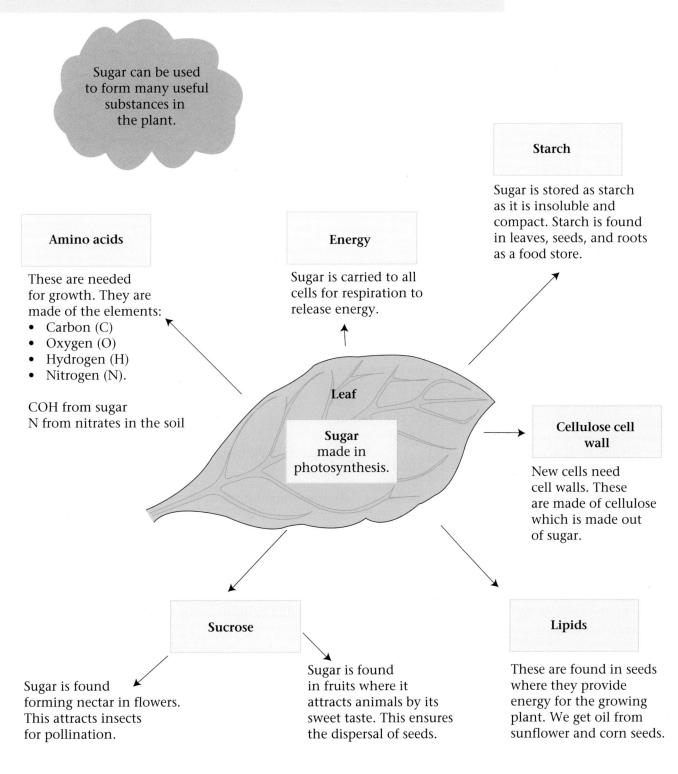

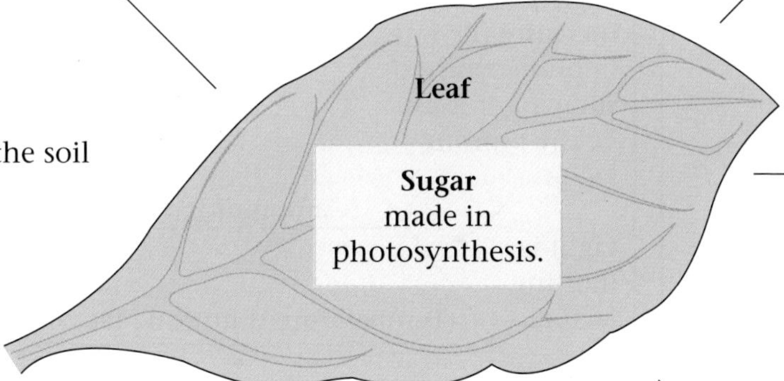

Amino acids

These are needed for growth. They are made of the elements:

- Carbon (C)
- Oxygen (O)
- Hydrogen (H)
- Nitrogen (N).

COH from sugar
N from nitrates in the soil

Energy

Sugar is carried to all cells for respiration to release energy.

Starch

Sugar is stored as starch as it is insoluble and compact. Starch is found in leaves, seeds, and roots as a food store.

Cellulose cell wall

New cells need cell walls. These are made of cellulose which is made out of sugar.

Sucrose

Sugar is found forming nectar in flowers. This attracts insects for pollination.

Sugar is found in fruits where it attracts animals by its sweet taste. This ensures the dispersal of seeds.

Lipids

These are found in seeds where they provide energy for the growing plant. We get oil from sunflower and corn seeds.

Questions:

1. What is translocation? In what tube does this occur?
2. Every cell in the plant needs sugar for respiration. What is released in this process?
3. Why is sucrose needed in the flower? Give two reasons.
4. How is sugar stored and why?
5. What two materials might sugar be changed to in the seeds and why?
6. Why is sugar easy to transport round the plant?
7. To grow, plants need amino acids. What element is added to sugar to provide this? How do plants take in this element?

LIMITING FACTORS IN PHOTOSYNTHESIS

Photosynthesis is a process that needs **light**, **warmth** and **carbon dioxide**. If one of these factors is in **short supply**, it will limit the rate of photosynthesis, and so is called a **limiting factor.**

1. Light as limiting factor

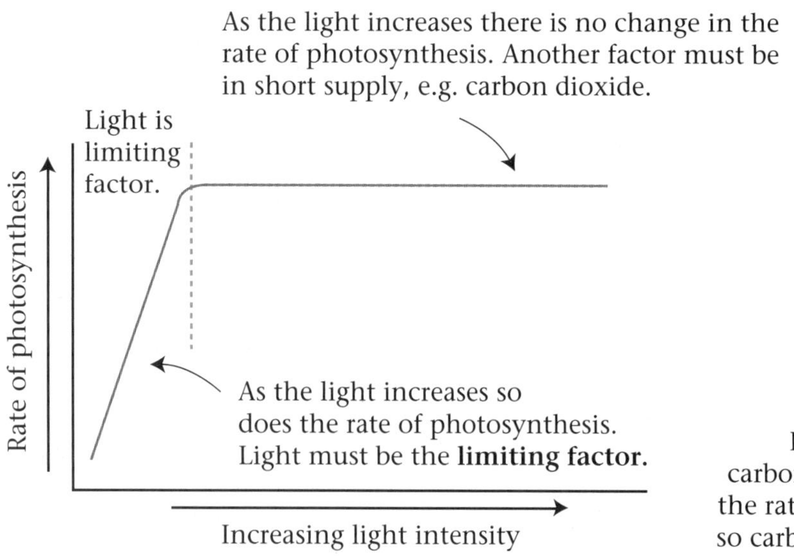

As the light increases there is no change in the rate of photosynthesis. Another factor must be in short supply, e.g. carbon dioxide.
Light is limiting factor.
Rate of photosynthesis
As the light increases so does the rate of photosynthesis. Light must be the **limiting factor.**
Increasing light intensity

There are three main limiting factors in photosynthesis
- Light
- Carbon dioxide
- Temperature (warmth)

Glasshouses are used commercially to produce crops as the conditions can be controlled to maximise photosynthesis. Paraffin burners are often used as they provide both heat and extra carbon dioxide for the plants to ensure that these are not limiting factors. Growers can produce crops out of season if additional, heat and light are provided. The grower must weigh-up whether the extra cost of heating and lighting will be covered by the extra profit of the crop out of season.

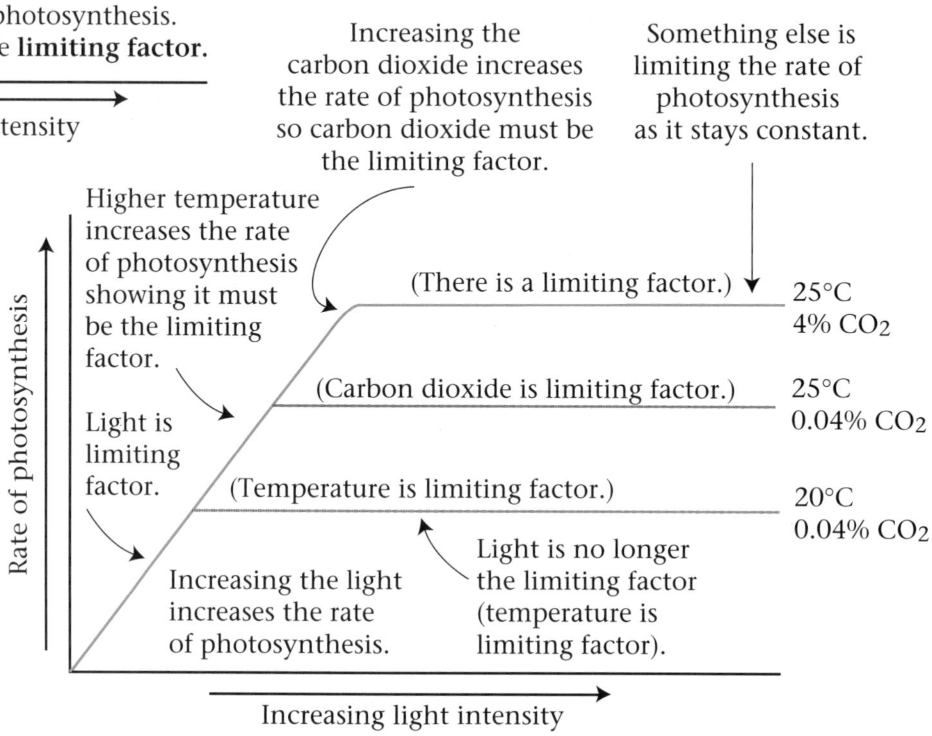

Increasing the carbon dioxide increases the rate of photosynthesis so carbon dioxide must be the limiting factor.
Something else is limiting the rate of photosynthesis as it stays constant.
Higher temperature increases the rate of photosynthesis showing it must be the limiting factor.
(There is a limiting factor.)
25°C
4% CO2
(Carbon dioxide is limiting factor.)
25°C
0.04% CO2
Light is limiting factor.
(Temperature is limiting factor.)
20°C
0.04% CO2
Rate of photosynthesis
Increasing the light increases the rate of photosynthesis.
Light is no longer the limiting factor (temperature is limiting factor).
Increasing light intensity

2. Carbon dioxide as limiting factor

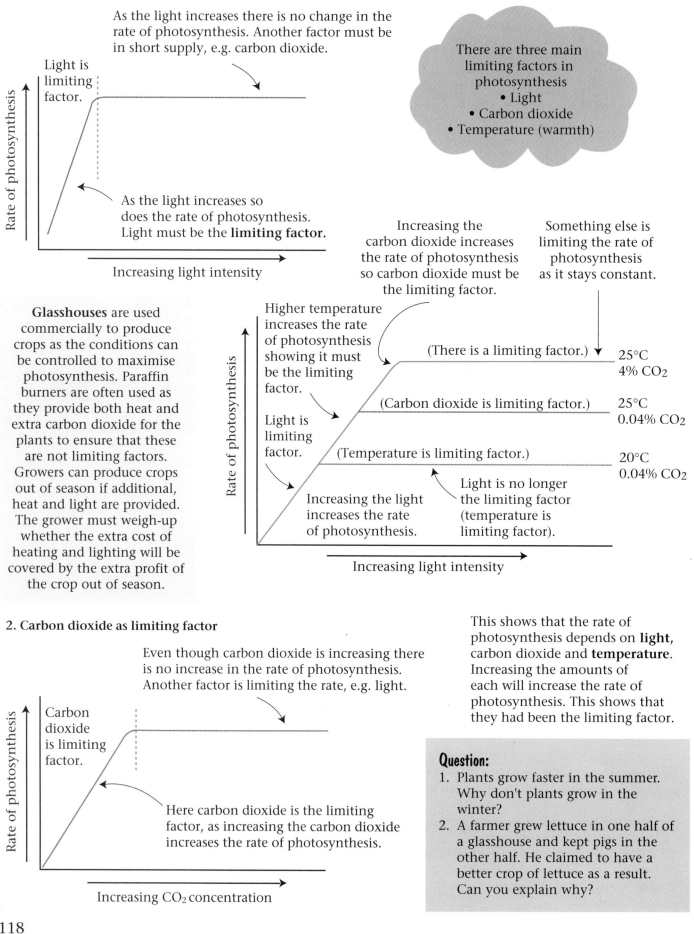

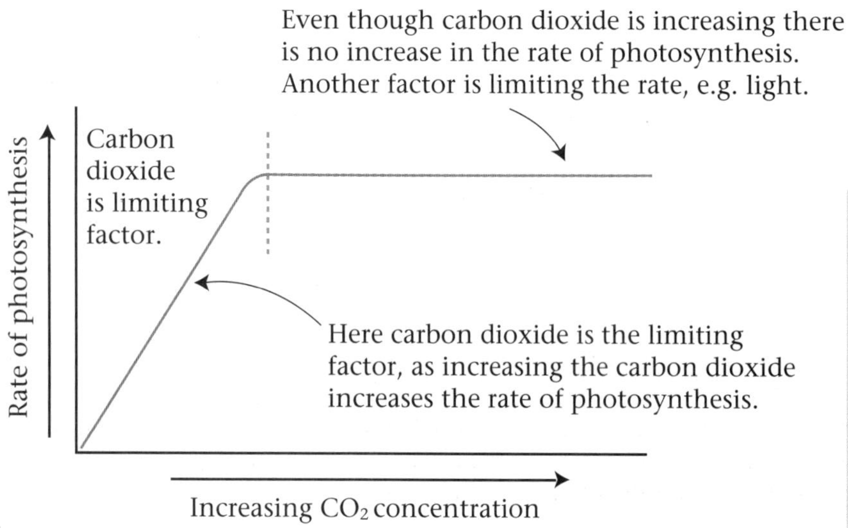

Even though carbon dioxide is increasing there is no increase in the rate of photosynthesis. Another factor is limiting the rate, e.g. light.
Carbon dioxide is limiting factor.
Rate of photosynthesis
Here carbon dioxide is the limiting factor, as increasing the carbon dioxide increases the rate of photosynthesis.
Increasing CO2 concentration

This shows that the rate of photosynthesis depends on **light**, carbon dioxide and **temperature**. Increasing the amounts of each will increase the rate of photosynthesis. This shows that they had been the limiting factor.

Question:
1. Plants grow faster in the summer. Why don't plants grow in the winter?
2. A farmer grew lettuce in one half of a glasshouse and kept pigs in the other half. He claimed to have a better crop of lettuce as a result. Can you explain why?

MINERALS AND PLANTS

Plants make sugar in photosynthesis. To stay healthy, they also need minerals.

These cuttings are all the same species, same age and started off the same size.

Experiment with cuttings of *Tradescantia* grown in solution.

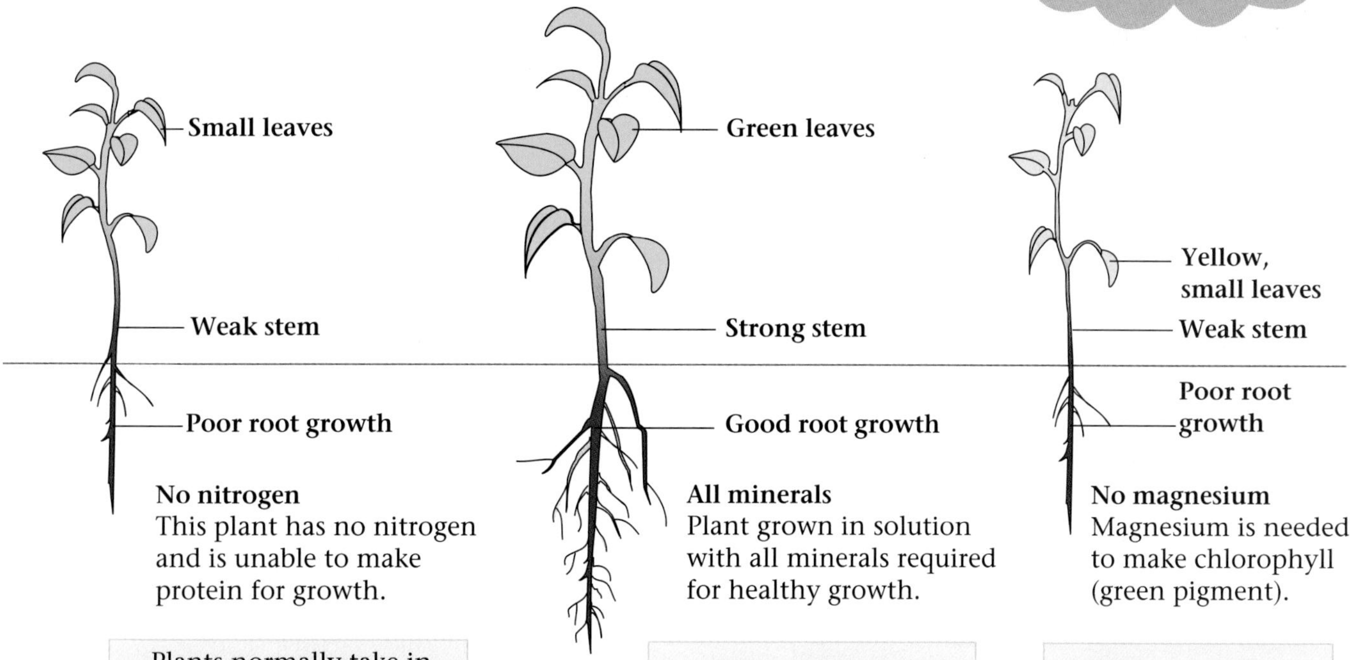

No nitrogen
This plant has no nitrogen and is unable to make protein for growth.

All minerals
Plant grown in solution with all minerals required for healthy growth.

No magnesium
Magnesium is needed to make chlorophyll (green pigment).

Plants normally take in nitrogen as nitrates which are soluble and enter roots dissolved in water.

These are taken in through the roots dissolved in water.

Without chlorophyll, little sugar can be made, so little growth.

Uptake of minerals

Minerals can enter by **diffusion** and **active transport** (low to high concentration).

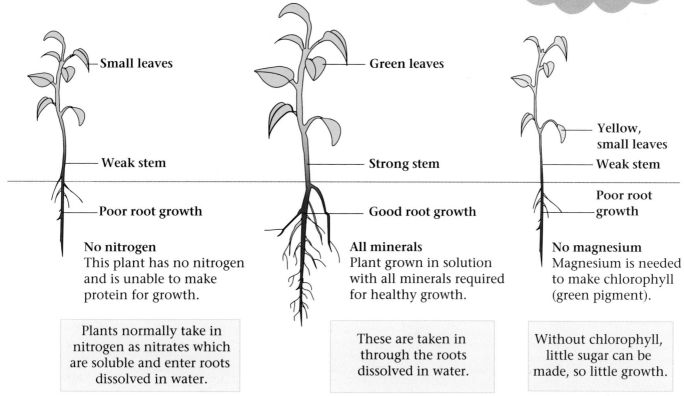

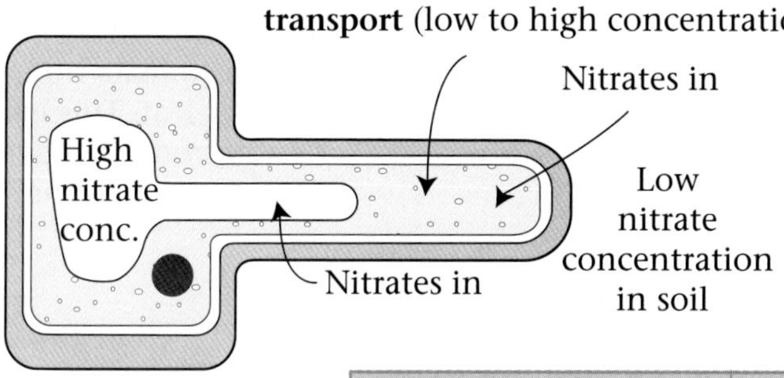

Root hair cell

Nitrates move from the lower concentration in the soil to the higher concentration in the root cells. This is the opposite of diffusion and requires energy. It is called active transport.

Element/Mineral	Why it is needed	Deficiency causes
Nitrates	For proteins and growth.	Poor growth.
Magnesium	To make chlorophyll.	Yellow leaves.
Iron	To make chlorophyll.	Yellow leaves.
Potassium	For photosynthesis and respiration.	Yellow-edged leaves.
Phosphate	To make protein and for cell membranes.	Stunted growth of roots.

Questions:
1. Where do minerals enter a plant?
2. By what two processes can minerals enter?
3. If a plant is grown without magnesium, how will it be affected?
4. Why do plants need nitrogen? How does it enter the plant, in what form?
5. Why is it important that the same plant was used in this experiment at the same size and age?
6. How does energy help in the uptake of minerals and how do plants release energy?

WATER MOVEMENT THROUGH A PLANT

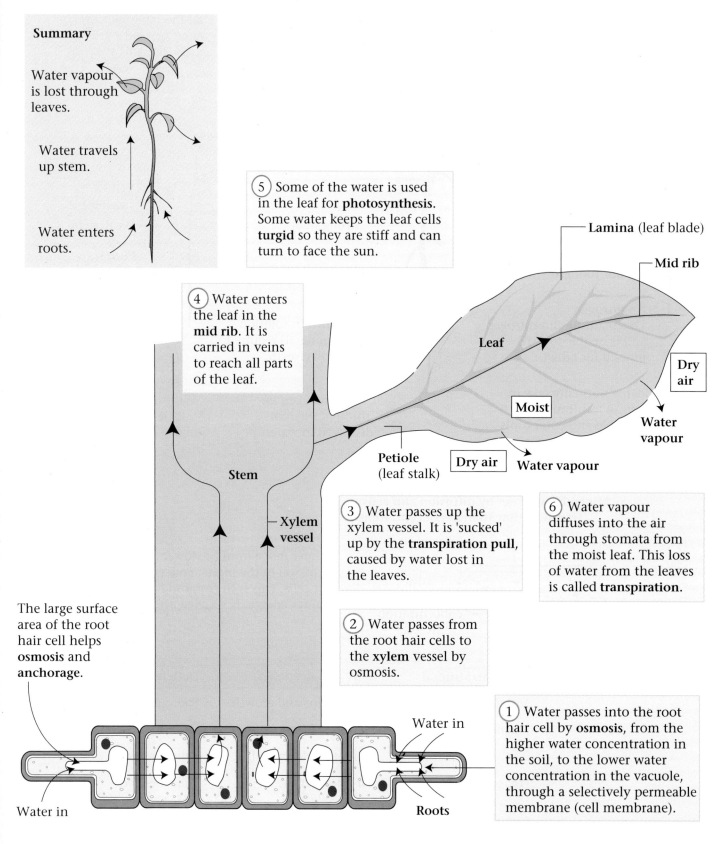

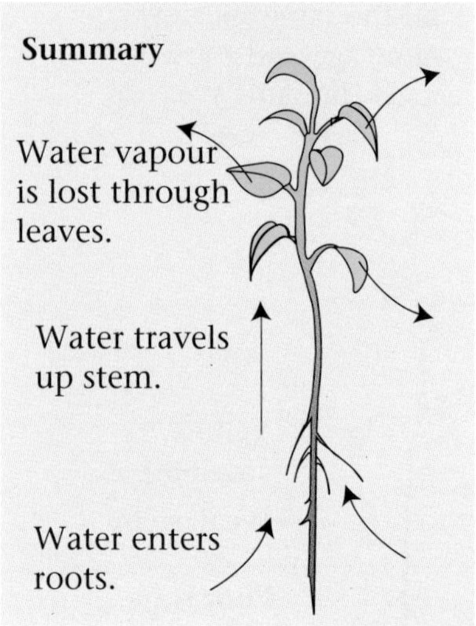

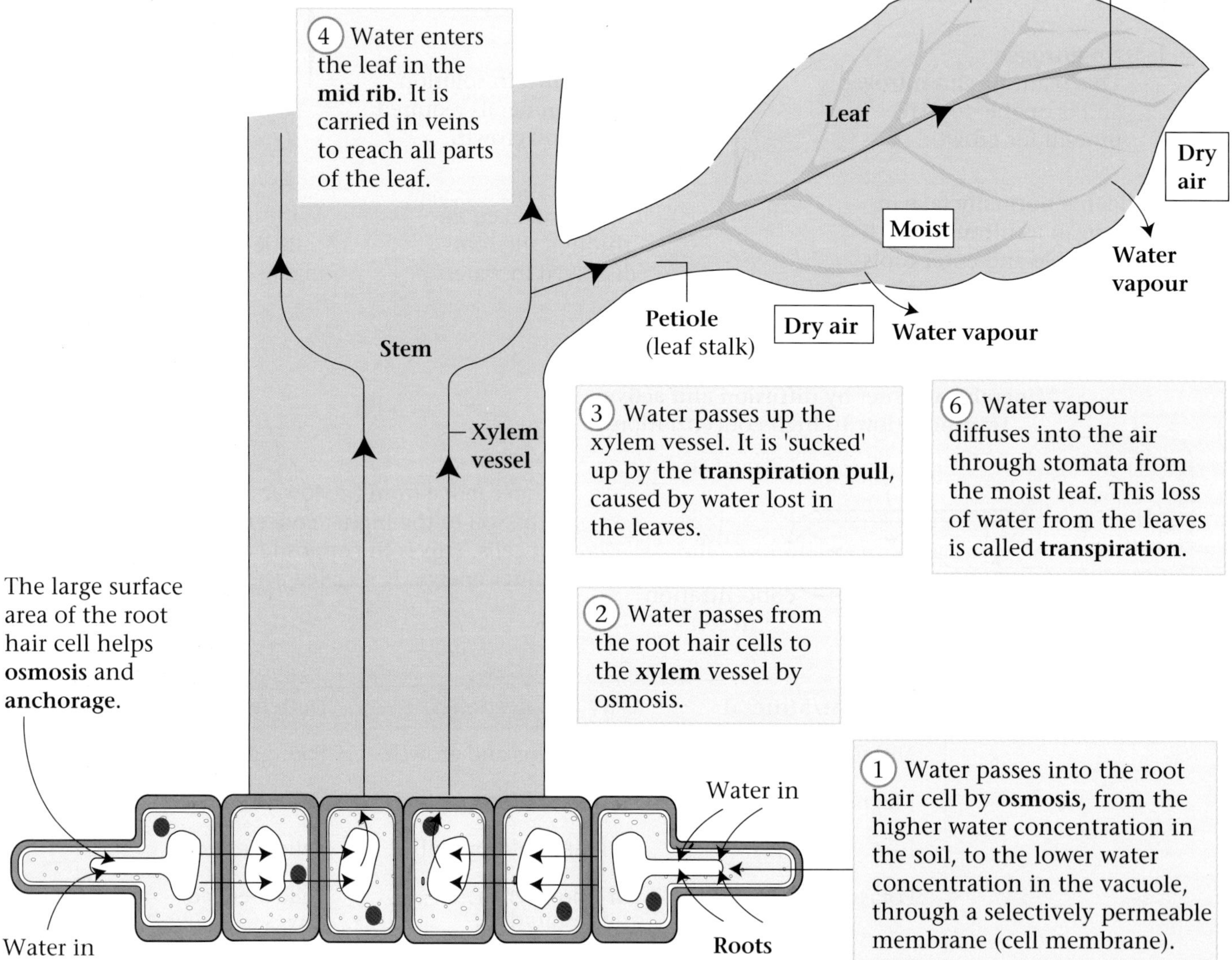

(1) Water passes into the root hair cell by **osmosis**, from the higher water concentration in the soil, to the lower water concentration in the vacuole, through a selectively permeable membrane (cell membrane).

(2) Water passes from the root hair cells to the **xylem** vessel by osmosis.

(3) Water passes up the xylem vessel. It is 'sucked' up by the **transpiration pull**, caused by water lost in the leaves.

(4) Water enters the leaf in the **mid rib**. It is carried in veins to reach all parts of the leaf.

(5) Some of the water is used in the leaf for **photosynthesis**. Some water keeps the leaf cells **turgid** so they are stiff and can turn to face the sun.

(6) Water vapour diffuses into the air through stomata from the moist leaf. This loss of water from the leaves is called **transpiration**.

Questions:

1. How does the shape of the root hair cell help in the uptake of water?
2. Why does water enter the plant roots and how does rain help this process?
3. Which vessel carries water up the plant?
4. How does water enter the leaf?
5. How is water used in the leaf?
6. How does water vapour pass out of the leaf and into the air?
7. What is this loss of water called?
8. What is the transpiration pull?

TRANSPIRATION The loss of water from a plant.

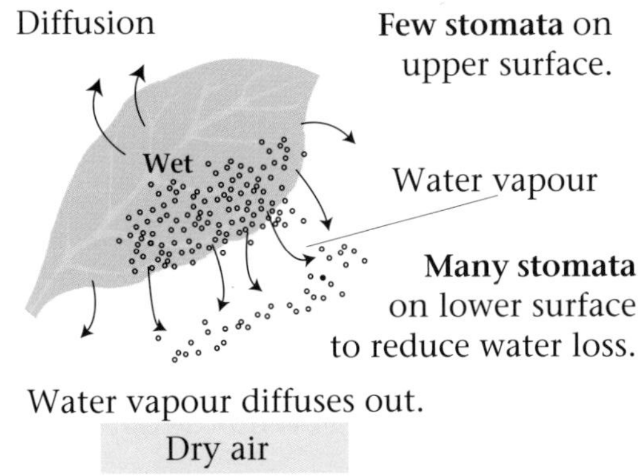

Importance of transpiration
Water is lost from the leaves in transpiration, through holes called stomata.

- Water *with minerals* enters roots to replace water lost through leaves.
- Nitrates and phosphates are vital for healthy plant growth.
- Water is needed for turgidity and support in plants; without water plants wilt.
- Water is needed for photosynthesis and translocation.
- Loss of water helps to cool plants down.

Why transpiration occurs
Stomata need to be open to let gases in and out for photosynthesis and respiration. Unfortunately, water vapour can escape from the leaf through these holes. On hot, dry and windy days many plants close some of their stomata to reduce this loss of water. The opening and closing of stomata is controlled by the **guard cells**.

Comparison of water loss in different conditions

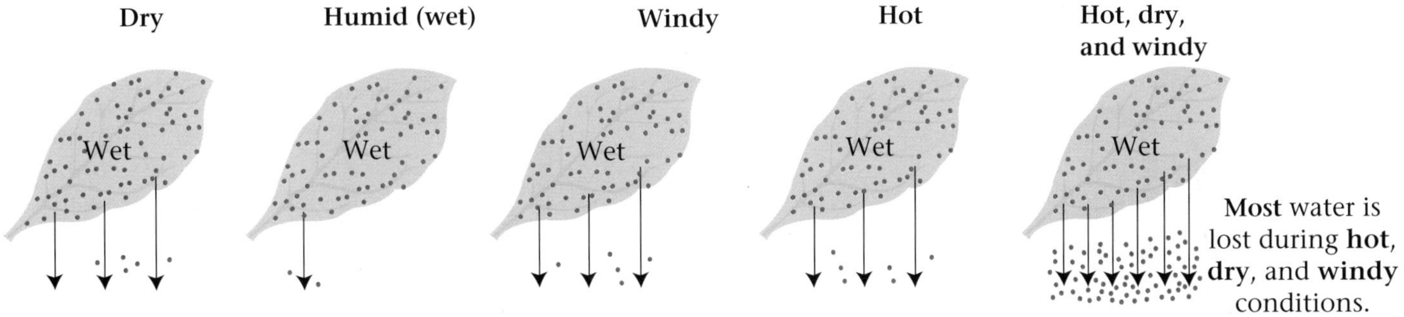

Holes in the leaf are called **stomata**. If it is moist in the leaf and dry outside, water vapour comes out by **diffusion. The greater the difference in concentration inside and outside the leaf, the faster the water loss by diffusion.** More water comes from the roots to replace water lost in the leaves.

Water vapour diffuses out and forms a **diffusion shell** of damp air. This layer slows the rate of diffusion as the difference in concentration of water vapour inside and outside the leaf is reduced. These **'diffusion shells'** build up only on still days, as the wind blows any diffusion shells away, causing a rapid rate of diffusion.

Most stomata are found on the **lower surface** of the leaf. Here it is cooler and less windy than on the upper surface. This reduces water loss from the leaf.

At night stomata close. This reduces water loss in transpiration.

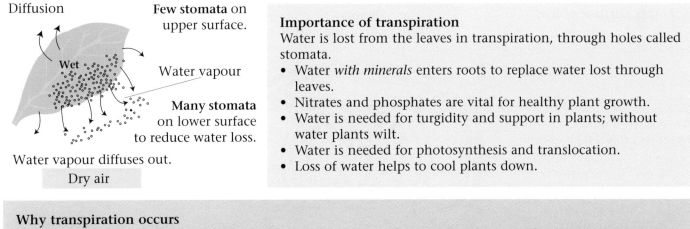

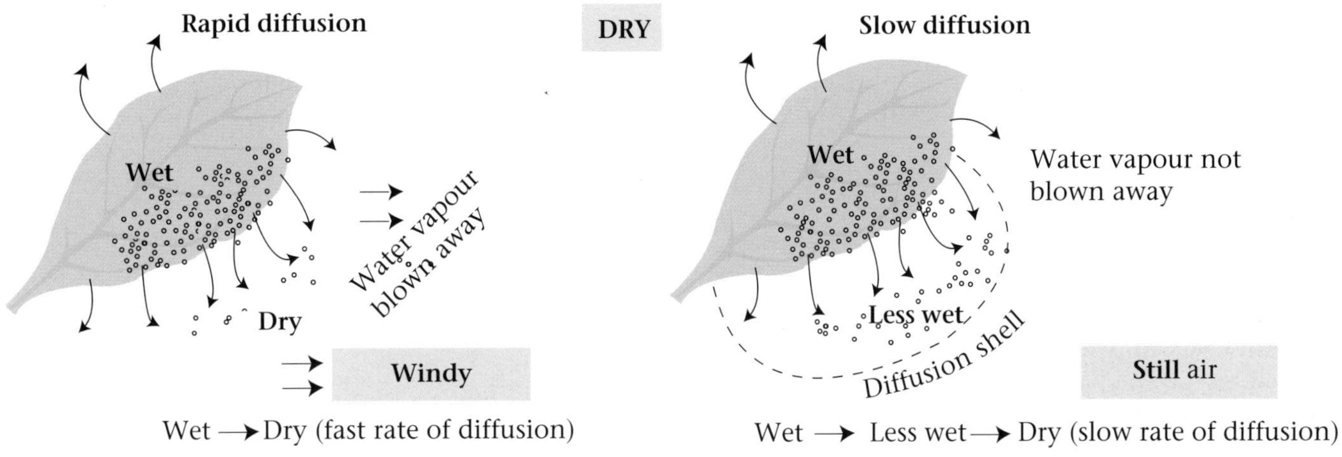

OPENING AND CLOSING OF STOMATA

1. *Lower surface of leaf*

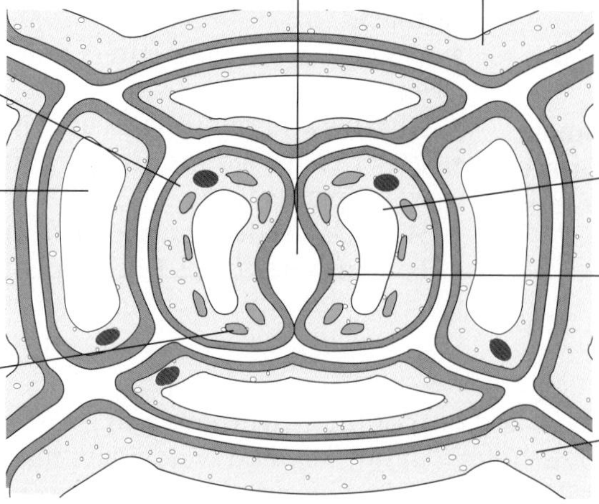

2. *Plant cells*

When water enters

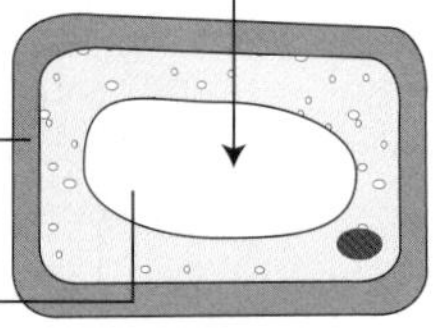

This is **turgid.**

When water leaves

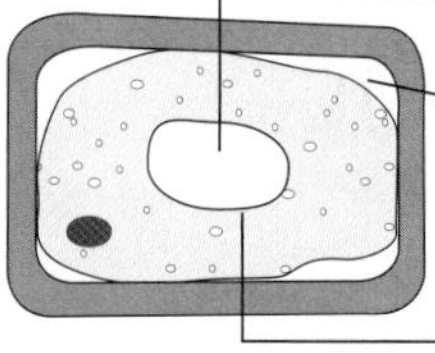

This is **flaccid.**

Plant cells full of water are firm and turgid. This gives them support.

3. *Guard cells*

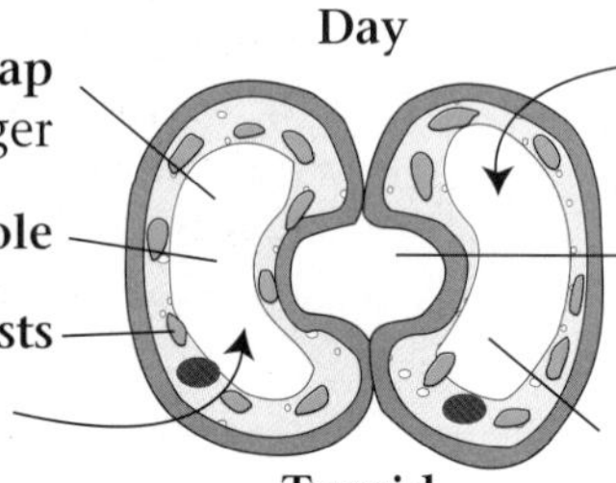

Turgid

Water enters guard cells from surrounding cells by osmosis and the vacuoles swells.

↓

Guard cells swell up and become turgid.

↓

This pulls the guard cells apart.

↓

Stomata open.

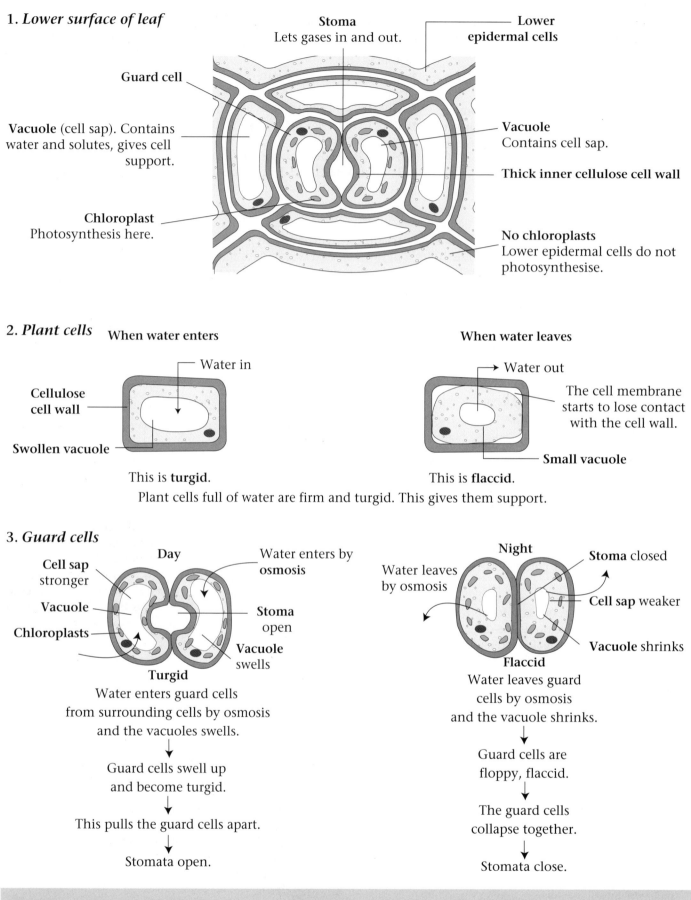

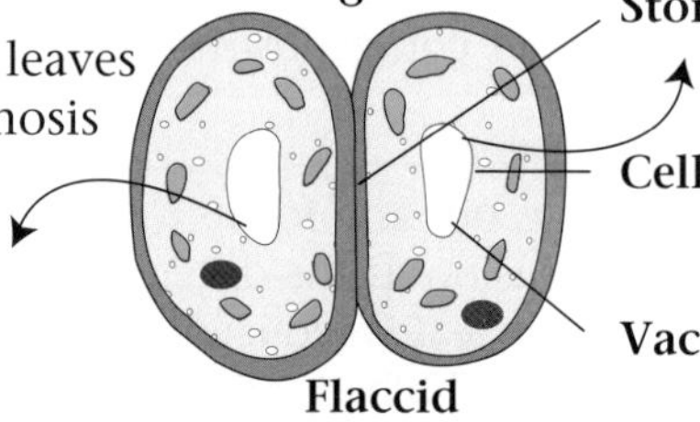

Flaccid

Water leaves guard cells by osmosis and the vacuole shrinks.

↓

Guard cells are floppy, flaccid.

↓

The guard cells collapse together.

↓

Stomata close.

Questions:

1. Which cells on the lower surface of the leaf are able to photosynthesise and why?
2. What do we call plant cells that contain a lot of water?
3. How does the entry of water to guard cells affect the stomata?
4. What stops plant cells bursting when water enters?

LEAVES

1. Marram grass – a **xerophyte** (a plant adapted to dry conditions). Found in **dry (arid)** conditions, i.e. sand dunes.

Leaf is **curled** to reduce area exposed to sun and wind, to reduce water loss.

Thick waxy cuticle reduces water loss here.

Water vapour trapped.

Sunken stomata less affected by drying action of wind.

Wind

Hairs reduce air movement in leaf, reducing the drying action of wind.

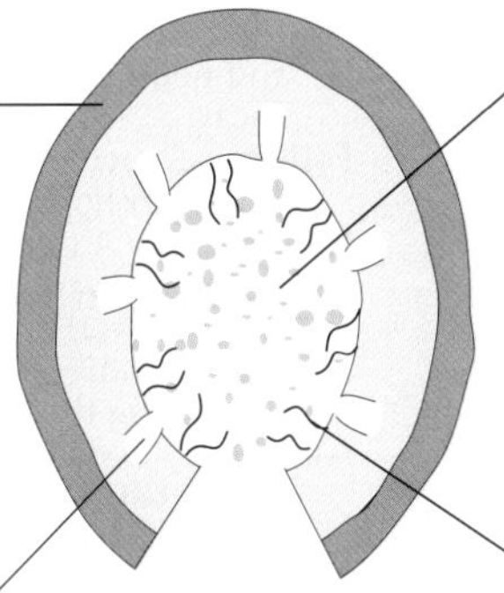

Contrasting leaves

Marram	Beech
Thick cuticle	Thin cuticle
Leaf curled	Leaf flat
Sunken stomata	Stomata at surface
Little surface area exposed	Large surface area exposed
Hairs to reduce air movement	No hairs
Designed to reduce water loss in dry conditions	Designed for maximum photosynthesis

Other adaptations to living in dry conditions
- Extensive roots to reach any available water.
- Storage of water in succulent stem.
- Stomata which close in dry conditions.
- Thick waxy cuticle in stem to reduce water loss.

1. Beech leaf – found in **damp** conditions, i.e. woodland.

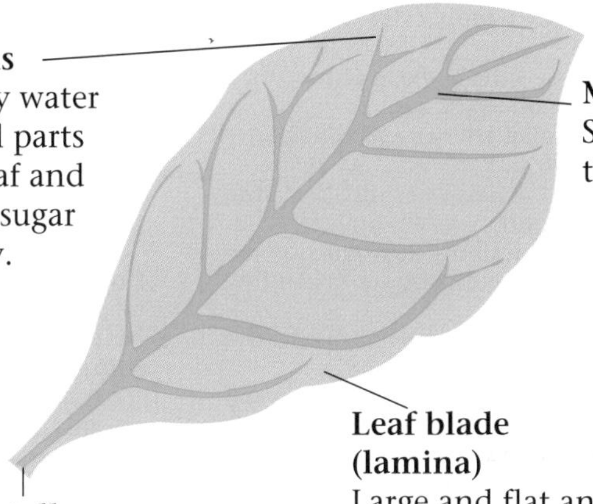

Leaves at night Only respiration as light is not available.

Respiration
Oxygen + sugar ⟶ Carbon dioxide + water + energy
(Oxygen – from air; sugar – from sugar made during day in photosynthesis)

Veins
Carry water to all parts of leaf and take sugar away.

Mid rib
Supports the leaf.

Leaf stalk
Contains veins bringing water in and taking sugar to rest of plant.

Leaf blade (lamina)
Large and flat and faces the sun to absorb light for photosynthesis. Has lots of stomata to let gases in and out.

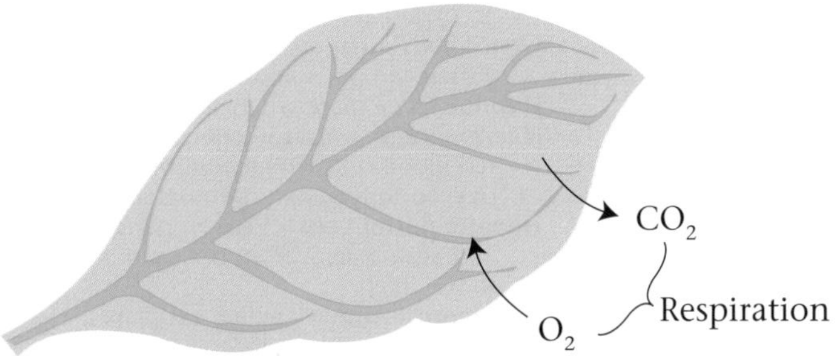

Leaves in day Both respiration and photosynthesis occur.

Respiration
Oxygen + sugar ⟶ Carbon dioxide + water + energy

Photosynthesis
Carbon dioxide + water $\xrightarrow[\text{Chlorophyll}]{\text{Light}}$ Sugar + oxygen

Overall exchange

CO_2 O_2
CO_2
O_2

CO_2
O_2

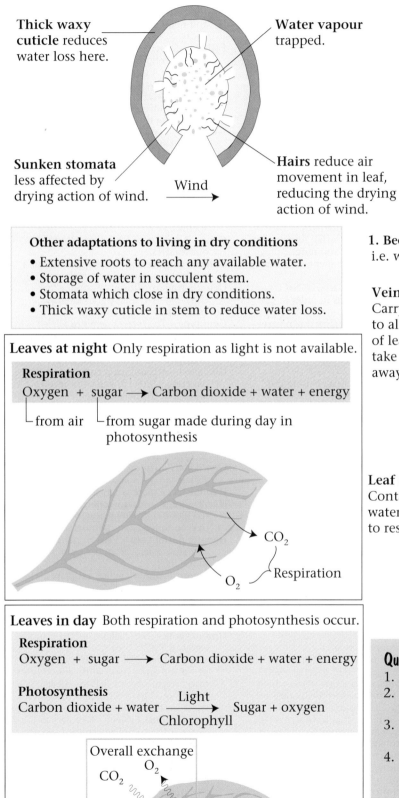

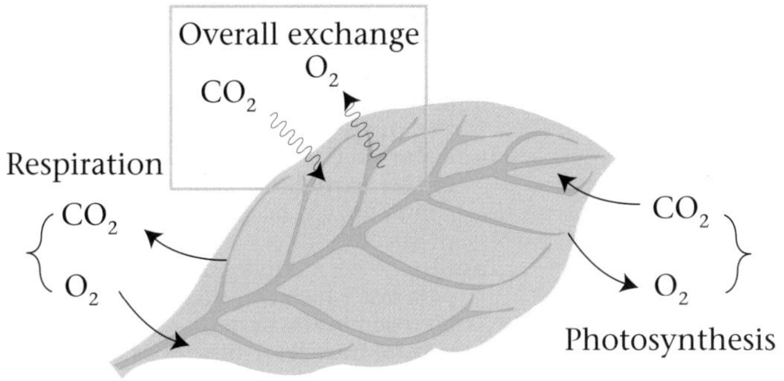

Questions:
1. Name a plant that lives in dry conditions.
2. What features of the plant reduce water loss from the leaves?
3. What other features are found in plants living in dry conditions?
4. Beech leaves are found in damp areas. What structures in the leaf allow water to reach every part?
5. What gas enters a leaf at night? Which process is taking place in the leaf.
6. What processes take place in a leaf during the day?
7. Overall, during the day, what gases a) enter the leaf, b) pass out of the leaf?

WORDS TO REMEMBER

Absorption – The movement of soluble food from the small intestine into the blood.
Acid rain – Acids formed from a combination of air pollutants and water which fall as rain.
Active transport – The movement of molecules from an area of low concentration to an area of higher concentration. This process requires energy.
ADH – Antidiuretic hormone secreted from the pituitary gland which controls the reabsorption of water in the nephron.
Adrenaline – Hormone produced during times of stress to increase the energy available.
Aerobic respiration – Respiration which requires oxygen.
Alcohol – A depressant drug found in drinks.
Algal bloom – The layer of algae covering a lake following eutrophication.
Allele – An alternative form of a gene e.g. T or t.
Alveoli – Air sacs in the lungs where gas exchange takes place.
Anaerobic respiration – Respiration which does not require oxygen.
Antibiotics – Chemicals used to kill bacteria.
Antibodies – Chemicals of specific shape produced by lymphocytes which destroy antigens.
Antigens – The surface shape of bacteria which antibodies attack.
Arterioles – Small arteries which link arteries to capillaries.
Artery – Blood vessels carrying blood away from the heart.
Asexual reproduction – Reproduction not involving gametes, leading to identical offspring.
Atria – The upper chambers of the heart where blood enters.
Auxins – Plant hormones produced in the shoot and root tip.
Benedict's solution – A chemical used to detect reducing sugars such as glucose.
Biological control – Using one species to control the numbers of another.
Biuret solution – A chemical used to detect the presence of protein.
Breathing – The inhaling and exhaling of air, also known as ventilation.
Camouflage – The blending of an animal into its habitat.
Capillaries – Tiny permeable blood vessels with walls only one cell thick where exchange of material takes place between the blood and the body cells.
Carbohydrate – A biological molecule including sugar, starch, glycogen and cellulose.
Cardiac muscle – The muscle making up the heart.
Carnivore – Animals which eat other animals.
Cell – The smallest unit of life.
Cell body – The swollen part of the nerve cell containing the nucleus.
Cellulose – Polysaccharide forming the cell wall of plants.
CFC – Chlorofluorocarbons produced by aerosols and fridges causing holes in the ozone layer.
Chloroplasts – Organelle containing chlorophyll where photosynthesis takes place.
Cholesterol – A substance made by the liver and found in the blood. High levels increase the risk of disease of the heart and blood vessels.
Chromosome – A strand of genes, found in the nucleus.
Cilia – Little hairs which beat in a regular way, e.g. in the trachea.
Ciliary muscle – The muscle that can alter the shape of the lens.
Clones – Genetically identical organisms.
CNS – Central nervous system made up of the brain and spinal cord.
Codominance – Two alleles of a pair are equally dominant and both are expressed.
Contraception pill – Pill containing oestrogen and/or progesterone to prevent ovulation and therefore pregnancy.
Coronary artery – The blood vessel which provides the heart muscle with oxygen and glucose.
Corpus luteum – Structure in the ovary which develops from the follicle, following ovulation.
Cuspid valves – Valves between the atria and the ventricles preventing the backflow of blood.
Deamination – The breakdown of amino acids to produce urea.
Decay – The breakdown and spoiling of food due to the action of microbes.
Decomposers – Organisms that feed on dead material, e.g. bacteria and fungi.
Deforestation – The permanent, large scale removal of forests.
Denitrifying bacteria – Bacteria that convert nitrates to nitrogen gas in order to get oxygen.
Depressants – Drugs that slow down nerve transmission and therefore reactions.
Diabetes – A condition caused by lack of insulin and detected by the presence of glucose in the urine.
Diaphragm – The muscle separating the thorax from the abdomen whose movements cause breathing.
Diffusion – The movement of molecules from an area of high concentration to an area of low concentration.
Digestion – The breakdown of large insoluble food to small soluble substances.
DNA – A molecule found in the nucleus which makes up the genes.
Dominant – The stronger allele.
Drugs – Chemicals which alter the way the body works.
Dry mass – The mass of plants once the water is removed by heating to 100°C.
Effector – A muscle or gland that brings about an effect.
Egestion – The removal of faeces from the anus.
Embryonic stem cells – Unspecialised cells in the embryo that can be used to replace damaged tissue.
Endangered species – Species whose numbers have declined to low levels.
Endocrine gland – A ductless gland which secretes a hormone into the blood.
Endometrium – The lining of the uterus.
Endotherm – Animals that have a constant body temperature, mammals and birds.
Enzymes – Proteins which speed up chemical reactions, e.g. they speed up the breakdown of food.
Eutrophication – A sudden increase in the nutrient content of lakes or rivers, often as a result of leaching.
Evaporation – The loss of water as water vapour. This leads to cooling.
Evolution – Changes that take place in organisms over a long period of time.
Excretion – The removal of waste made in cells from the body.
Extinct – Species that no longer exist.
Fertilisation – The joining of the ovum and sperm.
Fertility drugs – These help a woman to become pregnant by stimulating the release of ova.
Fish farming – The mass production of fish like salmon, in cages.
Flaccid – A plant cell which has lost water and is wilting.
Fluoride – A chemical that hardens enamel and neutralises the acid that causes decay.
Follicle – Structure in the ovary in which the ovum develops.
Food chain – A linear feeding relationship between organisms at each feeding or trophic level, e.g. grass → rabbit → fox.
Food preservation – A method of preventing the decay of food.
Food web – Feeding relationships shown by arrows between many organisms at each trophic level.
Fossils – Remains of animals and plants preserved in rock.
FSH – Follicle-stimulating hormone. This causes the development of the follicle.
Gametes – The sex cells e.g. the ovum and sperm.
Gene – A section of DNA that determines our features.
Genetic engineering – Altering genes for a particular purpose e.g. to make insulin.
Genetic fingerprinting – A unique 'bar code' of DNA, used e.g. to identify criminals.
Genotype – The type of genes or alleles present.
Glomerular filtrate – The fluid formed by ultrafiltration in the glomerulus.
Glucagon – A hormone secreted by the pancreas which converts glycogen to glucose in the liver.
Glycogen – A store of glucose in the liver.
Greenhouse effect – The warming effect of polluted air.
Growth – The permanent increase in size of an organism.
Guard cells – The specialised cells in leaves which can open and close stomata.
Habitat – The place where an organism lives.
Haemoglobin – An iron compound found in red blood cells which combines with oxygen to form oxyhaemoglobin.
Hepatic portal vein – Vein carrying glucose and amino acids from the small intestine to the liver.
Herbivore – Animals which eat plants.
Heterozygote – The 2 alleles of a pair are different e.g. Tt.
Homeostasis – The maintenance of a constant internal environment.
Homozygote – The 2 alleles are the same e.g. tt or TT.
Hormone – Chemical messenger produced in a gland and carried in the blood to its target organ where it has an effect, e.g. insulin produced in the pancreas has its effect in the liver.
Human Genome Project – A project to discover the complete genetic make-up of humans.
Immunity – The ability of an organism to resist an infection.
Implantation – Attachment of the zygote to the uterus wall.
Indicator species – Species whose presence or absence indicates a particular condition, e.g. lichens indicate low levels of SO_2.
Insulin – A hormone secreted by the pancreas which converts glucose to glycogen in the liver.
Intercostal muscles – Muscles between the ribs which move the ribs up and down during breathing.

Iodine solution – A chemical used to detect the presence of starch.
Iris muscle – Muscle in the eye which controls the size of the pupil and therefore controls the amount of light entering the eye.
Lactic acid – A product of anaerobic respiration in animals.
Leaching – The removal of soluble minerals from the soil by rain.
LH – Luteinising hormone. This causes ovulation.
Limiting factor – A factor in short supply which limits the rate of a reaction.
Lipid – A biological molecule used for insulation, waterproofing and as an energy store.
Lymph – Tissue fluid which drains into lymph vessels.
Lymphocytes – White blood cells which secrete antibodies.
Meiosis – A type of cell division that forms the gametes.
Menopause – The stopping of the menstrual cycle in middle-aged women.
Menstrual cycle – The 28 day cycle in females which results in an ovum being produced, controlled by hormones.
Microbes – Microorganisms e.g. bacteria, fungi and viruses.
Microvilli – The folded membrane of a cell that increases the surface area.
Minerals – Substances required by animals and plants for healthy growth.
Mitosis – A type of cell division that leads to growth.
Motor neurone – Nerve cell running from the CNS to an effector.
MRSA – A type of bacteria resistant to many antibiotics.
Mutation – A change in a gene or chromosome.
Myelin sheath – Fatty layer surrounding a nerve cell.
Natural selection – The process in which better adapted organisms survive to breed and pass on their useful features, whilst less well-adapted ones do not.
Negative feedback mechanism – A change from the normal level is detected, corrected and returned to the normal level.
Nephron – A tiny tube in the kidneys where the blood is filtered.
Nicotine – A stimulant found in cigarettes.
Nitrifying bacteria – Bacteria that convert ammonia to nitrates in order to gain energy.
Nitrogen-fixing bacteria – Bacteria living in the root nodules of leguminous plants that change nitrogen gas into ammonia.
Oestrogen – Ovarian hormone which repairs the uterus lining following the period.
Organ – A group of tissues working together for one function, e.g. the heart is an organ.
Osmoregulation – The control of water level in the body.
Osmosis – The movement of water from an area of high water concentration to an area of lower water concentration through a partially permeable membrane.
Ovaries – Reproductive glands in females producing ova, and the hormones controlling the menstrual cycle, oestrogen and progesterone.
Ovulation – The release of the ovum from the ovary.
Oxygen debt – The amount of oxygen needed to break down lactic acid formed during anaerobic respiration.
Ozone – A layer of gas in the upper atmosphere which absorbs harmful UV rays from the sun.
Palisade layer – A layer in leaves where most photosynthesis takes place.
Pancreas – A digestive gland secreting enzymes and an endocrine gland secreting the hormones insulin and glucagon.
Pathogens – Organisms that cause disease.
Penicillin – An antibiotic produced by the Penicillium mould.
Period – The monthly shedding of the uterus wall and some blood.
Peristalsis – The squeezing of food along the gut due to the contraction of muscles.
Pesticides – A chemical used to kill pests, such as insects on crops.
Phagocytes – White blood cells which surround and digest bacteria.
Phenotype – The appearance of the organism.
Phloem – Tubes in the stem which carry sugar and amino acids around a plant.
Photosynthesis – The process in which plants make sugar.
Pituitary gland – Structure in the brain producing the hormones ADH, FSH,LH and growth hormone.
Placenta – Structure which forms in the uterus wall of a pregnant female which attaches the fetus to the mother.
Plasma – The watery part of the blood.
Platelets – Cell fragments that are involved in blood clotting.
Population – A group of the same species together in a habitat.
Predator – Animals that hunt, catch and eat their prey, e.g. hawks.
Predator-prey relationship – The number of predators depends on the number of prey and vice versa. Each controls the size of the others population.
Primary consumer – Animals feeding on the producers, e.g. rabbits.
Producers – Plants producing their own food (sugar) in photosynthesis.
Progesterone – Ovarian hormone maintaining the uterus lining.
Protein – A biological molecule required for growth.
Pupil – The hole in the iris muscle through which light enters the eye.
Pyramid of biomass – A diagram to illustrate the mass of organisms at each trophic level. The size of the box depends on the mass of organisms.
Pyramid of energy flow – A diagram to illustrate the flow of energy between the trophic levels. The size of the box depends on the energy content.
Pyramid of numbers – A diagram to illustrate the number of organisms at each trophic level. The size of the box depends on the number of organisms.
Quadrat – A square frame used to estimate plant populations.
Receptor – A sense organ that responds to a stimulus, such as light.
Recessive – The weaker allele.
Red blood cells – Biconcave discs with haemoglobin that carry oxygen.
Respiration – The release of energy from food.
Rings of cartilage – Rings found in the trachea and bronchi to keep the air tubes open.
Saprobiotic – Organisms feeding on dead material.
Saturated fats – Hard fats that increase the risk of heart disease.
Secondary consumer – Animals feeding on primary consumers, e.g. foxes.
Selective breeding – A technique used by animal and plant breeders to combine required features and produce a desired variety.
Selective reabsorption – The reabsorption of required materials in the nephron.
Semilunar valves – Valves which close to prevent backflow of blood in veins.
Sensory neurone – Nerve cell running from a receptor to the CNS.
Sewage – Animal waste; faeces and urine.
Sex chromosomes – The pair of chromosomes that determine our sex e.g. XX (female) XY (male).
Sex-linked characteristics – Features caused by genes found on the sex chromosomes.
Sexual reproduction – Reproduction involving gametes joining, leading to variation.
Skeleton – The bony framework of the body.
Solvents – A depressant drug found in glue, lighter fuel and aerosol paints.
Species – Organisms which can interbreed and produce fertile offspring.
Stable population – A population whose numbers remain fairly constant.
Starch – A store of carbohydrate found in plants.
Stimulants – Drugs that speed up nerve transmission.
Stomata – Holes in the leaf, mainly on the under surface, through which gases can enter and leave a leaf.
Temperature regulation – Controlling body temperature.
Tertiary consumer – Animals feeding on secondary consumers.
Testes – Reproductive glands in males producing sperm and testosterone.
Thermal pollution – The reduction of oxygen in lakes caused by the influx of hot water.
Tissue – A group of similar cells working together e.g. muscle.
Tissue culture – Use of tiny pieces of tissue to artificially grow new plants.
Tissue fluid – The liquid produced by ultrafiltration at the capillary bed which bathes the cells.
Tooth decay – The dissolving of teeth by acid.
Transpiration – The loss of water from a plant, mainly as water vapour through the leaves.
Trophic level – A feeding level in a food chain and in ecology, e.g. producer trophic level.
Tropism – A growth movement in plants in response to a stimulus, e.g. phototropism is a growth movement in response to light.
Turgid – A plant cell full of water and swollen.
Ultrafiltration – The process pushing fluid out of the blood caused by high blood pressure.
Urine – Liquid waste excreted from the kidneys.
Vasoconstriction – The narrowing of arteries which reduces the volume of blood flowing through it.
Vasodilation – The widening of arteries which increases the volume of blood flowing through it.
Vein – Blood vessels carrying blood to the heart.
Ventricles – The lower chambers of the heart with thick muscle which pump blood out of the heart into arteries.
Venules – Small veins linking capillaries to veins.
Villi – Folds in the surface of the small intestine increasing the surface area for absorption.
Warfarin – A chemical that prevents the blood from clotting, used as rat poison.
Water potential – The ability of water molecules to move, a measure of their kinetic energy.
Waxy cuticle – Waxy layer covering a leaf which reduces the loss of water from a leaf.
Xerophyte – A plant adapted to dry conditions, e.g. marram grass.
Xylem – Tubes in the stem which transports water and minerals up a plant.

INDEX